AF389384

Relationship between Forest Ecophysiology and Environment

Relationship between Forest Ecophysiology and Environment

Editors

Roberto Tognetti
John D. Marshall

MDPI • Basel • Beijing • Wuhan • Barcelona • Belgrade • Manchester • Tokyo • Cluj • Tianjin

Editors
Roberto Tognetti
University of Molise
Italy

John D. Marshall
Swedish University of Agricultural Sciences (SLU)
Sweden

Editorial Office
MDPI
St. Alban-Anlage 66
4052 Basel, Switzerland

This is a reprint of articles from the Special Issue published online in the open access journal *Forests* (ISSN 1999-4907) (available at: https://www.mdpi.com/journal/forests/special_issues/Ecophysiology_Environment).

For citation purposes, cite each article independently as indicated on the article page online and as indicated below:

LastName, A.A.; LastName, B.B.; LastName, C.C. Article Title. *Journal Name* **Year**, *Volume Number*, Page Range.

ISBN 978-3-0365-0648-7 (Hbk)
ISBN 978-3-0365-0649-4 (PDF)

Contents

About the Editors

Roberto Tognetti (Ph.D.): Plants live in a wide range of environments and the conditions in these environments can fluctuate dramatically over the time scales of seconds to years and beyond. I use the combined potential of biometeorology and ecology to study the effects of disturbances on tree productivity and plant development. I focus on the basic environmental physiology of carbon, water, and nutrient cycling and strive to integrate these physiological processes to gain an understanding of plant functions and ecosystem processes, in a changing global environmental setting. Roberto Tognetti received an M.Sc. in forest science from the University of Firenze and a Ph.D. in plant physiological ecology from Trinity College Dublin. He is Full Professor of forest ecophysiology and silviculture at the Department of Agricultural, Environmental and Food Sciences, University of Molise.

John D. Marshall (Ph.D.): Trees and forests are like icebergs, where the part you can see is far less than the sum of everything that's going on. And as with an iceberg, the part you can't see can be important. I work with measurements of processes that are largely invisible, like photosynthesis and transpiration, but can be measured and can tell us important things about how a forest will respond to something like fertilization or climate change. The tools we use to measure these invisible processes and help us to build budgets for carbon and water, which allow us to explain and predict forest responses to change. John D. Marshall received an M.Sc. in forestry from Michigan State University and a Ph.D. in forest science from Oregon State University. He is Guest Professor and Senior Adviser at the Department of Forest Ecology and Management, Ecophysiology Unit, Swedish University of Agricultural Sciences.

Editorial

Relationship between Forest Ecophysiology and Environment

Roberto Tognetti [1,2,*] and John D. Marshall [3]

[1] Department of Agricultural, Environmental and Food Sciences, University of Molise, I-86100 Campobasso, CB, Italy

[2] Edmund Mach Foundation, EFI Project Center Mountain Forests (MOUNTFOR), I-38010 San Michele All'Adige, TN, Italy

[3] Department of Forest Ecology and Management, Swedish University of Agricultural Sciences (SLU), S-90183 Umeå, Sweden; john.marshall@slu.se

* Correspondence: tognetti@unimol.it; Tel.: +39-0874-404-735

Received: 2 January 2021; Accepted: 7 January 2021; Published: 9 January 2021

Although aspects of forest ecophysiology and forest environments have received considerable attention from research scientists in the last three decades, assessment of implications for meeting the climate targets and international agreements is still a matter of debate. The goal of limiting the increase in global temperature gives forests a prominent role, mainly because they can be significant carbon sinks, capable of removing carbon dioxide from the atmosphere. However, in addition to serving as carbon sinks, forest plants suffer from the negative impacts of climate change. These negative impacts include chronic stress (e.g., drought stress), acute responses to strengthened disturbance events (e.g., windstorms), and long-term ecological processes (e.g., species shifts). Detecting how the intensification of disturbances and stresses will affect the trajectory of plant vulnerability and adaptive capacity to stress is a major scientific challenge, because forests provide a number of ecosystem services.

The impact of disturbances may exceed a threshold that defines undesirable effects on ecosystem processes and services. Understanding the mechanisms of resilience and tipping points in forest environments would guide better climate-smart management practices. This Special Issue explores recent advances in research on relationships between plant traits and the environment in forest ecosystems, as well as physiological thresholds associated with ecological changes.

Ecophysiological measurements are needed to reveal both subtle changes and abrupt deviations at appropriate spatial and temporal scales. Nevertheless, comparing static trait correlations does not provide sufficient information on the functional and structural acclimation potential of individual leaves to rapidly changing environmental conditions: this acclimation is especially notable for light, as Deguchi and Koyama observed in their experiment [1]. However, light affects the dynamics (i.e., production, growth, and death) of organs that, in turn, influence the crown environment. Variation in the light environment within the crown of tree saplings was shown to be affected by the redeployment of shoots and leaves by Koyama et al. [2], with differences between fast- and slow-growing species because of their distinct crown dynamics. These species differences can result in pronounced differences in plant carbon gain. Differences in carbon pools and fluxes were investigated by Diao et al. [3], in pure conifers vs. mixed broadleaves, through continuous monitoring of CO_2 isotopes in situ. Not just the tree species (or forest type) but also the tree size mattered. Indeed, growing from seedlings in the understory to adult trees in the canopy layer required acclimation to different light environments along the way. This was illustrated by Deng et al. [4], who investigated leaf morphological and physiological traits across a vertical ambient light and tree height gradient in a tropical seasonal rainforest, highlighting the role of leaf functional traits in the adaptation to changing conditions during plant ontogeny.

In Mediterranean environments, water availability is the most important driver limiting tree growth and plant health. Szymczak et al. [5] used electronic dendrometers to analyze the responses of two native pine species to drought and precipitation events, observing different adaptation strategies to deal with the Mediterranean precipitation regime. Plants have developed different adaptation strategies to cope with abiotic stresses, and the response mechanism is regulated by multiple signaling pathways. Han et al. [6] examined transcription factors and their role in the interaction of these signaling pathways, particularly in response to cold and salt stress. Among functional traits, wood density is directly associated with mechanical support, carbon and nutrient storage, drought tolerance, water transport, and pathogen defense. Nelson et al. [7] tested whether canopy structure, climatic niche, geofloristic history, and habitat specialization related to wood density variation among members of a regional woody plant community in Mediterranean climatic conditions; their findings challenge classic assumptions about the adaptive significance of high wood density as a drought tolerance trait.

Increasing nitrogen deposition adds to warming temperature as a major driver, affecting the carbon sequestration potential, productivity, and water-use efficiency of forests. Giammarchi et al. [8] reported on long-term, manipulative experiments to improve the understanding of the carbon sequestration and mitigation ability of forests in response to increased nitrogen deposition. Atmospheric pollution (e.g., nitrogen oxides) adds to warming temperature in affecting plant metabolism and growth, the consequence being negative above certain thresholds. Wang et al. [9] fumigated leaves of an important agroforestry tree with nitrogen oxides and assessed the impact on nitrogen metabolism and photosynthetic efficiency, concluding that at a low concentration of fumigant, nitrogen oxides can be absorbed by leaves, protecting the photosynthetic apparatus and reducing the haze effect.

Regardless of environmental stress, forests are expected to deliver a number of functions and services, and mixed forests are considered more resistant to disturbances than pure stands. Russo et al. [10] found that mixed conifer and broadleaf forest stands had higher stand productivity and improved wood quality in comparison with corresponding monocultures, in a Mediterranean setting. In agroforestry systems, in addition to environmental disturbance, autotoxicity may strongly affect natural regeneration. Huang et al. [11] conducted an experiment on seedling growth, with different concentration gradients, using associated aqueous litter extracts, and concluded that autotoxicity inhibited seedling growth and natural regeneration of the studied tree species.

Warmer and drier environmental conditions will also affect future boreal and mountain forests, with an impact on photosynthetic rate and plant productivity. Ruiz-Pérez et al. [12] applied a mechanistic model of energy and mass transfer in the soil–plant–atmosphere continuum to explore the role of plant traits and projected environmental conditions on leaf temperature and photosynthetic rate, suggesting that trait selection should consider specifically the warmest period within the growing season. Understanding the responses of mature trees to global change is fundamental for predicting the potential or limitation of forest productivity. Liu et al. [13] used elevation gradients as proxy for changes in environmental factors to test whether the responses of carbon assimilation and nutrient allocation to environmental changes vary with tree ontogeny and leaf habit, noticing that young trees perform similarly to adult individuals. Stable isotope techniques provide insights into plant responses to environmental factors, notably leaf gas exchange and water use efficiency, though variation during plant development stages warrants further investigation. Li et al. [14] tested whether the $\delta^{13}C$ and $\delta^{15}N$ signature changed with leaf or tree age in a mountain conifer and identified that leaf age compared to tree age plays a dominant role, with leaf nitrogen concentration also being an important determinant.

The effect of disturbance on reproductive biology is particularly critical for endangered plant species with small population size and low genetic diversity. Zhang et al. [15] conducted field fixed-point observations and laboratory experiments to study the reproductive system and seed germination of an endangered plant species with an extremely small population, in a tropical setting, concluding that this population can be expanded through indoor seedling breeding for salinity resistance.

This Special Issue comprises a selection of papers reporting recent advances in research on relationships between plant functions and the environment. Natural settings, manipulated environments, and modeling exercises have been used to investigate abrupt deviations from the normal mode at diverse spatial and temporal scales. These reports offer insights into ecological dynamics and physiological responses to climatic perturbations and are useful to assess threshold-type relationships for tree decline and forest dieback, as well as to monitor consequences associated with ecological change. Integrating research results into innovative monitoring and decision-making may help forest managers to face climate change. We would like to thank all the authors and the reviewers of the papers published in this Special Issue for their great contributions and efforts. We are also grateful to the editorial board members and to the staff of the journal for their kind support in the preparation of this Special Issue.

Funding: R.T. and J.D.M. were supported by the *Progetto bilaterale di Grande Rilevanza* Italy-Sweden "Natural hazards in future forests: how to inform climate change adaptation" (MAECI) and the COST (European Cooperation in Science and Technology) Action CA15226 (Climate-Smart Forestry in Mountain Regions—CLIMO). J.D.M. was supported in part by the Knut and Alice Wallenberg Foundation (#2015.0047).

Conflicts of Interest: The authors declare no conflict of interest.

References

1. Deguchi, R.; Koyama, K. Photosynthetic and Morphological Acclimation to High and Low Light Environments in *Petasites japonicus* subsp. *giganteus*. *Forests* **2020**, *11*, 1365. [CrossRef]
2. Koyama, K.; Shirakawa, H.; Kikuzawa, K. Redeployment of Shoots into Better-Lit Positions within the Crowns of Saplings of Five Species with Different Growth Patterns. *Forests* **2020**, *11*, 1301. [CrossRef]
3. Diao, H.; Wang, A.; Yuan, F.; Guan, D.; Dai, G.; Wu, J. Environmental Effects on Carbon Isotope Discrimination from Assimilation to Respiration in a Coniferous and Broad-Leaved Mixed Forest of Northeast China. *Forests* **2020**, *11*, 1156. [CrossRef]
4. Deng, Y.; Deng, X.; Dong, J.; Zhang, W.; Hu, T.; Nakamura, A.; Song, X.; Fu, P.; Cao, M. Detecting Growth Phase Shifts Based on Leaf Trait Variation of a Canopy Dipterocarp Tree Species (*Parashorea chinensis*). *Forests* **2020**, *11*, 1145. [CrossRef]
5. Szymczak, S.; Häusser, M.; Garel, E.; Santoni, S.; Huneau, F.; Knerr, I.; Trachte, K.; Bendix, J.; Bräuning, A. How Do Mediterranean Pine Trees Respond to Drought and Precipitation Events along an Elevation Gradient? *Forests* **2020**, *11*, 758. [CrossRef]
6. Han, D.; Han, J.; Yang, G.; Wang, S.; Xu, T.; Li, W. An ERF Transcription Factor Gene from *Malus baccata* (L.) Borkh, *MbERF11*, Affects Cold and Salt Stress Tolerance in *Arabidopsis*. *Forests* **2020**, *11*, 514. [CrossRef]
7. Nelson, R.A.; Francis, E.J.; Berry, J.A.; Cornwell, W.K.; Anderegg, L.D.L. The Role of Climate Niche, Geofloristic History, Habitat Preference, and Allometry on Wood Density within a California Plant Community. *Forests* **2020**, *11*, 105. [CrossRef]
8. Giammarchi, F.; Panzacchi, P.; Ventura, M.; Tonon, G. Tree Growth and Water-Use Efficiency Do Not React in the Short Term to Artificially Increased Nitrogen Deposition. *Forests* **2020**, *11*, 47. [CrossRef]
9. Wang, Y.; Jin, W.; Che, Y.; Huang, D.; Wang, J.; Zhao, M.; Sun, G. Atmospheric Nitrogen Dioxide Improves Photosynthesis in Mulberry Leaves via Effective Utilization of Excess Absorbed Light Energy. *Forests* **2019**, *10*, 312. [CrossRef]
10. Russo, D.; Marziliano, P.A.; Macrì, G.; Zimbalatti, G.; Tognetti, R.; Lombardi, F. Tree Growth and Wood Quality in Pure Vs. Mixed-Species Stands of European Beech and Calabrian Pine in Mediterranean Mountain Forests. *Forests* **2020**, *11*, 6. [CrossRef]

11. Huang, X.; Chen, J.; Liu, J.; Li, J.; Wu, M.; Tong, B. Autotoxicity Hinders the Natural Regeneration of *Cinnamomum migao* H. W. Li in Southwest China. *Forests* **2019**, *10*, 919. [CrossRef]

12. Ruiz-Pérez, G.; Launiainen, S.; Vico, G. Role of Plant Traits in Photosynthesis and Thermal Damage Avoidance under Warmer and Drier Climates in Boreal Forests. *Forests* **2019**, *10*, 398. [CrossRef]

13. Liu, J.-F.; Jiang, Z.-P.; Schaub, M.; Gessler, A.; Ni, Y.-Y.; Xiao, W.-F.; Li, M.-H. No Ontogenetic Shifts in C-, N- and P-Allocation for Two Distinct Tree Species along Elevational Gradients in the Swiss Alps. *Forests* **2019**, *10*, 394. [CrossRef]

14. Li, C.; Wang, B.; Chen, T.; Xu, G.; Wu, M.; Wu, G.; Wang, J. Leaf Age Compared to Tree Age Plays a Dominant Role in Leaf δ^{13}C and δ^{15}N of Qinghai Spruce (*Picea crassifolia* Kom.). *Forests* **2019**, *10*, 310. [CrossRef]

15. Zhang, M.; Yang, X.; Long, W.; Li, D.; Lv, X. Reasons for the Extremely Small Population of putative hybrid *Sonneratia* × *hainanensis* W.C. Ko (Lythraceae). *Forests* **2019**, *10*, 526. [CrossRef]

Publisher's Note: MDPI stays neutral with regard to jurisdictional claims in published maps and institutional affiliations.

 *forests*

Article

Photosynthetic and Morphological Acclimation to High and Low Light Environments in *Petasites japonicus* subsp. *giganteus*

Ray Deguchi and Kohei Koyama *

Laboratory of Plant Ecology, Department of Life Science and Agriculture, Obihiro University of Agriculture and Veterinary Medicine, Inada-cho, Obihiro, Hokkaido 080-8555, Japan
* Correspondence: koyama@obihiro.ac.jp

Received: 31 October 2020; Accepted: 17 December 2020; Published: 19 December 2020

Abstract: Within each species, leaf traits such as light-saturated photosynthetic rate or dark respiration rate acclimate to local light environment. Comparing only static physiological traits, however, may not be sufficient to evaluate the effects of such acclimation in the shade because the light environment changes diurnally. We investigated leaf photosynthetic and morphological acclimation for a perennial herb, butterbur (*Petasites japonicus* (Siebold et Zucc.) Maxim. subsp. *giganteus* (G.Nicholson) Kitam.) (Asteraceae), in both a well-lit clearing and a shaded understory of a temperate forest. Diurnal changes in light intensity incident on the leaves were also measured on a sunny day and an overcast day. Leaves in the clearing were more folded and upright, whereas leaves in the understory were flatter. Leaf mass per area (LMA) was approximately twofold higher in the clearing than in the understory, while light-saturated photosynthetic rate and dark respiration rate per unit mass of leaf were similar between the sites. Consequently, both light-saturated photosynthetic rate and dark respiration rate per unit area of leaf were approximately twofold higher in the clearing than in the understory, consistent with previous studies on different species. Using this experimental dataset, we performed a simulation in which sun and shade leaves were hypothetically exchanged to investigate whether such plasticity increased carbon gain at each local environment. As expected, in the clearing, the locally acclimated sun leaves gained more carbon than the hypothetically transferred shade leaves. By contrast, in the understory, the daily net carbon gain was similar between the simulated sun and shade leaves on the sunny day due to the frequent sunflecks. Lower LMA and lower photosynthetic capacity in the understory reduced leaf construction cost per area rather than maximizing net daily carbon gain. These results indicate that information on static photosynthetic parameters may not be sufficient to evaluate shade acclimation in forest understories.

Keywords: phenotypic plasticity; shade tolerance; shade acclimation; light acclimation; light regime; sunfleck; leaf thickness; leaf angle; leaf three-dimensional structure

---

1. Introduction

In forests, individual plants from a single species often experience various light environments, from well-lit clearings or large gaps to shaded understories [1–4]. For plants, as sessile organisms, phenotypic plasticity is essential for survival in such heterogeneous environments [3,5–8]. This phenotypic plasticity and the consequent intraspecific variation also greatly influence community-level plant traits and productivity [9–15], highlighting the importance of the quantification of phenotypic plasticity of plant traits under different light environments.

In shaded understories, maximizing net carbon gain [3,6,16–18] and maximizing stress tolerance [19–22] are two major determinants of plant survival [18,23]. For maximizing net photosynthetic carbon gain, acclimation of leaf physiological traits [24,25] and biomass allocation patterns [18,24,25] are both important strategies. Within a species, plants grown in shaded places have leaves with a lower light-saturated photosynthetic rate [1–7,15,18–20,26–31] and a lower dark respiration rate [1–7,18], have thinner leaves with a lower leaf mass per unit area associated with their lower biomass investment per unit area [4,15,18–20,27–29,32–34], and have a higher leaf mass ratio (i.e., leaf mass relative to whole-plant mass) [5,18,25,28] than plants grown in well-lit places. Analogous leaf acclimation to different light environments has also been reported for sunlit and shaded leaves within a single canopy or within a single plant [26,29,34–47]. A low dark respiration rate of a shade leaf leads to a lower photosynthetic light compensation point (LCP) [3,5,6,17,18,24,30,48]. It has been frequently suggested that the net daily carbon gain would increase by lowering the LCP in the shade [3,17,24]. Such a simple consideration, however, has limitations because it only evaluates static photosynthetic parameters. In the forest understory, light intensity changes diurnally due to the diurnal elevation of the sun and fluctuates dynamically due to sunflecks [2,49–60]. A comparison of only static photosynthetic parameters, such as light-saturated photosynthetic rates and dark respiration rates, may therefore poorly reflect actual daily photosynthesis in field environments [42,51,56,61,62]. Given this, it has been questioned whether simple sun vs. shade acclimation can be understood based on the steady-state photosynthetic rate [23,42]. Additionally, the results of laboratory experiments under controlled low light environments [6,20] or those of field shading experiments using shade cloths [19,30,33] may not provide an accurate estimate of carbon gain in the understory, because they do not take into consideration sunflecks. To understand the effect of shade acclimation on daily net carbon gain, therefore, the effects of sunflecks also should be considered.

Here, we investigated the shade acclimation of *Petasites japonicus* subsp. *giganteus* that naturally grew in either well-lit or shaded places in a temperate forest. In a previous study on the same species [32], the phenotypic plasticity of some leaf traits under different light environments was reported. However, because the authors did not measure photosynthetic parameters and local light intensity, they did not clarify whether such plasticity contributed to maximizing carbon gain under each light environment. The objectives of this study, therefore, were (1) to quantify the photosynthetic and morphological acclimation to different light environments for this species, and (2) to test whether leaf physiological acclimation contributed to maximizing leaf-level carbon gain under diurnally changing light environment due to sunflecks.

2. Materials and Methods

2.1. Study Species

Butterbur (*Petasites japonicus* (Siebold et Zucc.) Maxim. subsp. *giganteus* (G.Nicholson) Kitam.) (Asteraceae) is a perennial herb distributed in Northeast Asia [63]. This species is found naturally in environments of varying amounts of light, such as roadsides, well-lit forest gaps, and in shaded forest understories. This species also is grown as a vegetable in eastern Asia, including Japan, Korea [64], and Taiwan [65]. Large radical leaves (often reaching 1–2 m in height) elongate from an underground shoot in this species (Figure 1a–c). Therefore, investigating the leaves is equivalent to investigating the entire above-ground part (ramet) for this clonal species. These leaves are usually horizontally arranged on the ground so as to prevent overtopping others, but small immature leaves that are not fully expanded often exist below fully expanded leaves.

2.2. Study Site and Sampling

We performed the study at two sites in the same forest (clearing [C] and understory [U]), which were approximately 100 m apart, in the Forest of Obihiro (Obihironomori) (42°53′ N, 143°09′ E, altitude: 86 m a.s.l.). This secondary forest comprises a mixture of planted and regenerated trees and is located in Obihiro City in eastern Hokkaido in a cool-temperate region in Japan. The mean annual temperature and precipitation at the Japan Meteorological Agency Obihiro Weather Station (6 km from the study site) between 1998 and 2017 were 7.2 °C and 937 mm, respectively [66]. In the clearing site (approximately

30 × 30 m), few trees were taller than the investigated leaves (Figure 2a). The understory site (Figure 2b) (approximately 15 × 10 m) was located under a young birch forest (*Betula platyphylla* Sukaczev var. *japonica* (Miq.) H.Hara; DBH: 17.5–25.1 cm), in which some walnut (*Juglans mandshurica* Maxim. var. *sachalinensis* (Komatsu) Kitam.) grew as subcanopy trees. Within each plot, the investigated leaves were selected along a transect. Although we found multiple separate patches of leaves at each site, the number of genets was unknown. Therefore, the investigated leaves were selected as evenly as possible along the entire length of each transect.

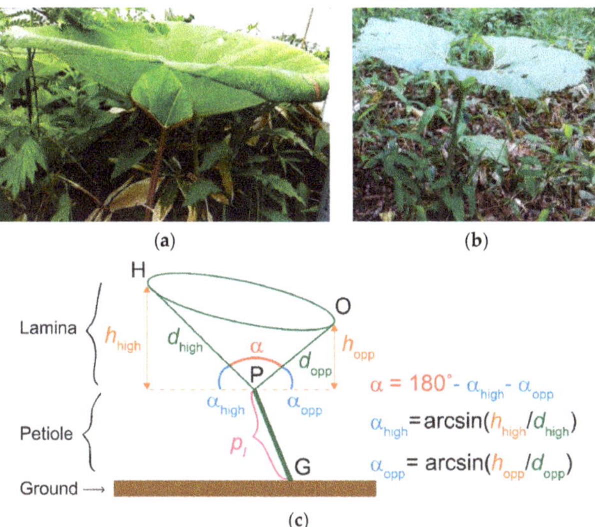

Figure 1. The measured morphological parameters of the leaves of the butterburs (*Petasites japonicus* (Siebold et Zucc.) Maxim. subsp. *giganteus* (G.Nicholson) Kitam.) investigated in this study. Leaves in (**a**) the clearing and (**b**) the understory, in addition to (**c**) the measured leaf parameters, are shown. H: The highest point on the leaf lamina. O: The point located at the opposite side of H on the lamina edge. P: The point of attachment of the lamina to the petiole. G: The point of attachment of the petiole to the ground. d_{high}: The distance between H and P. h_{high}: The vertical distance between H and P. d_{opp}: the distance between O and P. h_{opp}: the vertical distance between O and P. p_l: Above-ground petiole length (the distance between P and G). α: Lamina openness angle. Photographs were taken in June 2020 by Kohei Koyama.

Figure 2. The study sites, (**a**) the clearing and (**b**) the understory, in the Forest of Obihiro. Photographs were taken on (**a**) June 28 and (**b**) 3 July 2020 by Kohei Koyama.

In June 2020, we marked 62 leaves (32 from plants in the clearing and 30 from plants in the understory). Leaf three-dimensional structure was measured using measuring tapes on 24–25 June 2020, and the lamina openness angle [67,68] was calculated (Figure 1c). Photosynthetic light response curves were measured for a total of 12 leaves (6 at each site) on 21, 22 and 24 June 2020 with a portable photosynthesis system (LI-6400; LI-COR, Lincoln, NE, USA) equipped with an LED light source (LI-6400-02B) (Figure 3a). Due to the amount of rainfall prior to the measurement days (June 18 (3 mm), June 19 (7.5 mm), June 20 (4 mm), and June 23 (1 mm), data from [66]), the soil in the fields was wet during the measurements. Measurements were taken in the morning (7:30–12:00) each day under cloudy and humid conditions, and the environment inside the chamber showed favorable conditions for photosynthesis: leaf temperature (measured by a thermocouple inside the chamber) ranged from 17.93 to 24.48 °C, and the vapor pressure deficit (VPD) based on leaf temperature was always less than 0.9 kPa. In the understory, we first induced the leaves by keeping incident photosynthetic photon flux density (PPFD) on the leaf surface at 1000–1500 µmol m^{-2} s^{-1} until equilibration. This process was omitted for most of the leaves in the clearing, which showed a very quick response under PPFD 2000 µmol m^{-2} s^{-1}. Then, we progressively lowered the incident PPFD on the leaf surface (2000, 1500, 1000, 750, 500, 250, 125, 63, 32, and 0 µmol m^{-2} s^{-1}). On each occasion of changing light intensity, we kept the PPFD constant until the equilibration of the leaves. The CO$_2$ concentration of the air entering the leaf chamber was controlled at 400 ppm. All the data recorded by the LI-6400 (e.g., photosynthesis, stomatal conductance, transpiration, humidity, temperature at each PPFD, etc.) are available in the Supplementary Materials.

(**a**)　　　　　　　　　　　　　　　(**b**)

Figure 3. Measurements of (**a**) photosynthesis and (**b**) the diurnal course of incident light in the clearing site. The two panels show the same leaf. The position on the leaf lamina, at which photosynthetic traits and incident light were measured, was marked with a light-resistant ink pen (red box). In the cases when that part of the lamina was inclined, the light sensor was inclined such that the lamina and the top of the sensor were parallel to one another. Leaf mass per area (LMA) was subsequently measured by sampling the lamina part within the same red box. Photographs were taken on (**a**) June 21 and (**b**) 1 July 2020 by Kohei Koyama.

Net photosynthetic rate per unit area of each leaf (P_{net} µmol m^{-2} s^{-1}) was assumed to be expressed by the non-rectangular hyperbola (NRH) [69]:

$$P_{net} = \frac{\Phi I + P_{g\_max\_area} - \sqrt{(\Phi I + P_{g\_max\_area})^2 - 4\theta \Phi I P_{g\_max\_area}}}{2\theta} - R_{area}, \tag{1}$$

where I (µmol m^{-2} s^{-1}) indicates the incident PPFD for each leaf at each moment, and $P_{g\_max\_area}$ (µmol m^{-2} s^{-1}) indicates the maximum gross photosynthetic rate when I approaches infinity. Φ (mol CO$_2$ mol^{-1} quanta) and θ (dimensionless) indicate the initial slope and the convexity, respectively. R_{area} (µmol m^{-2} s^{-1}) indicates the dark respiration rate. These parameters were fitted for each leaf with the R function *nls*. Light compensation point (LCP) was calculated by solving the quadratic form of NRH [68] for I on the condition that $P_{net} = 0$ [70] using the software Maxima (Maxima project, USA) [71]:

$$\theta(P_{net} + R_{area})^2 - \left(\phi I + P_{g\_max\_area}\right)(P_{net} + R_{area}) + \phi I P_{g\_max\_area} = 0$$
$$P_{net} = 0 \Rightarrow I \equiv LCP = \frac{R_{area}\left(R_{area}\theta - P_{g\_max\_area}\right)}{\left(R_{area} - P_{g\_max\_area}\right)\Phi} \tag{2}$$

2.3. Measurement of PPFD

2.3.1. Diurnal Course of Incident PPFD on the Leaves

We measured a time-series of incident PPFD on the selected leaves on two days: an overcast day (June 24; clearing, $n = 4$; understory, $n = 4$) and a sunny day (July 3; clearing, $n = 3$; understory: $n = 4$) in 2020. The sample size difference between these two days was due to a measurement failure caused by an operational error. On the days between June 24 and July 3, PPFD data were not obtained due to intermittent disruptions by rain. Those leaves were selected from the samples for which photosynthetic light response curves were measured. PPFDs were measured for the same parts of the leaves as for the photosynthetic parameters (the red box, Figure 3). For each target leaf, we set one quantum sensor (MIJ-14PAR Type2/K2; Environmental Measurement Japan, Fukuoka, Japan) on the pole. If the leaf part was inclined, the sensor was inclined to measure the incident PPFD on the leaf surface (Figure 3b). Each sensor was connected to a voltage logger (LR5041; HIOKI, Ueda, Japan). Voltage was recorded every 10 min for 24 h. These voltage values were transformed into PPFD using sensor-specific coefficients.

2.3.2. Instantaneous Measurement of rPPFD

In addition to measuring the diurnal course of PPFD for the selected leaves, we measured instantaneous PPFDs for all 62 leaves to estimate the light environment at the two sites. The measurements were conducted around midday on an overcast day (20 June 2020). We used two quantum sensors (MIJ-14PAR) and simultaneously measured the PPFD incident on each leaf and on top of an approximately 2 m tripod placed at the open place. One sensor at the tripod was fixed horizontally. The other was placed at a tip of a hand-held measuring bar [4,72] and was inclined for each leaf to measure the PPFD at the bottom surface of the cone- or funnel-shaped leaf (Figure 4). Each sensor was connected to a voltage logger (LR5041), and the voltages of the two sensors were recorded simultaneously. We calculated the relative PPFD (rPPFD) as the ratio of the PPFD on each leaf to the PPFD at the top of the tripod.

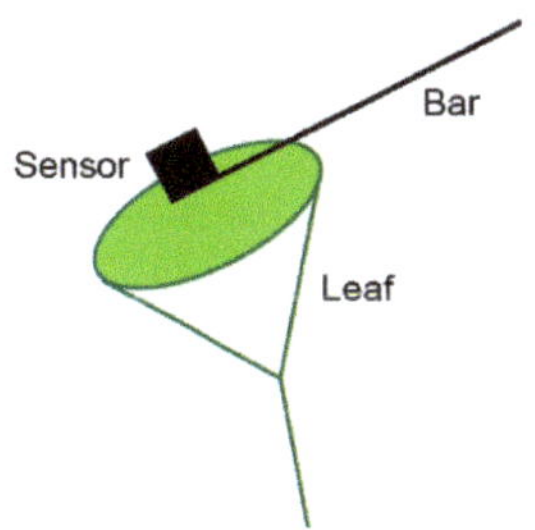

Figure 4. A schematic diagram of instantaneous relative photosynthetic photon flux density (rPPFD) measurement with a hand-held measuring bar (illustration by Kohei Koyama).

2.4. Leaf Thickness and LMA

On 4 and 7 July 2020, we sampled pieces (ca. 40 cm^2) from the lamina edges of 41 leaves (clearing, $n = 20$; understory, $n = 21$), measured their thickness with a digital caliper (CD-15PSX (resolution: 0.01 mm); Mitutoyo Corp, Kawasaki, Japan), and scanned them immediately after sampling with an A4 flatbed scanner (CanoScan LiDE 220; Canon, Tokyo, Japan). These samples included all 12 leaves for which photosynthetic rates were measured, and the lamina samples were taken at the same positions on the leaves as the photosynthetic measurements (Figure 3). The lamina parts were selected to avoid the thickest leaf veins. The laminae were then oven-dried at 70 °C for at least one week, and their dry mass was measured with a precision balance. The projected area of each piece of lamina was measured with the Image J software (NIH, Bethesda, MD, USA) [73]. We calculated leaf mass per area (LMA, g m^{-2}) as the ratio between the dry mass and the area of one side of each sampled piece of lamina [74]. We calculated mass-based values by dividing the area-based values by the LMA of that leaf [2].

2.5. Calculation of Daily Photosynthesis

We calculated the instantaneous net photosynthetic rate for each target leaf every 10 min for 24 h using the estimated light response curves (Equation (1)) and the diurnal course of incident PPFD. At night, the dark respiration rate (R_{area}) was used as the nighttime respiration rate. By integrating these, we calculated the daily net photosynthesis for each target leaf.

2.6. Simulation 1: Exchanged Leaves

We performed a simulation in which we hypothetically exchanged the photosynthetic light response curves between the two sites. We expected that if the difference in the photosynthetic traits between the sites was the acclimation to a local light environment, then the exchange of leaves would reduce the daily net photosynthesis for both sites. For each understory leaf, we replaced the photosynthetic light response parameters ($P_{g\_max\_area}$, R_{area}, Φ, and θ in Equation (1)) with the mean photosynthetic parameters obtained in the clearing and vice versa. Next, we calculated the daily net photosynthesis for each hypothetical leaf by using the actual incident PPFD on each leaf. We also calculated the critical PPFD for each understory leaf. This value was defined as the lowest PPFD, such that if instantaneous PPFD exceeded that value, the net photosynthetic rate at that moment would be higher for the hypothetically set clearing leaf than for the actual understory leaf in question. This value was used to investigate how often PPFD exceeded this value due to sunflecks.

2.7. Simulation 2: Understory without Sunflecks

To evaluate the significance of sunflecks in the understory, we performed another simulation in which sunflecks were hypothetically removed from the original dataset of the diurnal course of PPFD on the sunny day (July 3). If PPFD at a given moment exceeded 200 μmol m^{-2} s^{-1}, that PPFD value was replaced by a fixed value of 200 μmol m^{-2} s^{-1}. This value was approximately equal to the maximum PPFD (199.72) observed in the understory on the overcast day (June 24). We also observed that the diel cycle of PPFD in the understory did not exceed 200 μmol m^{-2} s^{-1} on the sunny day except during sunflecks (see Results), so that the background diel cycle of PPFD was retained by this simulation. We then calculated the daily net photosynthesis in the understory without sunflecks, either with (1) actual understory leaves, or (2) the hypothetically set clearing leaf (described above).

2.8. Statistical Analysis

All statistical analyses were performed with the statistical software R v4.0.3 (Vienna, Austria) [75] and packages ("cowplot" [76], "ggbeeswarm" [77], "ggplot2" [78], and "lme4" [79]). The results were compared between sites by fitting generalized linear models (GLM) using the function *glm* (family = Gamma (link = "log")), except for the simulation results. The differences in the simulation results obtained under different scenarios were tested using a generalized linear mixed model (GLMM), treating individual leaves (i.e., diurnal courses of PPFD on different leaves) as random intercepts, and using the function *glmer* (family = Gamma (link = "log")) [79], except in one case (simulated clearing leaves in the understory on the overcast day), in which simulated net daily photosynthesis values contained a negative value. For that case, a linear mixed model (LMM) was fit with the function *lmer* [79]. In all cases, the significance of the fixed effect was tested using the likelihood ratio test with the function *anova* (test = "Chisq").

3. Results

3.1. Leaf Shape

The differences in the light environment were quantified by large differences in rPPFD and daily light integral between the two sites (Table 1). Lamina openness angle was significantly larger in the understory (U) than in the clearing (C) ($p < 0.01$, Table 1; Figure 5a,b), indicating that the three-dimensional arrangement of leaf laminae was flatter in the understory (see photographs in Figure 1a,b). Compared with the difference in shape, the difference in lamina size was relatively small: d_{high} was slightly larger in the understory than in the clearing ($p = 0.025$), and neither d_{opp} ($p = 0.37$) nor the petiole length (p_l) ($p = 0.28$) significantly differed between the sites (Table 1; Figure 5c–e).

Table 1. Leaf traits in the clearing (C) and in the understory (U).

Symbol	Definition	Unit	Sample Size (n)		Mean (C)	Mean (U)	Ratio C/U	C vs. U p-Value
			C	U				
rPPFD	-	-	32	30	87.7%	10.8%	8.10	<0.01
α	Lamina openness angle	Degree (°)	32	30	116.3	136.5	0.85	<0.01
d_{high}	Lamina length (high)	cm	32	30	31.3	35.2	0.89	0.025
d_{opp}	Lamina length (opposite)	cm	32	30	29.7	31.1	0.96	0.38 ns
p_l	Petiole length	cm	32	30	55.1	57.7	0.95	0.28 ns
LMA	Leaf mass per area	$g\ m^{-2}$	20	21	44.3	23.7	1.87	<0.01
-	Leaf thickness	mm	20	21	0.428	0.348	1.23	<0.01
P_{net_2000}	Net photosynthetic rate at PPFD = 2000 mol m^{-2} s^{-1}	$\mu mol\ m^{-2}\ s^{-1}$	6	6	23.2	11.3	2.04	<0.01
$P_{g\_max\_area}$	Maximum gross photosynthetic rate per unit leaf area	$\mu mol\ m^{-2}\ s^{-1}$	6	6	27.2	12.6	2.16	<0.01
R_{area}	Dark respiration rate per unit leaf area	$\mu mol\ m^{-2}\ s^{-1}$	6	6	1.85	0.87	2.14	<0.01
Φ	Initial slope	mol CO$_2$ mol^{-1} quanta	6	6	0.075	0.076	0.98	0.48 ns
θ	Convexity	-	6	6	0.562	0.600	0.94	0.49 ns
LCP	Light compensation point	μmol quanta m^{-2} s^{-1}	6	6	25.7	11.8	2.18	<0.01
$P_{g\_max\_mass}$	Maximum gross photosynthetic rate per unit leaf mass	nmol g^{-1} s^{-1}	6	6	545	523	1.04	0.72 ns
R_{mass}	Dark respiration rate per unit leaf mass	nmol g^{-1} s^{-1}	6	6	37.1	36.7	1.01	0.97 ns
I_{day}	Daily light integral—sunny day	mol quanta m^{-2} day^{-1}	3	4	45.7	5.96	7.67	<0.01
	—overcast day		4	4	26.5	2.74	9.69	<0.01
P_{n_day}	Net daily photosynthesis per area—sunny day	mol m^{-2} day^{-1}	3	4	0.745	0.145	5.14	<0.01
	—overcast day		4	4	0.682	0.075	9.05	<0.01
R_{day}	Daily respiration per area	mol m^{-2} day^{-1}	6	6	0.160	0.075	-[1]	-[1]
P_{g_day}	Gross daily photosynthesis per area—sunny day	mol m^{-2} day^{-1}	3	4	0.906	0.237	3.83	<0.01
	—overcast day		4	4	0.841	0.167	5.03	<0.01

[1] Same values as R_{area}.

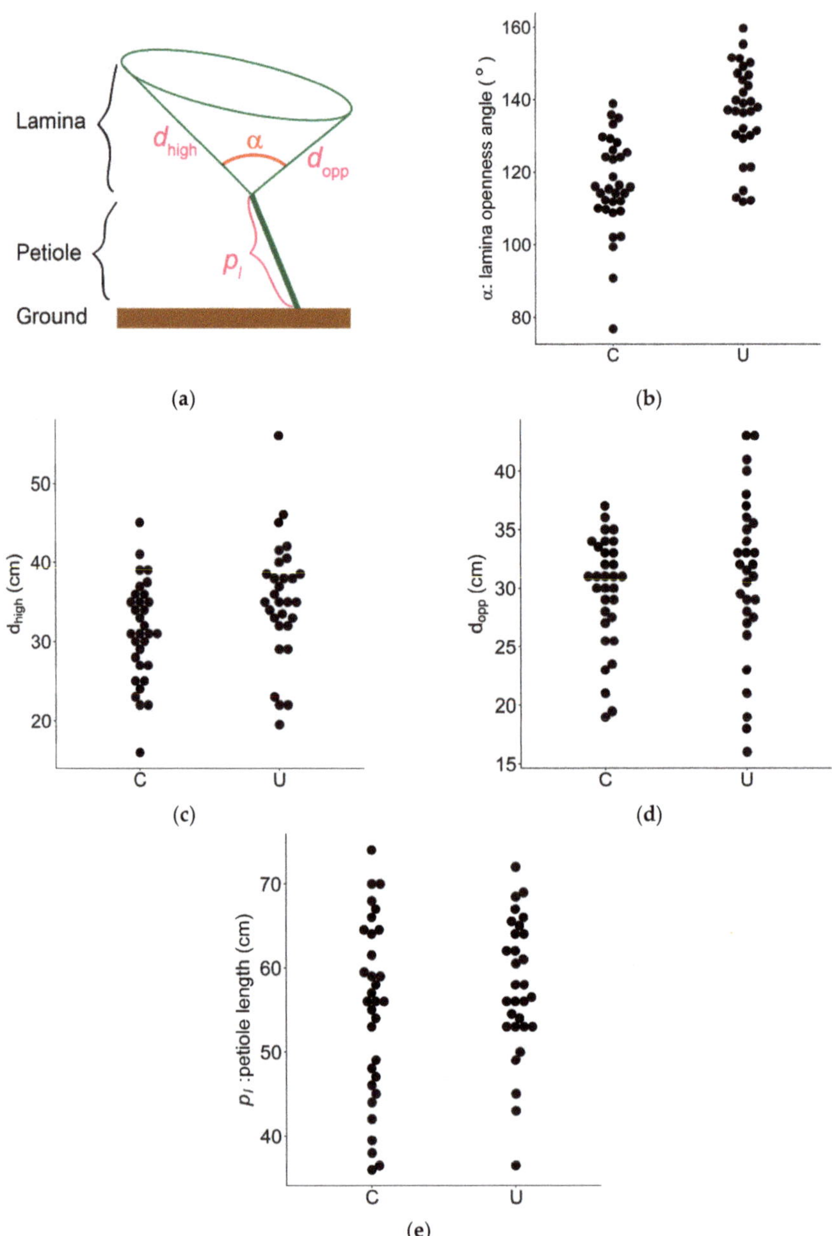

Figure 5. (**a**–**e**) Leaf shape parameters in the understory (U) and in the clearing (C). Each closed circle indicates one leaf (i.e., bee swarm plot).

3.2. Area-Based Photosynthetic Traits

The area-based net photosynthetic rate at PPFD = 2000 μmol m^{-2} s^{-1} (P_{net_2000}), the maximum gross photosynthetic rate ($P_{g\_max\_area}$), dark respiration rate (R_{area}), and light compensation point (LCP) were all significantly higher in the clearing than in the understory ($p < 0.01$) (Table 1; Figure 6a–d).

Neither the initial slope (Φ) nor the convexity (θ) significantly differed between the sites ($p = 0.48$–0.49) (Table 1; Figure 6e,f).

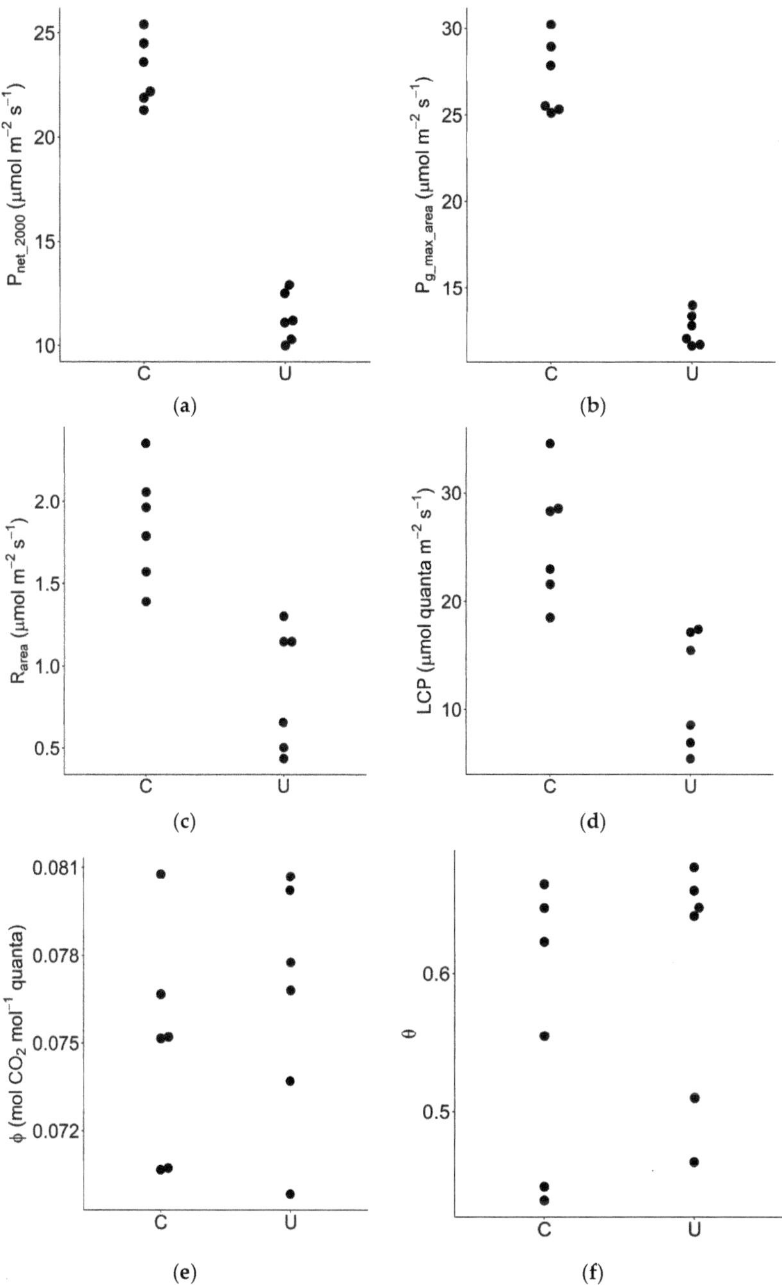

Figure 6. (a–f) Area-based photosynthetic parameters of the sun leaves in the clearing (C) and the shade leaves in the understory (U). Each closed circle indicates one leaf. LCP: light compensation point.

3.3. LMA and Leaf Thickness

The mean LMA of the leaves in the clearing leaf was 1.87 times larger than that of the leaves in the understory ($p < 0.01$) (Table 1; Figure 7a). Additionally, the mean leaf thickness was 1.23 times larger than that of the leaves in the understory ($p < 0.01$) (Table 1; Figure 7b).

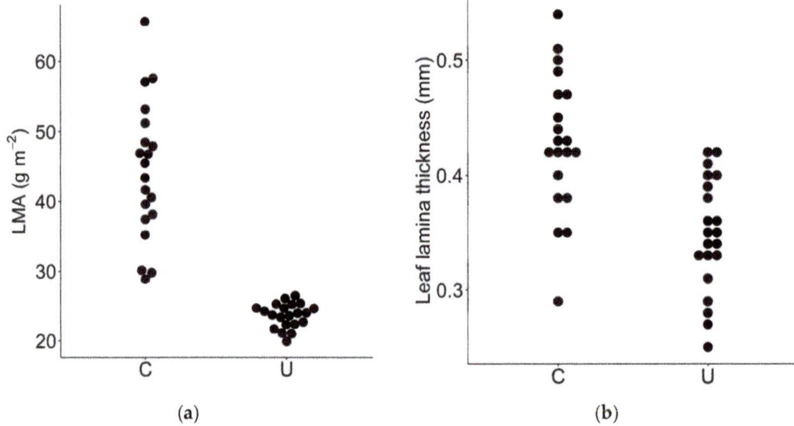

(a)   (b)

Figure 7. (**a**) Leaf mass per unit area (LMA) and (**b**) lamina thickness of the sun leaves in the clearing (C) and the shade leaves in the understory (U). Each closed circle indicates one leaf. The thicknesses were measured with a resolution of 0.01 mm.

3.4. Mass-Based Photosynthetic Traits

In contrast to the high plasticity in the area-based rates, no significant difference was found for mass-based photosynthetic and respiration rates between the sites ($p = 0.69$–0.96) (Table 1; Figure 8a,b).

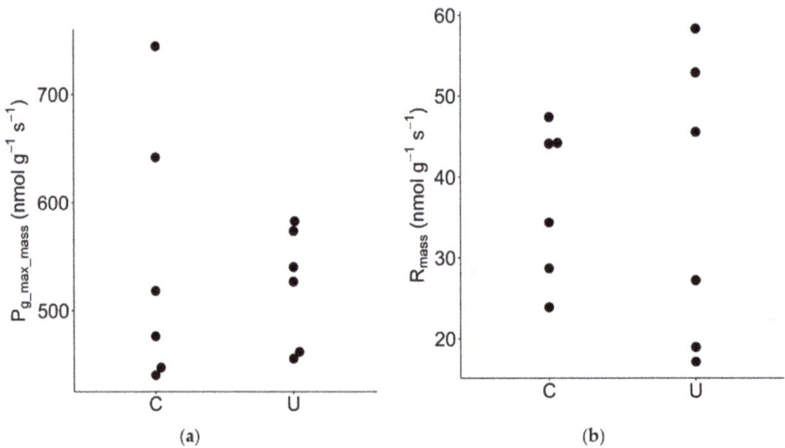

(a)   (b)

Figure 8. Mass-based (**a**) photosynthetic rates and (**b**) dark respiration rates of the sun leaves in the clearing (C) and the shade leaves in the understory (U). Each closed circle indicates one leaf.

3.5. Diurnal Courses of PPFD

Figure 9 shows the diurnal courses of PPFD incident on the leaves. The estimated critical values were 59–161 μmol m^{-2} s^{-1} (median: 111.5; these values were calculated for all the six understory leaves for which photosynthetic parameters were measured, and the diurnal course of PPFD was measured for four of them, as shown in Figure 9b,d). On the sunny day in the understory, instantaneous PPFD often exceeded the critical values due to sunflecks (Figure 9b).

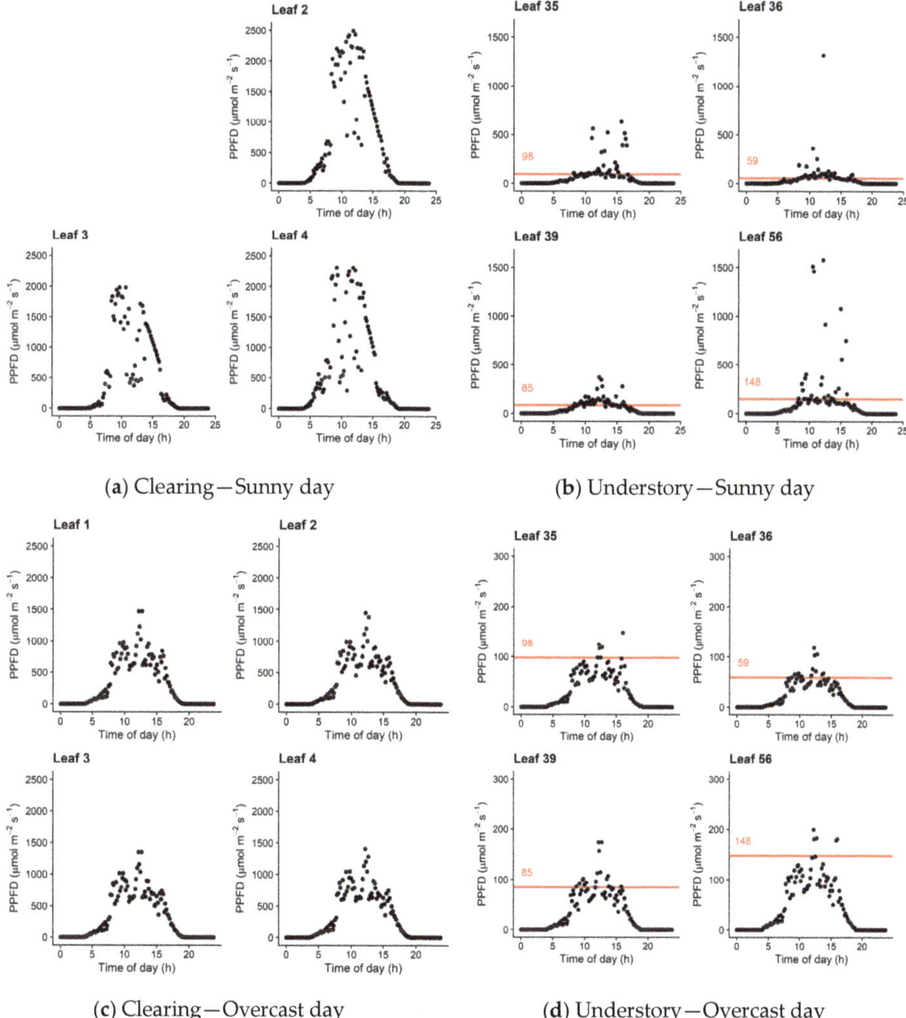

Figure 9. Diurnal course of light intensity (PPFD) incident on each leaf at each site (the clearing and understory) on a sunny day (3 July 2020) and on an overcast day (24 June 2020). (**a**) Clearing on the sunny day, (**b**) understory on the sunny day, (**c**) clearing on the overcast day, and (**d**) understory on the overcast day. Data for Leaf 1 on the panel (**a**) were not obtained owing to a measurement failure. The red horizontal lines on the understory panels indicate the critical values of PPFD; if PPFD at one moment exceeded that value, the instantaneous net photosynthetic rate would be higher for a typical clearing leaf than for the understory leaf in question.

3.6. Daily Carbon Gain under Actual and Simulated Conditions

The obtained results differed between the two sites and between weather conditions (Table 2). In the clearing, the simulated change from the actual clearing leaves into understory leaves greatly

(>40%) reduced the daily net photosynthesis on both the sunny day (from A to B in Figure 10a; mean: (A) 0.745 and (B) 0.430 mol m^{-2} day^{-1}; A vs. B: $p < 0.01$) and on the overcast day (from A to B in Figure 10b; (A) 0.682 and (B) 0.415 mol m^{-2} day^{-1}; A vs. B: $p < 0.01$). This was because on both days, the reduction in gross photosynthesis (from C into D in Figure 10a,b) was much larger in magnitude than that of respiratory loss (from E into F in Figure 10a,b).

Table 2. Daily carbon exchange per unit area of leaf. The mean value for each item is shown.

Daily Carbon Exchange	Simulation	Clearing (C)		Understory (U)	
		Sunny	Overcast	Sunny	Overcast
Daily net photosynthesis (mol m^{-2} day^{-1})	Sample size	$n = 3$	$n = 4$	$n = 4$	$n = 4$
	Actual leaves	0.745	0.682	0.145	0.075
	Exchanged leaves (between C and U)	0.430	0.415	0.138	0.024
	Without sunflecks (actual leaves)	-		0.134	-
	Without sunflecks (exchanged leaves)	-		0.106	-
Daily gross photosynthesis (mol m^{-2} day^{-1})	Sample size	$n = 3$	$n = 4$	$n = 4$	$n = 4$
	Actual leaves	0.906	0.803	0.237	0.167
	Exchanged leaves	0.505	0.471	0.298	0.184
Daily respiration (mol m^{-2} day^{-1})	Sample size [1]	$n = 6$		$n = 6$	
	Actual leaves	0.160		0.075	
	Exchanged leaves	0.075		0.160	

[1] Daily leaf respiration rates were calculated for all the 12 leaves (6 in each site) for which dark respiration rates were measured, and the mean value at each site is shown in Table 2 and used in the leaf-exchange simulation. The same daily respiration rates were used for the two days. Among these 12 leaves, daily net- and gross-photosynthetic rates were measured or simulated for seven or eight selected leaves.

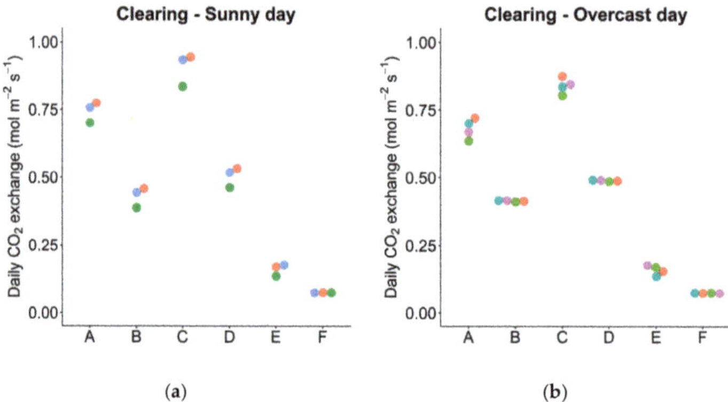

Figure 10. Daily carbon exchange per unit area of leaf in the clearing on (**a**) a sunny day (3 July 2020) and (**b**) an overcast day (24 June 2020). Results of the same leaf simulated in different scenarios appear in the same color. A: Estimated actual net photosynthesis. B: Simulated net photosynthesis when the photosynthetic parameters were hypothetically changed to those of the understory. C: Estimated actual gross photosynthesis. D: Simulated gross photosynthesis when the photosynthetic parameters were hypothetically changed to those of the understory. E: Estimated actual daily respiration. F: Simulated daily respiration when the leaves were changed to the understory leaves. In simulation F, all three or four leaves were assumed to have the same respiration rate as the mean value of the understory leaves.

By contrast, in the understory, the simulated change from actual understory leaves to the clearing leaves did not greatly reduce the daily net carbon gain on the sunny day (Table 2; from A to B, Figure 11a; (A) 0.145 and (B) 0.138 mol m^{-2} day^{-1}; A vs. B: $p = 0.082$). This is because on the sunny day, during which frequent sunflecks were observed (Figure 9b), the increment of gross daily photosynthesis (from C into D in Figure 11a) had approximately the same magnitude as the increment of respiratory loss (from E to F in Figure 11a). Those two effects offset each other. On the overcast day, during which few sunflecks were observed (Figure 9d), the simulated change from the actual understory leaves to the clearing leaves reduced the daily net photosynthesis (from A to B, Figure 11b; mean: (A) 0.075 and (B) 0.024 mol m^{-2} day^{-1}; A vs. B: $p < 0.01$).

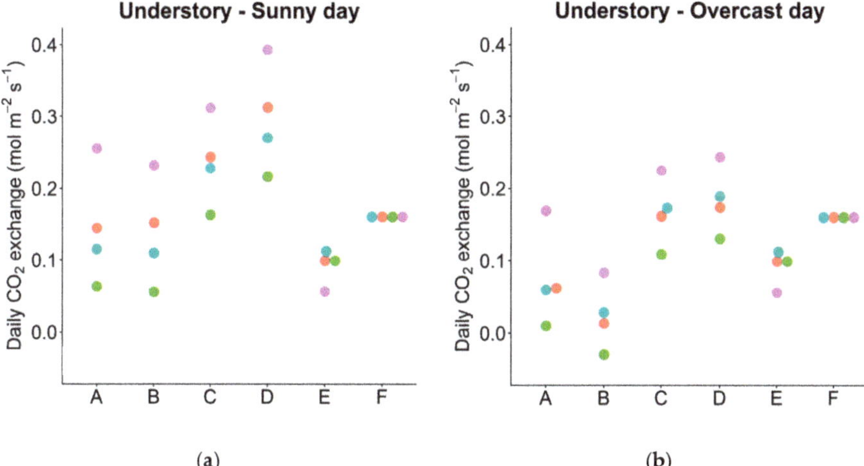

(a)   (b)

Figure 11. Daily carbon gain per unit area of leaf in the understory on (**a**) a sunny day (3 July 2020) and (**b**) an overcast day (24 June 2020). Results of the same leaf simulated in different scenarios appear in the same color. A: Estimated actual net photosynthesis. B: Simulated net photosynthesis when the leaves were hypothetically changed to clearing leaves. C: Estimated actual gross photosynthesis. D: Simulated gross photosynthesis when the leaves were changed to the clearing leaves. E: Estimated actual daily respiration. F: Simulated daily respiration when the leaves were changed to the clearing leaves. In simulation F, all four leaves were assumed to have the same respiration rate as the mean value of the clearing leaves.

3.7. Simulation: Understory without Sunflecks

We further examined whether the obtained differences could be explained by the effect of sunflecks. We simulated the net daily carbon gain in the understory on the same sunny day (July 3) under a hypothetical situation in which all sunflecks' PPFD values (>200 μmol m^{-2} s^{-1}) were replaced by a fixed value 200 μmol m^{-2} s^{-1}. Without sunflecks, the actual understory leaves indeed performed slightly better than the simulated clearing leaves in the understory (Figure 12; mean: (A) 0.134 and (B) 0.106 mol m^{-2} day^{-1}; A vs. B: $p < 0.01$).

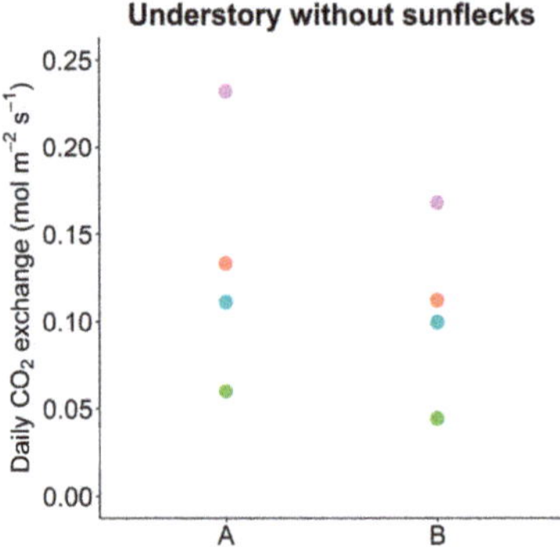

Figure 12. Simulated daily carbon gain without sunflecks on the sunny day (3 July 2020), in which sunflecks (>200 μmol m^{-2} s^{-1}) were replaced by the fixed value 200 μmol m^{-2} s^{-1}. A: Net daily photosynthesis calculated with actual leaves in the understory. B: Net daily photosynthesis when the leaves were hypothetically changed to the clearing leaves. The same leaf simulated under the two different scenarios appears as the same color.

4. Discussion

4.1. Carbon Gain or Saving via Acclimation

It is frequently discussed that photosynthetic acclimation increases daily net photosynthesis [3,6,17,24,48]. Supporting this theory, net daily carbon photosynthesis per unit leaf area in the clearing was higher for the actual sun leaves than for the simulated shade leaves (Figure 10). This was because during the daytime hours, PPFD on the leaves was always higher than the critical values, irrespective of the diurnal changes or weather (Figure 9a,c). By contrast, in the understory, our results did not always support the same theory; when sunflecks are present, photosynthetic shade acclimation may not always increase net daily photosynthesis. The understory leaves performed better than the clearing leaves in the understory on the overcast day, but not on the sunny day (Figure 11). Lower dark respiration rates in shade-acclimated leaves indeed resulted in lower LCPs (Table 1), but this does not necessarily imply that the shade leaves had higher net carbon gain than the sun leaves in the understory. In the understory, the simulated sun leaves frequently outperformed the shade leaves during the sunflecks (Figure 9b). When these sunflecks were hypothetically removed (Figure 12), or on the overcast day during which few sunflecks were observed (Figure 9d), such shade-acclimated understory leaves indeed had higher net carbon gain than the clearing leaves in the understory light environment. These results suggest that the observed difference was caused by sunflecks. Our results therefore support the notion [23] that under a diurnally fluctuating light environment, information on the static photosynthetic parameters and daily averaged light environment may not be sufficient to evaluate shade acclimation in forest understories. The implication may therefore be that laboratory experiments under controlled light [6,20] or field experiments using shade cloths [19,30,33], in which sunflecks were not taken into consideration, may not provide an accurate estimate of carbon gain in the understory. In this regard, it is possible to obtain better estimates through experiments using plants grown in natural conditions, as observed in [2,60,62] and the present study; those using natural canopy shading [18]; or those using advanced techniques that allow the rapid fluctuation of artificial light intensity in the case of a laboratory experiment [51,80].

For the present case, the observed reduced LMA in the understory can instead be interpreted as an effective cost-saving strategy [24,25,81] rather than as maximizing net daily photosynthesis in low-light environments. LMA was approximately twofold larger in the clearing than in the understory, whereas mass-based photosynthetic capacity ($P_{g\_max\_mass}$) and respiration rate (R_{mass}) were similar between the sites. Consequently, both light-saturated photosynthetic capacity and respiration rate per unit area of the leaves ($P_{g\_max\_area}$ and R_{area}) were approximately twofold higher in the clearing. Higher investment of photosynthetic apparatus per unit area results in a higher LMA and $P_{g\_max\_area}$ [24]. Additionally, a greater leaf thickness increases the internal surface area available for the diffusive transfer of CO_2 within a leaf [24,28,82–85]. Our results therefore confirm the findings of previous studies on other species that within-species variation in LMA and leaf thickness explain the variation of area-based photosynthetic traits across different light environments [4,28,29,34,35,45,86]. The lower LMA and lower photosynthetic capacity of shade-acclimated leaves incur a lower carbon cost [23–25,27,29,82] and lower nitrogen costs [31,37,41,42,87–89] to produce a unit area of leaf. In the case of this species, having lower LMA associated with lower $P_{g\_max\_area}$ in the understory may therefore have reduced leaf construction cost per unit area, in support of the cost-saving hypotheses [24,25,81]. Such reduced LMA, or equivalently, increased leaf area per unit mass (specific leaf area, SLA), increases whole-plant leaf area and light capture with a given amount of resources as a method of acclimation to low-light environments [5,6,18,19].

4.2. Morphological Acclimation

Leaf laminae in the shaded understory were flatter and therefore more horizontally displayed, whereas laminae in the clearing were more upright to decrease excessive irradiance and maximize leaf area per unit ground (Figure 5b). Similar changes in leaf three-dimensional structures (i.e., flatter in the shade) to maximize light capture in low light environments have been reported for a different species of a forest herb [55], for other within-canopy variation of lamina morphology for broadleaved trees [36,67,68], and for the three-dimensional arrangement of conifer needles [90–92]. This result is consistent with several previous findings that the leaves in well-lit places are more vertically upright, while leaves in shaded places are more horizontally displayed to maximize light capture [36,45–47,60]. However, we did not focus on the consequence of the morphological acclimation in the present study. As predicted by the game theory [93], leaf angle is determined not only on the basis of optimal light capture but also on the competition [93] and/or the contact [94] with neighboring plants. Further study is therefore needed to evaluate the consequence of morphological acclimation by taking the existing competition into consideration. Additionally, in the present study, we made a simplified assumption that the leaf 3D structure was approximated by a cone (Figure 1c). However, the shape of the actual leaves was much more complex and was trumpet-shaped; a lamina was more horizontal at the edge of each leaf and gradually was more vertical towards the center (see photographs in Figure 1). In the present study, we measured photosynthetic rate only at the edge of each leaf with the LI-6400 (Figure 3); environmental heterogeneity within a single leaf [95] was not investigated. More detailed studies that model complex 3D structure [68,95] are needed for this species.

Our study had several additional limitations. First, we examined only leaves. Although investigating leaves is equivalent to investigating the entire above-ground part of individual ramets for this species (Figure 1), the importance of the whole-plant carbon economy, including roots, has long been recognized [25]. Although leaf respiration rate is positively correlated with the respiration rates of roots [96] and the entire plant [97], further studies on whole-plant respiration rates [97,98] and whole-plant biomass allocation patterns [5,6,18,20,99] are needed for this species. Second, we ignored photosynthetic induction time and instead estimated the instantaneous photosynthetic rate using steady-state photosynthetic light response curves. Efficiency of photosynthesis may differ between steady-state and short-sunfleck conditions [49,56,58] due to stomatal [42,100–106], mesophyll [42], and biochemical [102,104,107] limitations. In our dataset, however, understory leaves frequently received sunflecks during the day (Figure 9b). Leaves of forest understory plants that are induced once maintain an induced condition for a relatively long time [49,58], so the magnitude of this overestimation might not be very large. Induction times were reported to be similar between shade-tolerant and shade-intolerant species [108]. Currently, however, there is little information on within-species differences in induction time between sunlit and shaded leaves. Third, the diurnal course of the photosynthetic rate depends not only on light but also on other environmental factors (i.e., humidity, temperature, VPD, etc.) [42,109,110] in addition to whole-plant water availability [4,111–115]. Therefore, the effect of midday depression due to stomatal closure [60,110,112,115] and photoinhibition [60,112] also would significantly alter the daily carbon gain of leaves. Furthermore, the strength of such effects may differ between sun and shade leaves [60,112,115]. Further detailed studies are therefore needed to reconfirm our findings before generation.

5. Conclusions

Petasites japonicus subsp. *giganteus* had a high capacity for acclimation to different light environments. In this species, having lower LMA associated with a lower photosynthetic rate in the understory did not increase net daily photosynthesis on the sunny day due to frequent sunflecks, but instead reduced construction costs per unit leaf area. These results indicate that when sunflecks were present, information on static leaf photosynthetic traits may not be sufficient to evaluate shade acclimation in forest understories.

Supplementary Materials: The following are available online at http://www.mdpi.com/1999-4907/11/12/1365/s1. Supplementary Materials: data and PPFD_and_P_net.

Author Contributions: Conceptualization, K.K.; methodology, K.K.; formal analysis, K.K.; investigation, R.D. and K.K.; writing—original draft preparation, R.D. and K.K; writing—review and editing, K.K.; supervision, K.K. All authors have read and agreed to the published version of the manuscript.

Funding: This work was funded by the Japan Society for the Promotion of Science (KAKENHI Grant Number 18K06406).

Acknowledgments: We thank Yasuyo Nagase and Miro Harada for performing preliminary studies. We thank staff members of Obihiro City Office and Obihiro Forest Hagukumu for permitting us to perform the fieldwork at the study sites.

Conflicts of Interest: The authors declare no conflict of interest.

Data Availability Statement: The datasets used in this article, including all the LI-6400 data, are available in the Supplementary Materials.

References

1. Walters, M.B.; Field, C.B. Photosynthetic light acclimation in two rainforest *Piper* species with different ecological amplitudes. *Oecologia* **1987**, *72*, 449–456. [CrossRef]
2. Ellsworth, D.S.; Reich, P.B. Leaf mass per area, nitrogen content and photosynthetic carbon gain in *Acer saccharum* seedlings in contrasting forest light environments. *Funct. Ecol.* **1992**, *6*, 423–435. [CrossRef]
3. Noda, H.; Muraoka, H.; Washitani, I. Morphological and physiological acclimation responses to contrasting light and water regimes in *Primula sieboldii*. *Ecol. Res.* **2004**, *19*, 331–340. [CrossRef]
4. Muraoka, H.; Tang, Y.; Koizumi, H.; Washitani, I. Combined effects of light and water availability on photosynthesis and growth of *Arisaema heterophyllum* in the forest understory and an open site. *Oecologia* **1997**, *112*, 26–34. [CrossRef]
5. Gruntman, M.; Segev, U.; Tielbörger, K.; Gange, A. Shade-induced plasticity in invasive *Impatiens glandulifera* populations. *Weed Res.* **2019**, *60*, 16–25. [CrossRef]
6. Muraoka, H.; Tang, Y.H.; Koizumi, H.; Washitani, I. Effects of light and soil water availability on leaf photosynthesis and growth of *Arisaema heterophyllum*, a riparian forest understorey plant. *J. Plant Res.* **2002**, *115*, 419–427. [CrossRef] [PubMed]
7. Wang, A.-Y.; Hao, G.-Y.; Guo, J.-J.; Liu, Z.-H.; Zhang, J.-L.; Cao, K.-F. Differentiation in leaf physiological traits related to shade and drought tolerance underlies contrasting adaptations of two *Cyclobalanopsis* (Fagaceae) species at the seedling stage. *Forests* **2020**, *11*, 844. [CrossRef]
8. Valladares, F.; Wright, S.J.; Lasso, E.; Kitajima, K.; Pearcy, R.W. Plastic phenotypic response to light of 16 congeneric shrubs from a panamanian rainforest. *Ecology* **2000**, *81*, 1925–1936. [CrossRef]
9. Niu, K.; Zhang, S.; Lechowicz, M.J.; Perez Carmona, C. Harsh environmental regimes increase the functional significance of intraspecific variation in plant communities. *Funct. Ecol.* **2020**, *34*, 1666–1677. [CrossRef]
10. Niu, K.; He, J.-S.; Lechowicz, M.J.; Souza, L. Grazing-induced shifts in community functional composition and soil nutrient availability in Tibetan alpine meadows. *J. Appl. Ecol.* **2016**, *53*, 1554–1564. [CrossRef]

11. Volf, M.; Redmond, C.; Albert, A.J.; Le Bagousse-Pinguet, Y.; Biella, P.; Gotzenberger, L.; Hrazsky, Z.; Janecek, S.; Klimesova, J.; Leps, J.; et al. Effects of long- and short-term management on the functional structure of meadows through species turnover and intraspecific trait variability. *Oecologia* **2016**, *180*, 941–950. [CrossRef] [PubMed]

12. Jung, V.; Violle, C.; Mondy, C.; Hoffmann, L.; Muller, S. Intraspecific variability and trait-based community assembly. *J. Ecol.* **2010**, *98*, 1134–1140. [CrossRef]

13. Enquist, B.J.; Norberg, J.; Bonser, S.P.; Violle, C.; Webb, C.T.; Henderson, A.; Sloat, L.L.; Savage, V.M. Scaling from traits to ecosystems. *Adv. Ecol. Res.* **2015**, *52*, 249–318. [CrossRef]

14. Violle, C.; Enquist, B.J.; McGill, B.J.; Jiang, L.; Albert, C.H.; Hulshof, C.; Jung, V.; Messier, J. The return of the variance: Intraspecific variability in community ecology. *Trends Ecol. Evol.* **2012**, *27*, 244–252. [CrossRef]

15. dos Santos, V.A.H.F.; Ferreira, M.J. Are photosynthetic leaf traits related to the first-year growth of tropical tree seedlings? A light-induced plasticity test in a secondary forest enrichment planting. *For. Ecol. Manag.* **2020**, *460*, 117900. [CrossRef]

16. Baltzer, J.L.; Thomas, S.C. Determinants of whole-plant light requirements in Bornean rain forest tree saplings. *J. Ecol.* **2007**, *95*, 1208–1221. [CrossRef]

17. Craine, J.M.; Reich, P.B. Leaf-level light compensation points in shade-tolerant woody seedlings. *New Phytol.* **2005**, *166*, 710–713. [CrossRef]

18. Ntawuhiganayo, E.B.; Uwizeye, F.K.; Zibera, E.; Dusenge, M.E.; Ziegler, C.; Ntirugulirwa, B.; Nsabimana, D.; Wallin, G.; Uddling, J. Traits controlling shade tolerance in tropical montane trees. *Tree Physiol.* **2020**, *40*, 183–197. [CrossRef]

19. Kitajima, K. Relative importance of photosynthetic traits and allocation patterns as correlates of seedling shade tolerance of 13 tropical trees. *Oecologia* **1994**, *98*, 419–428. [CrossRef]

20. Pons, T.L.; Poorter, H. The effect of irradiance on the carbon balance and tissue characteristics of five herbaceous species differing in shade-tolerance. *Front. Plant Sci.* **2014**, *5*, 12. [CrossRef]

21. Kupers, S.J.; Wirth, C.; Engelbrecht, B.M.J.; Hernández, A.; Condit, R.; Wright, S.J.; Rüger, N. Performance of tropical forest seedlings under shade and drought: An interspecific trade-off in demographic responses. *Sci. Rep.* **2019**, *9*, 18784. [CrossRef] [PubMed]

22. Bartholomew, D.C.; Bittencourt, P.R.L.; da Costa, A.C.L.; Banin, L.F.; de Britto Costa, P.; Coughlin, S.I.; Domingues, T.F.; Ferreira, L.V.; Giles, A.; Mencuccini, M.; et al. Small tropical forest trees have a greater capacity to adjust carbon metabolism to long-term drought than large canopy trees. *Plant Cell Environ.* **2020**, *43*, 2380–2393. [CrossRef] [PubMed]

23. Valladares, F.; Niinemets, Ü. Shade tolerance, a key plant feature of complex nature and consequences. *Annu. Rev. Ecol. Evol. S.* **2008**, *39*, 237–257. [CrossRef]

24. Björkman, O. Responses to different quantum flux densities. In *Physiological Plant Ecology I: Responses to the Physical Environment*; Lange, O.L., Nobel, P.S., Osmond, C.B., Ziegler, H., Eds.; Springer: Berlin/Heidelberg, Germany, 1981; pp. 57–107. [CrossRef]

25. Givnish, T.J. Adaptation to sun and shade: A whole-plant perspective. *Funct. Plant Biol.* **1988**, *15*, 63–92. [CrossRef]

26. Yoshimura, K. Irradiance heterogeneity within crown affects photosynthetic capacity and nitrogen distribution of leaves in *Cedrela sinensis*. *Plant Cell Environ.* **2010**, *33*, 750–758. [CrossRef]

27. Rozendaal, D.M.A.; Hurtado, V.H.; Poorter, L. Plasticity in leaf traits of 38 tropical tree species in response to light; relationships with light demand and adult stature. *Funct. Ecol.* **2006**, *20*, 207–216. [CrossRef]

28. Poorter, H.; Niinemets, U.; Ntagkas, N.; Siebenkas, A.; Maenpaa, M.; Matsubara, S.; Pons, T. A meta-analysis of plant responses to light intensity for 70 traits ranging from molecules to whole plant performance. *New Phytol.* **2019**, *223*, 1073–1105. [CrossRef]

29. Poorter, H.; Pepin, S.; Rijkers, T.; de Jong, Y.; Evans, J.R.; Körner, C. Construction costs, chemical composition and payback time of high- and low-irradiance leaves. *J. Exp. Bot.* **2006**, *57*, 355–371. [CrossRef]

30. Wan, Y.; Zhang, Y.; Zhang, M.; Hong, A.; Yang, H.; Liu, Y. Shade effects on growth, photosynthesis and chlorophyll fluorescence parameters of three *Paeonia species*. *PeerJ* **2020**, *8*, e9316. [CrossRef]

31. Vincent, G. Leaf photosynthetic capacity and nitrogen content adjustment to canopy openness in tropical forest tree seedlings. *J. Trop. Ecol.* **2001**, *17*, 495–509. [CrossRef]

32. Lei, T.T.; Tabuchi, R.; Kitao, M.; Koike, T. Functional relationship between chlorophyll content and leaf reflectance, and light-capturing efficiency of Japanese forest species. *Physiol. Plant.* **1996**, *96*, 411–418. [CrossRef]

33. Onoda, Y.; Schieving, F.; Anten, N.P. Effects of light and nutrient availability on leaf mechanical properties of Plantago major: A conceptual approach. *Ann. Bot.* **2008**, *101*, 727–736. [CrossRef] [PubMed]

34. Poorter, H.; Niinemets, U.; Poorter, L.; Wright, I.J.; Villar, R. Causes and consequences of variation in leaf mass per area (LMA): A meta-analysis. *New Phytol.* **2009**, *182*, 565–588. [CrossRef] [PubMed]

35. Ellsworth, D.S.; Reich, P.B. Canopy structure and vertical patterns of photosynthesis and related leaf traits in a deciduous forest. *Oecologia* **1993**, *96*, 169–178. [CrossRef] [PubMed]

36. Niinemets, Ü. Within-canopy variations in functional leaf traits: Structural, chemical and ecological controls and diversity of responses. In *Canopy Photosynthesis: From Basics to Applications*; Hikosaka, K., Niinemets, Ü., Anten, N.P.R., Eds.; Springer: Dordrecht, The Netherlands, 2016; pp. 101–141.

37. Anten, N.P.R.; During, H.J. Is analysing the nitrogen use at the plant canopy level a matter of choosing the right optimization criterion? *Oecologia* **2011**, *167*, 293–303. [CrossRef] [PubMed]

38. Koyama, K.; Kikuzawa, K. Geometrical similarity analysis of photosynthetic light response curves, light saturation and light use efficiency. *Oecologia* **2010**, *164*, 53–63. [CrossRef] [PubMed]

39. Koyama, K.; Kikuzawa, K. Is whole-plant photosynthetic rate proportional to leaf area? A test of scalings and a logistic equation by leaf demography census. *Am. Nat.* **2009**, *173*, 640–649. [CrossRef]

40. Koyama, K.; Kikuzawa, K. Reduction of photosynthesis before midday depression occurred: Leaf photosynthesis of *Fagus crenata* in a temperate forest in relation to canopy position and a number of days after rainfall. *Ecol. Res.* **2011**, *26*, 999–1006. [CrossRef]

41. Muryono, M.; Chen, C.P.; Sakai, H.; Tokida, T.; Hasegawa, T.; Usui, Y.; Nakamura, H.; Hikosaka, K. Nitrogen distribution in leaf canopies of high-yielding rice cultivar takanari. *Crop. Sci.* **2017**, *57*, 2080–2088. [CrossRef]

42. Campany, C.E.; Tjoelker, M.G.; Von Caemmerer, S.; Duursma, R.A. Coupled response of stomatal and mesophyll conductance to light enhances photosynthesis of shade leaves under sunflecks. *Plant Cell Environ.* **2016**, *39*, 2762–2773. [CrossRef]

43. Bhusal, N.; Han, S.-G.; Yoon, T.-M. Summer pruning and reflective film enhance fruit quality in excessively tall spindle apple trees. *Hortic. Environ. Biotechnol.* **2017**, *58*, 560–567. [CrossRef]

44. Kusi, J.; Karsai, I. Plastic leaf morphology in three species of Quercus: The more exposed leaves are smaller, more lobated and denser. *Plant Spec. Biol.* **2020**, *35*, 24–37. [CrossRef]

45. Hollinger, D.Y. Optimality and nitrogen allocation in a tree canopy. *Tree Physiol.* **1996**, *16*, 627–634. [CrossRef] [PubMed]

46. Posada, J.M.; Lechowicz, M.J.; Kitajima, K. Optimal photosynthetic use of light by tropical tree crowns achieved by adjustment of individual leaf angles and nitrogen content. *Ann. Bot.* **2009**, *103*, 795–805. [CrossRef] [PubMed]

47. Hagemeier, M.; Leuschner, C. Functional crown architecture of five temperate broadleaf tree species: Vertical gradients in leaf morphology, leaf angle, and leaf area density. *Forests* **2019**, *10*, 265. [CrossRef]

48. Walters, M.B.; Reich, P.B. Trade-offs in low-light CO_2 exchange: A component of variation in shade tolerance among cold temperate tree seedlings. *Funct. Ecol.* **2000**, *14*, 155–165. [CrossRef]

49. Chazdon, R.L.; Pearcy, R.W. Photosynthetic responses to light variation in rainforest species. i. induction under constant and fluctuating light conditions. *Oecologia* **1986**, *69*, 517–523. [CrossRef]

50. Chazdon, R.L. Sunflecks and their importance to forest understorey plants. In *Advances in Ecological Research*; Begon, M., Fitter, A.H., Ford, E.D., Macfadyen, A., Eds.; Academic Press: Cambridge, MA, USA, 1988; Volume 18, pp. 1–63.

51. Morales, A.; Kaiser, E. Photosynthetic acclimation to fluctuating irradiance in plants. *Front. Plant Sci.* **2020**, *11*, 268. [CrossRef]

52. Parker, G.G.; Fitzjarrald, D.R.; Gonçalves Sampaio, I.C. Consequences of environmental heterogeneity for the photosynthetic light environment of a tropical forest. *Agr. For. Meteorol.* **2019**, *278*, 107661. [CrossRef]

53. Hartikainen, S.M.; Pieristè, M.; Lassila, J.; Robson, T.M. Seasonal patterns in spectral irradiance and leaf UV-A absorbance under forest canopies. *Front. Plant Sci.* **2020**, *10*. [CrossRef]

54. Muraoka, H.; Koizumi, H.; Pearcy, R.W. Leaf display and photosynthesis of tree seedlings in a cool-temperate deciduous broadleaf forest understorey. *Oecologia* **2003**, *135*, 500–509. [CrossRef] [PubMed]

55. Muraoka, H.; Takenaka, A.; Tang, Y.; Koizumi, H.; Washitani, I. Flexible leaf orientations of *Arisaema heterophyllum* maximize light capture in a forest understorey and avoid excess irradiance at a deforested site. *Ann. Bot.* **1998**, *82*, 297–307. [CrossRef]

56. Kaiser, E.; Morales, A.; Harbinson, J. Fluctuating light takes crop photosynthesis on a rollercoaster ride. *Plant Physiol.* **2018**, *176*, 977–989. [CrossRef] [PubMed]

57. Slattery, R.A.; Walker, B.J.; Weber, A.P.M.; Ort, D.R. The impacts of fluctuating light on crop performance. *Plant Physiol.* **2018**, *176*, 990. [CrossRef]

58. Way, D.A.; Pearcy, R.W. Sunflecks in trees and forests: From photosynthetic physiology to global change biology. *Tree Physiol.* **2012**, *32*, 1066–1081. [CrossRef]

59. Miyashita, A.; Sugiura, D.; Sawakami, K.; Ichihashi, R.; Tani, T.; Tateno, M. Long-term, short-interval measurements of the frequency distributions of the photosynthetically active photon flux density and net assimilation rate of leaves in a cool-temperate forest. *Agric. For. Meteorol.* **2012**, *152*, 1–10. [CrossRef]

60. Valladares, F.; Pearcy, R.W. Drought can be more critical in the shade than in the sun: A field study of carbon gain and photo-inhibition in a Californian shrub during a dry El Nino year. *Plant Cell Environ.* **2002**, *25*, 749–759. [CrossRef]

61. Poorter, H.; Fiorani, F.; Pieruschka, R.; Wojciechowski, T.; van der Putten, W.H.; Kleyer, M.; Schurr, U.; Postma, J. Pampered inside, pestered outside? Differences and similarities between plants growing in controlled conditions and in the field. *New Phytol.* **2016**, *212*, 838–855. [CrossRef]

62. Valladares, F.; Zaragoza-Castells, J.; Sanchez-Gomez, D.; Matesanz, S.; Alonso, B.; Portsmuth, A.; Delgado, A.; Atkin, O.K. Is shade beneficial for Mediterranean shrubs experiencing periods of extreme drought and late-winter frosts? *Ann. Bot.* **2008**, *102*, 923–933. [CrossRef]

63. Ohashi, H.; Kadota, Y.; Murata, J.; Yonekura, K.; Kihara, H. *Wild Flowers of Japan*; Heibonsha: Tokyo, Japan, 2015.

64. Sok, D.E.; Oh, S.H.; Kim, Y.B.; Kang, H.G.; Kim, M.R. Neuroprotection by extract of *Petasites japonicus* leaves, a traditional vegetable, against oxidative stress in brain of mice challenged with kainic acid. *Eur. J. Nutr.* **2006**, *45*, 61–69. [CrossRef]

65. Xu, J.; Ji, F.; Cao, X.; Ma, J.; Ohizumi, Y.; Lee, D.; Guo, Y. Sesquiterpenoids from an edible plant *Petasites japonicus* and their promoting effects on neurite outgrowth. *J. Funct. Food.* **2016**, *22*, 291–299. [CrossRef]

66. Japan Meteorological Agency. Available online: http://www.jma.go.jp (accessed on 14 September 2020).

67. Fleck, S.; Niinemets, U.; Cescatti, A.; Tenhunen, J.D. Three-dimensional lamina architecture alters light-harvesting efficiency in Fagus: A leaf-scale analysis. *Tree Physiol.* **2003**, *23*, 577–589. [CrossRef] [PubMed]

68. Chambelland, J.-C.; Dassot, M.; Adam, B.; Donès, N.; Balandier, P.; Marquier, A.; Saudreau, M.; Sonohat, G.; Sinoquet, H. A double-digitising method for building 3D virtual trees with non-planar leaves: Application to the morphology and light-capture properties of young beech trees (*Fagus sylvatica*). *Funct. Plant Biol.* **2008**, *35*, 1059–1069. [CrossRef] [PubMed]

69. Johnson, I.R.; Thornley, J.H.M. A model of instantaneous and daily canopy photosynthesis. *J. Theor. Biol.* **1984**, *107*, 531–545. [CrossRef]

70. Liu, Y.Y.; Li, J.; Liu, S.C.; Yu, Q.; Tong, X.J.; Zhu, T.T.; Gao, X.X.; Yu, L.X. Sugarcane leaf photosynthetic light responses and their difference between varieties under high temperature stress. *Photosynthetica* **2020**, *58*, 1009–1018. [CrossRef]

71. Maxima.sourceforge.net. Maxima, a Computer Algebra System. Version 5.43.0 (2020). Available online: http://maxima.sourceforge.net/ (accessed on 14 September 2020).

72. Koyama, K.; Shirakawa, H.; Kikuzawa, K. Redeployment of shoots into better-lit positions within the crowns of saplings of five species with different growth patterns. *Forests* **2020**, *11*, 1301. [CrossRef]

73. Schneider, C.A.; Rasband, W.S.; Eliceiri, K.W. NIH Image to ImageJ: 25 years of image analysis. *Nat. Methods* **2012**, *9*, 671. [CrossRef]

74. John, G.P.; Scoffoni, C.; Buckley, T.N.; Villar, R.; Poorter, H.; Sack, L. The anatomical and compositional basis of leaf mass per area. *Ecol. Lett.* **2017**, *20*, 412–425. [CrossRef]

75. R Core Team. *R: A Language and Environment for Statistical Computing. R Foundation for Statistical Computing*; R Core Team: Vienna, Austria, 2020.

76. Wilke, C.O. *cowplot*: Streamlined Plot Theme and Plot Annotations for 'ggplot2'. *CRAN Repos.* **2016**, *2*, R2.

77. Clarke, E.; Sherrill-Mix, S. *ggbeeswarm*: Categorical Scatter (Violin Point) Plots. 2017. Available online: https://cran.R-project.org (accessed on 14 September 2020).

78. Wickham, H. *ggplot2: Elegant Graphics for Data Analysis*; Springer: New York, NY, USA, 2016.

79. Bates, D.; Mächler, M.; Bolker, B.; Walker, S. Fitting linear mixed-effects models using lme4. *J. Stat. Softw.* **2015**, *67*, 48. [CrossRef]

80. Pattison, P.M.; Tsao, J.Y.; Brainard, G.C.; Bugbee, B. LEDs for photons, physiology and food. *Nature* **2018**, *563*, 493–500. [CrossRef] [PubMed]

81. Tateno, M.; Taneda, H. Photosynthetically versatile thin shade leaves: A paradox of irradiance-response curves. *Photosynthetica* **2007**, *45*, 299–302. [CrossRef]

82. Oguchi, R.; Hikosaka, K.; Hiura, T.; Hirose, T. Costs and benefits of photosynthetic light acclimation by tree seedlings in response to gap formation. *Oecologia* **2008**, *155*, 665–675. [CrossRef] [PubMed]

83. Oguchi, R.; Onoda, Y.; Terashima, I.; Tholen, D. Leaf Anatomy and Function. In *The Leaf: A Platform for Performing Photosynthesis*; Adams Iii, W.W., Terashima, I., Eds.; Springer International Publishing: Cham, Switzerland, 2018; pp. 97–139. [CrossRef]

84. Terashima, I.; Miyazawa, S.-I.; Hanba, Y.T. Why are sun leaves thicker than shade leaves?—Consideration based on analyses of CO_2 diffusion in the leaf. *J. Plant Res.* **2001**, *114*, 93–105. [CrossRef]

85. Bhusal, N.; Bhusal, S.J.; Yoon, T.-M. Comparisons of physiological and anatomical characteristics between two cultivars in bi-leader apple trees (*Malus* × *domestica* Borkh.). *Sci. Hortic.* **2018**, *231*, 73–81. [CrossRef]

86. Wyka, T.P.; Oleksyn, J.; Żytkowiak, R.; Karolewski, P.; Jagodziński, A.M.; Reich, P.B. Responses of leaf structure and photosynthetic properties to intra-canopy light gradients: A common garden test with four broadleaf deciduous angiosperm and seven evergreen conifer tree species. *Oecologia* **2012**, *170*, 11–24. [CrossRef]

87. Mooney, H.A.; Gulmon, S.L. Environmental and evolutionary constraints on the photosynthetic characteristics of higher plants. In *Topics in Plant Population Biology*; Solbrig, O.T., Jain, S., Johnson, G.B., Raven, P.H., Eds.; Macmillan Education: London, UK, 1979; pp. 316–337.

88. Tanaka, T.; Oikawa, S.; Kurokawa, C. Leaf shedding increases the photosynthetic rate of the canopy in N_2-fixing and non-N_2-fixing woody species. *Tree Physiol.* **2018**, *38*, 1903–1911. [CrossRef]

89. Oikawa, S.; Hikosaka, K.; Hirose, T. Does leaf shedding increase the whole-plant carbon gain despite some nitrogen being lost with shedding? *New Phytol.* **2008**, *178*, 617–624. [CrossRef]

90. Kubínová, Z.; Janáček, J.; Lhotáková, Z.; Šprtová, M.; Kubínová, L.; Albrechtová, J. Norway spruce needle size and cross section shape variability induced by irradiance on a macro- and microscale and CO_2 concentration. *Trees* **2017**, *32*, 231–244. [CrossRef]

91. Dörken, V.M.; Lepetit, B. Morpho-anatomical and physiological differences between sun and shade leaves in *Abies alba* Mill. (Pinaceae, Coniferales): A combined approach. *Plant Cell Environ.* **2018**, *41*, 1683–1697. [CrossRef]

92. Niinemets, Ü.; Tobias, M.; Cescatti, A.; Sparrow, A. Size-dependent variation in shoot light-harvesting efficiency in shade-intolerant conifers. *Int. J. Plant Sci.* **2006**, *167*, 19–32. [CrossRef]

93. Anten, N.P.R. Optimization and game theory in canopy models. In *Canopy Photosynthesis: From Basics to Applications*; Hikosaka, K., Niinemets, Ü, Anten, N.P.R., Eds.; Springer: Dordrecht, The Netherlands, 2016; pp. 355–377.

94. De Wit, M.; Kegge, W.; Evers, J.B.; Vergeer-van Eijk, M.H.; Gankema, P.; Voesenek, L.A.; Pierik, R. Plant neighbor detection through touching leaf tips precedes phytochrome signals. *Proc. Natl. Acad. Sci. USA* **2012**, *109*, 14705–14710. [CrossRef] [PubMed]

95. Saudreau, M.; Ezanic, A.; Adam, B.; Caillon, R.; Walser, P.; Pincebourde, S. Temperature heterogeneity over leaf surfaces: The contribution of the lamina microtopography. *Plant Cell Environ.* **2017**, *40*, 2174–2188. [CrossRef] [PubMed]

96. Reich, P.B.; Walters, M.B.; Tjoelker, M.G.; Vanderklein, D.; Buschena, C. Photosynthesis and respiration rates depend on leaf and root morphology and nitrogen concentration in nine boreal tree species differing in relative growth rate. *Funct. Ecol.* **1998**, *12*, 395–405. [CrossRef]

97. Walters, M.B.; Reich, P.B. Low-light carbon balance and shade tolerance in the seedlings of woody plants: Do winter deciduous and broad-leaved evergreen species differ? *New Phytol.* **1999**, *143*, 143–154. [CrossRef]

98. Kurosawa, Y.; Mori, S.; Wang, M.; Ferrio, J.P.; Yamaji, K.; Koyama, K.; Haruma, T.; Doyama, K. Initial burst of root development with decreasing respiratory carbon cost in *Fagus crenata* Blume seedlings. *Plant Spec. Biol.* **2020**. [CrossRef]

99. Miyashita, A.; Tateno, M. A novel index of leaf RGR predicts tree shade tolerance. *Funct. Ecol.* **2014**, *28*, 1321–1329. [CrossRef]

100. Tatsumi, K.; Kuwabara, Y.; Motobayashi, T. Photosynthetic light-use efficiency of rice leaves under fluctuating incident light. *Agrosyst. Geosci. Environ.* **2020**, *3*, e20030. [CrossRef]

101. Matthews, J.S.A.; Vialet-Chabrand, S.; Lawson, T. Role of blue and red light in stomatal dynamic behaviour. *J. Exp. Bot.* **2020**, *71*, 2253–2269. [CrossRef]

102. Zhang, S.B.; Hao, Y.J.; Deng, Q.L. Photosynthetic induction is slower in young leaves than in mature leaves in a tropical invader, *Chromolaena Odorata*. *Photosynth.* **2019**, *57*, 1044–1052. [CrossRef]

103. Kimura, H.; Hashimoto-Sugimoto, M.; Iba, K.; Terashima, I.; Yamori, W. Improved stomatal opening enhances photosynthetic rate and biomass production in fluctuating light. *J. Exp. Bot.* **2020**, *71*, 2339–2350. [CrossRef] [PubMed]

104. Sakoda, K.; Yamori, W.; Groszmann, M.; Evans, J.R. Stomatal, mesophyll conductance, and biochemical limitations to photosynthesis during induction. *Plant Physiol.* **2020**. [CrossRef]

105. Shimadzu, S.; Seo, M.; Terashima, I.; Yamori, W. Whole irradiated plant leaves showed faster photosynthetic induction than individually irradiated leaves via improved stomatal opening. *Front. Plant Sci.* **2019**, *10*. [CrossRef] [PubMed]

106. Yamori, W.; Kusumi, K.; Iba, K.; Terashima, I. Increased stomatal conductance induces rapid changes to photosynthetic rate in response to naturally fluctuating light conditions in rice. *Plant Cell Environ.* **2020**, *43*, 1230–1240. [CrossRef] [PubMed]

107. Alter, P.; Dreissen, A.; Luo, F.L.; Matsubara, S. Acclimatory responses of *Arabidopsis* to fluctuating light environment: Comparison of different sunfleck regimes and accessions. *Photosynth. Res.* **2012**, *113*, 221–237. [CrossRef]

108. Naumburg, E.; Ellsworth, D.S. Photosynthetic sunfleck utilization potential of understory saplings growing under elevated CO_2 in FACE. *Oecologia* **2000**, *122*, 163–174. [CrossRef]

109. Duursma, R.A.; Payton, P.; Bange, M.P.; Broughton, K.J.; Smith, R.A.; Medlyn, B.E.; Tissue, D.T. Near-optimal response of instantaneous transpiration efficiency to vapour pressure deficit, temperature and CO_2 in cotton (*Gossypium hirsutum* L.). *Agric. For. Meteorol.* **2013**, *168*, 168–176. [CrossRef]

110. Koyama, K.; Takemoto, S. Morning reduction of photosynthetic capacity before midday depression. *Sci. Rep.* **2014**, *4*, 4389. [CrossRef]

111. Scoffoni, C.; Sack, L.; Ort, D. The causes and consequences of leaf hydraulic decline with dehydration. *J. Exp. Bot.* **2017**, *68*, 4479–4496. [CrossRef]

112. Muraoka, H.; Tang, Y.H.; Terashima, I.; Koizumi, H.; Washitani, I. Contributions of diffusional limitation, photoinhibition and photorespiration to midday depression of photosynthesis in *Arisaema heterophyllum* in natural high light. *Plant Cell Environ.* **2000**, *23*, 235–250. [CrossRef]

113. Espadafor, M.; Orgaz, F.; Testi, L.; Lorite, I.J.; González-Dugo, V.; Fereres, E. Responses of transpiration and transpiration efficiency of almond trees to moderate water deficits. *Sci. Hortic.* **2017**, *225*, 6–14. [CrossRef]

114. Paillassa, J.; Wright, I.J.; Prentice, I.C.; Pepin, S.; Smith, N.G.; Ethier, G.; Westerband, A.C.; Lamarque, L.J.; Wang, H.; Cornwell, W.K.; et al. When and where soil is important to modify the carbon and water economy of leaves. *New Phytol.* **2020**, *228*, 121–135. [CrossRef] [PubMed]

115. Bhusal, N.; Lee, M.; Reum Han, A.; Han, A.; Kim, H.S. Responses to drought stress in *Prunus sargentii* and *Larix kaempferi* seedlings using morphological and physiological parameters. *For. Ecol. Manag.* **2020**, *465*, 118099. [CrossRef]

Publisher's Note: MDPI stays neutral with regard to jurisdictional claims in published maps and institutional affiliations.

 *forests*

Article

Redeployment of Shoots into Better-Lit Positions within the Crowns of Saplings of Five Species with Different Growth Patterns

Kohei Koyama [1,2,*,†], **Hiroyuki Shirakawa** [2,†,‡] **and Kihachiro Kikuzawa** [2]

[1]   Department of Life Science and Agriculture, Obihiro University of Agriculture and Veterinary Medicine, Inada-cho, Obihiro, Hokkaido 080-8555, Japan

[2]   Laboratory of Forest Biology, Graduate School of Agriculture, Kyoto University, Kyoto 606-8502, Japan; kkikuzawa@kyoto.zaq.jp

*   Correspondence: koyama@obihiro.ac.jp

†   These authors made equal contributions.

‡   Present address: Japan Broadcasting Corporation (NHK), Tokyo 150-8001, Japan.

Received: 16 August 2020; Accepted: 30 November 2020; Published: 3 December 2020

Abstract: Research Highlights: We demonstrate the first quantitative evidence that the shoot shedding of fast-growing species growing in a high-light environment is part of the process of shoot redeployment into better-lit outer parts of the crown. Background and Objectives: Light foraging by redeploying organs from shaded regions of a tree crown into better-lit regions is considered to apply to both leaves and shoots. To date, however, this hypothesis has never been tested for shoots. Materials and Methods: We investigated the shoot dynamics of saplings of five deciduous woody species. We included fast-growing and slow-growing species (*Alnus sieboldiana* Matsum., *Castanea crenata* Siebold & Zucc., *Betula ermanii* Cham., *Acer distylum* Siebold & Zucc., and *Fagus crenata* Blume). Results: Shoots in the shaded regions of the crowns of the fast-growing trees showed higher mortality rates than those at better-lit positions. Because of the selective shedding of the shaded shoots, at the end of the growth period the light environment experienced by the shoots that survived until the following spring was similar to that at the early stage of the same growth period. By contrast, the slow-growing trees displayed slow and determinate growth, with a very low mortality rate of shoots at all positions in the crown. Conclusions: The rapid shoot turnover of the fast-growing species resulted in the redeployment of shoots into better-lit positions within the tree crown in a manner similar to the redeployment of leaves.

Keywords: branch lifespan; shoot lifespan; stem lifespan; branch shedding; shoot shedding; stem shedding; canopy; crown development; tree architecture; light foraging

---

1. Introduction

A plant canopy is a dynamically changing system due to the continuous production, growth, and death of leaves and shoots [1–15]. The light environment within a canopy therefore changes temporally during plant growth [1,9,10,14,16]. Although studies on static plant form and function, such as allometric scaling approaches [17–24], are useful for analyzing time-averaged plant characteristics, investigations of plant organ dynamics are essential to understanding the development of individual plants and stands.

The dynamics of the aboveground parts of plants are driven by three processes: the production, growth, and death of leaves and shoots. Plants in different light environments differ in their shoot dynamics. In low-light environments (e.g., forest understories), plants may accelerate the shedding of their shaded, lower-positioned leaves or shoots because of their negative carbon balances [1] (but see [6]). For plants in open habitats, the continuous production of leaves into better-lit, higher positions in a canopy is an essential strategy to outcompete neighbors [25–27]. New leaves in the higher or outer positions of a canopy shade the leaves in the lower or inner part of the canopy [14,28–30]. Senescence and death of the shaded leaves of plants in high-light environments are a part of organ redeployment into better-lit positions [9,27,28,31–39]. Light foraging by organ redeployment from shaded regions into better-lit regions of a tree crown is considered to apply to shoots in addition to leaves [40,41]. To the best of our knowledge, however, no empirical study has tested the applicability of this hypothesis to shoots.

The temporal changes of a crown are determined by two-way interactions between the light environment and the dynamics of the leaves and shoots; light affects the dynamics (i.e., production, growth, and death) of organs, and the dynamics of organs affect the environment [1,4,5,8,10,14,42]. Shoot production rates are higher and shoot mortality rates are lower in better-lit, outer parts than in shaded, inner parts of the crown [13,40,43]. Furthermore, shoot growth and death depend on whether the shoots are shaded relative to other shoots throughout the entire crown, a phenomenon known as correlative inhibition [41,44–47]. Although these previous studies have successfully elucidated how the light environment affects shoot dynamics, they have clarified only the one-way effects of environmental factors on shoot dynamics. By contrast, only a few empirical studies (e.g., [6]) have quantified the effect of shoot dynamics on light environments, despite the importance of these effects having long been recognized in theoretical studies [1,4,5,8]. Therefore, further empirical studies of this two-way interaction are needed.

In this study, we investigated the two-way interactions between the within-crown light environment and shoot dynamics of saplings of five temperate, winter-deciduous woody plants that included fast- and slow-growing species grown in open, well-lit places. We tested the hypothesis that light foraging by organ redeployment from shaded regions into better-lit regions of a canopy, an idea well supported for leaves, also applies to shoots. We also expected that the interaction between shoot dynamics and light environments would be greater for fast-growing species than for slow-growing species because of their difference in shoot production and survival rates.

2. Materials and Methods

2.1. Study Species

We investigated saplings of five temperate, winter-deciduous tree species, all of which are native to Japan. The alder (*Alnus sieboldiana* Matsum.) is a nitrogen-fixing [48] pioneer species that grows in open habitats after disturbances [49]. Its higher photosynthetic rates than those of typical slow-growing species [50] indicate that this is a fast-growing species because the photosynthetic rates of leaves and whole-tree growth rates are positively correlated [51–53]. The Japanese chestnut (*Castanea crenata* Siebold & Zucc.) is an early-successional species that regenerates in sunny places [54]. In this study, we classified *A. sieboldiana* and *C. crenata* as fast-growing species based on their growth rates (see Results). The Russian rock birch (*Betula ermanii* Cham.) is an early-successional species, and its leaves have higher photosynthetic rates than those of typical late-successional species [55]. Thus, we classified *B. ermanii* as a medium-growth-rate species. *Acer distylum* Siebold & Zucc. is a shade-tolerant sub-canopy species [56] found in climax beech forests [57]. Although it can tolerate a shaded understory, this species also can grow rapidly in gaps [58]. Siebold's beech (*Fagus crenata* Blume) is a shade-tolerant [59] and late-successional [55] species that dominates in cool-temperate climax forests in Japan [60]. The leaves of this species have lower photosynthetic rates than those of typical early-successional species [55,61–64]. *A. distylum* and *F. crenata* were classified herein as slow-growing species.

2.2. Study Sites and Sample Trees

The current study was conducted from 2002 to 2003 at two study sites in Japan. To ensure that the number of current-year shoots measured for each species exceeded 300, which we believed to be sufficient to describe species-specific patterns, we examined three to six saplings of each species (Table 1). Saplings of *A. sieboldiana*, *C. crenata*, and *F. crenata* were investigated at the experimental garden of, or a plantation adjacent to, the Center for Ecological Research (CER), Kyoto University (34°58′ N, 135°57′ E, 150 m asl). The mean annual temperature and precipitation at this location are 14.6 °C and 1618 mm, respectively (years 2006–2010; data from CER). Saplings of *B. ermanii* and *A. distylum* were investigated in a nursery at the Kamigamo Experimental Forest Station (KEFS), Kyoto University (35°04′ N, 135°46′ E, 140 m asl), which is approximately 21 km from CER. The mean annual temperature and precipitation at KEFS are 14.6 °C and 1580 mm, respectively (years 1971–2000; data from KEFS). Therefore, the two study sites had very similar temperatures and precipitation. All of the saplings investigated in the study were grown in open, well-lit places.

Table 1. Sample sizes of current-year shoots from the saplings of different species of trees.

	Alnus sieboldiana	*Castanea crenata*	*Betula ermanii*	*Acer distylum*	*Fagus crenata*
No. of saplings	3	3	3	6	4
Tree height (m)	0.81–1.03	0.95–1.13	0.61–0.81	0.86–1.72	0.81–1.02
No. of shoots for the shoot census [1]	581	425	318	325	343
No. of shoots for the leaf census [1]	189	251	259	288	339

[1] Total number of current-year shoots from all saplings of each species that we investigated.

2.3. Shoot Census

For each sapling, we identified and periodically investigated all the current-year shoots that appeared in 2002 (Table 1, Figure 1). In 2002, we measured the length of each shoot after it had completely elongated. Shoot lengths (i.e., the annual shoot elongation) and the number of vegetative buds (i.e., the production of daughter shoots) were used as indices of the growth rates of each shoot. For *A. sieboldiana*, *C. crenata*, and *A. distylum*, we observed sylleptic shoots: lateral buds that grew from current-year shoots within the same year. All of these were regarded as part of their parent (proleptic) shoots, and the lengths and numbers of vegetative buds of the sylleptic shoots were summed with those of their parent to calculate the annual shoot elongation and bud production of the parent shoot. In April 2003, we evaluated the survival of each shoot. A shoot was defined as alive if it had at least one living bud (vegetative or flower bud). A shoot was defined as dead if it had been shed before April 2003 or if it lacked any living bud in April 2003.

(**a**)   (**b**)

Figure 1. *Cont.*

Figure 1. Young shoots of (**a**) *Alnus sieboldiana*, (**b**) *Castanea crenata*, (**c**) *Betula ermanii*, (**d**) *Acer distylum*, and (**e**) *Fagus crenata* (photographs captured in April 2002 by Hiroyuki Shirakawa).

2.4. Leaf Census

From bud break (late March 2002) to completion of leaf fall (late January 2003), we periodically investigated all leaves of the selected shoots of each species (Table 1) and recorded the numbers of leaves that appeared from these shoots every one or two weeks. The date of leaf emergence was postulated to be the midpoint between the date of a previous observation and the date that a new leaf was observed [65]. For each shoot, we calculated the shoot elongation period as the time between the emergence of the first and last leaf (i.e., the leaf emergence period calculated for each shoot). All leaves on the monitored shoots of *F. crenata* emerged as a flush within seven days between two successive census days (8 and 15 April 2002). The shoot elongation period of *F. crenata* was therefore estimated at 3.5 days.

2.5. Measurement of Light Environments

We measured the light environment within each crown for all the current-year shoots of the studied saplings. For the two fast-growing species (*A. sieboldiana* and *C. crenata*), which had indeterminate growth patterns, we investigated seasonal changes of the light environment within each canopy by conducting the same measurements twice during the same growth period in 2002 for all the shoots. The first measurement was taken in June or July, when almost all of the current-year proleptic shoots had emerged, but their elongation (including elongation of their sylleptic shoots) was not yet complete. The second measurement was conducted in September, when the elongation of almost all of the shoots was either complete or nearly so. For the other three medium- or slow-growing species, the measurements were conducted once, between June and September, when the elongation of most of the shoots was complete or nearly so (the dates of the measurements for each species are shown in the Results). We conducted the measurements from 11:00 to 13:00 local time on overcast days. On each measurement day, we used two quantum sensors (IKSX-7/101; Koito Manufacturing, Tokyo, Japan) to measure the photosynthetic photon flux density (PPFD) simultaneously. One sensor was fixed horizontally to a pole at the top of each tree. Light conditions at the tops of the trees in the open experimental gardens were nearly the same as those in the open full-lit location. Another sensor was fixed horizontally at the top of each shoot [46,66] with a hand-held measuring bar, as described in Muraoka et al. [67,68] (Figure 2). The investigators took care so as to not shade the sensor. We calculated the relative PPFD (rPPFD) as the ratio of the PPFD at the tip of each shoot to the PPFD at the top of the tree [46]. A previous study demonstrated that rPPFD measured on the horizontal surface at the tip of a branch represents the average light environment of that branch because it is highly correlated with the mean rPPFD of distal-to-proximal secondary shoots on that branch [46]. Therefore, we used the rPPFD at the tip of each shoot as a representative value for the average light environment of that shoot.

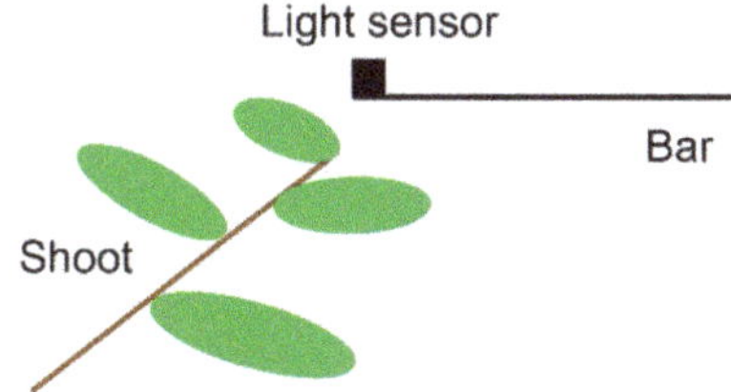

Figure 2. A schematic diagram demonstrating the measurement of light environments of different shoots in this study. A sensor was fixed horizontally with a hand-held measuring bar. (Illustration by Kohei Koyama).

Some sylleptic shoots emerged after the first measurements (June–July) and were considered part of their parent proleptic shoots. Therefore, if a proleptic shoot had sylleptic shoots, the continuum of proleptic–sylleptic shoots of the same age (i.e., those shoots that emerged within the same growth period) was considered as one current-year shoot. In that case, the top of that current-year shoot was measured in the second measurement. In the case of a shoot that had been shed prior to the second measurements, we measured the PPFD at the approximate prior position of the shoot tip (estimated from its length).

We analyzed the effect of rPPFD that was obtained at the first measurement on each shoot for each species on the subsequent survival of that shoot in April 2003 using a generalized linear mixed model (GLMM) with the function *glmer* (binomial (link = "logit")) [69] and R software v4.0.3 (Vienna, Austria) [70]:

$$\ln\left(\frac{y}{1-y}\right) = a + bx + r \Leftrightarrow y = \frac{\exp(a+bx+r)}{1+\exp(a+bx+r)} \Leftrightarrow y = \frac{1}{1+\exp(-a-bx-r)}, \tag{1}$$

where y is the survival of that shoot in April 2003 (0 = dead, 1 = survived), x is the fixed effect (rPPFD), a and b are the coefficients estimated by the *glmer*, and r is the random intercept (individual tree). The significance of the fixed effect of rPPFD (coefficient b) was tested by a likelihood ratio test with the R function *anova* (test = "Chisq") by comparing Equation (1) with the null model with no rPPFD dependence ($b = 0$).

As mentioned, for the two fast-growing species, we conducted the PPFD measurements twice (i.e., the early (June or July) and the late (September) stages) in 2002 for the same set of shoots. The survival of the shoots was monitored until April 2003. The shoots were then assigned to one of two groups: (1) survivor group, those that survived until April 2003, or (2) dead group, those that had died by April 2003. We performed two types of comparisons (A and B) as follows. (A) Using all of the shoots from the survivor and dead groups, we compared the early versus late environments within the same growth period in 2002 within each tree crown. We expected that because of the rapid and indeterminate growth of the fast-growing species, the later-stage shoots would experience an inferior light environment compared with the early-stage shoots. This expected result would indicate that rapid growth causes an increment of self-shading. (B) We compared all the shoots at the early stage with the rPPFD of the survivor group shoots at the late stage. The survivor group shoots at the late stage were considered to be the shoots for the next year. Finally, we compared the difference in the results between (A) and (B). We hypothesized that any difference between (A) and (B) would indicate an effect of shoot shedding on the within-crown light environment. We expected that shoot deaths would counter the development of self-shading by the selective shedding of shaded shoots. For these two types of comparisons, we tested the differences in the two distributions of rPPFD (early versus late) with two-sample Kolmogorov–Smirnov tests using R software. All of the datasets used in this article are available online as Supplementary Materials.

3. Results

3.1. Growth Rate and Pattern

Table 2 summarizes the results of the growth and survival rates of the shoots. The shoot elongation period was longest for the fastest-growing species (*A. sieboldiana*), which showed typical indeterminate growth and successive emergence of leaves (Figure 3). By contrast, the elongation period was shortest for the slowest-growing species (*F. crenata*), which displayed typical determinate growth and a flush of leaf emergence during a very short time. The remaining species showed intermediate patterns. *C. crenata* experienced three flushes of leaf emergence, while *B. ermanii* and *A. distylum* showed one flush of emergence of early leaves and the successive emergence of late leaves.

Table 2. Growth and survival rates of the shoots.

	A. sieboldiana	C. crenata	B. ermanii	A. distylum	F. crenata
Species category	Fast	Fast	Medium	Slow	Slow
Annual shoot elongation [1,2] (cm year^{-1})	26.9	8.0	4.4	1.5	2.2
Shoot elongation period (days) [1]	156	33	18	8	3.5
Number of daughter buds per shoot [1,2,3]	5.1	2.7	1.9	1.1	1.1
Shoot survival rate	0.54	0.78	0.88	0.93	0.98

[1] Mean values for each species. [2] The lengths of sylleptic shoots and the number of their vegetative buds were summed with those of their parent proleptic shoots. [3] Shoots without vegetative buds were not included.

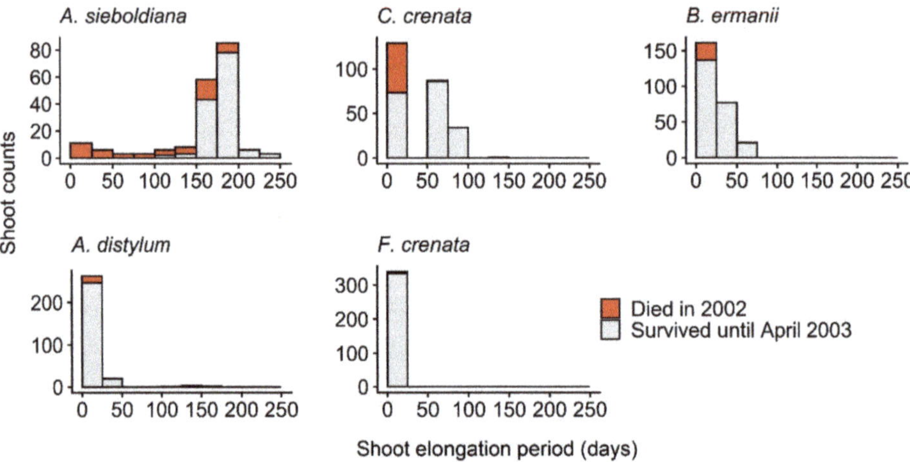

Figure 3. Histogram of shoot elongation period. The shoot elongation period is defined as the time interval between the emergence of the first and last leaves.

3.2. Shoot Survival

The shoot survival rate was lower for the fast-growing species than for the slow-growing species (Table 2). The shoots in the shaded part of the canopy of the two fast-growing species (*A. sieboldiana* and *C. crenata*) experienced a high mortality rate (Figure 4, Table 2). By contrast, the shoots of the slow-growing species (*A. distylum* and *F. crenata*) experienced a low mortality rate at all positions within the canopy, regardless of the light environment (Figure 4, Table 2). The mortality rate of the shoots of the medium-growth-rate species (*B. ermanii*) was intermediate between these two extremes (Figure 4, Table 2). Logistic regression analysis demonstrated that those shoots located in a low-rPPFD environment were significantly less likely to survive than shoots located in a high-rPPFD environment in fast-growing (*A. sieboldiana* and *C. crenata*, $p < 0.001$) and medium-growth-rate species (*B. ermanii*, $p < 0.001$), but not in slow-growing species (*A. distylum*, $p = 0.54$; *F. crenata*, $p = 0.45$).

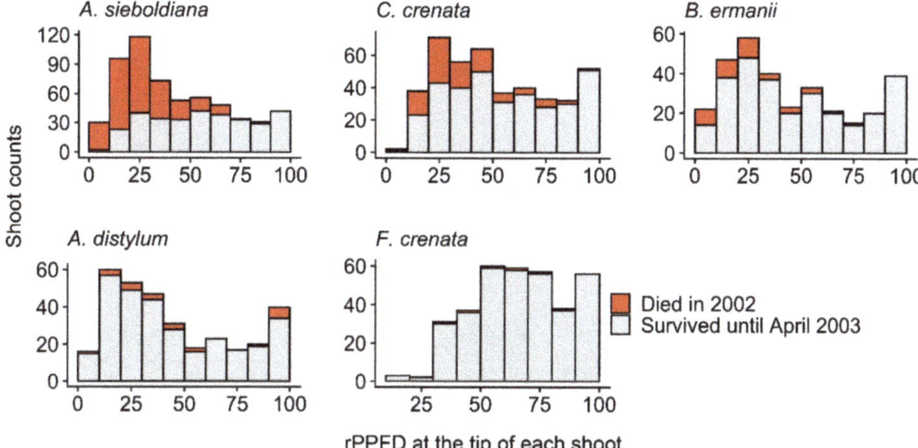

Figure 4. Relative photosynthetic photon flux density (rPPFD) at the shoot tips in summer and autumn in 2002 and number of surviving or dead shoots in April 2003. Dates of rPPFD measurements in 2002 are as follows: *A. sieboldiana*, 15–24 June; *C. crenata*, 19 July; *B. ermanii*, 17 June; *A. distylum*, 4 August–1 September; *F. crenata*, 27 August–26 September.

3.3. Light Environment

The long shoot elongation periods observed for the two fast-growing species caused the development of self-shading during a single growth period. Shoots of the two fast-growing species (*A. sieboldiana* and *C. crenata*) experienced significantly more self-shading in the late stage than the early stage of the same growth period ($p < 0.001$; Figure 5a). Nonetheless, the light environment in the later stage of growth experienced by the shoots that survived until April 2003 (i.e., the shoots for the next year) was similar to the one experienced by all the shoots in the early stage (Figure 5b). For both of the fast-growing species, the light distributions during the two stages did not differ significantly (*A. sieboldiana*, $p = 0.48$; *C. crenata*, $p = 0.10$; Figure 5b).

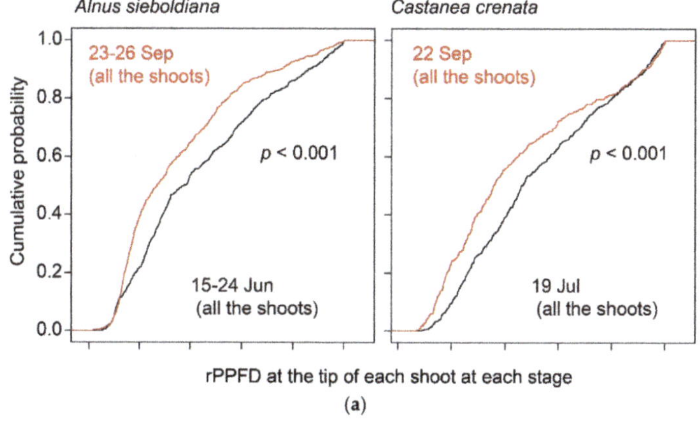

Figure 5. *Cont.*

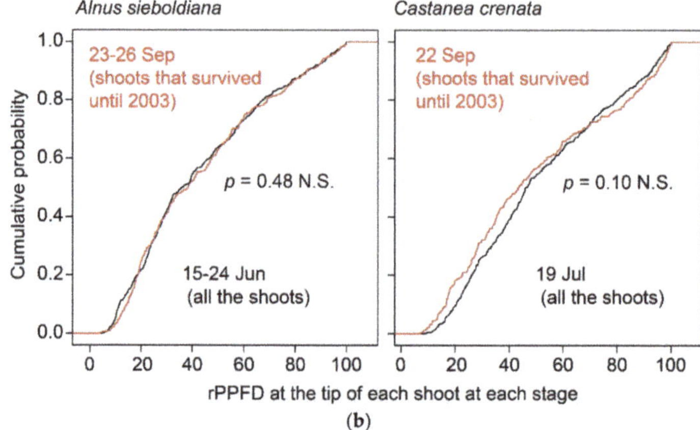

Figure 5. Comparison of the light environments between the first (early stage of the growth period, black curves) and the second (late stage, red curves) measurements illustrated by empirical cumulative distribution functions. (**a**) All the shoots were analyzed for both stages. (**b**) All the shoots were analyzed for the early stage, and only the shoots that survived until April 2003 were analyzed for the late stage. *p*-values are the results of two-sample Kolmogorov–Smirnov tests, which were used to determine whether the difference between each pair of distributions (early vs. late) was significant. N.S., not significant.

4. Discussion

The high metabolic activities of plant organs are generally associated with their fast turnover rates and short lifespans [71]. We observed that the shoot growth and mortality rates were higher for the fast-growing species than for the slow-growing species. Recent advances in plant ecophysiology have led to the concept of a "plant economics spectrum," a covariation of a suite of traits that can be largely explained by a position on the single axis of fast versus slow strategies [71]. The core relationships in the economics spectrum are the negative correlations between trait values associated with productivity (e.g., higher growth rate, higher leaf photosynthetic rate) and those associated with persistence (e.g., slow turnover rate, mechanical stability, longer lifespan). This growth–persistence trade-off has been reported both at the organ level (e.g., "leaf economics spectrum" [72], "wood economics spectrum" [73], "root economics spectrum" [74]) and at the whole-plant level (fast vs. slow strategies or "plant economics spectrum") [51,71,75–77], based on the rationale that more active organs afford individuals with better whole-plant growth rates at the expense of lower survival rates [1,51,73,75,78,79]. Such trade-offs have been found across different taxa, including different vascular plant groups [51,71–78], ferns [80], and mosses [81]. Similar trade-offs have also been observed in interspecific comparisons globally [72,73,82] and locally [55,83–85]. Analogous trade-offs have been quantified at intraspecific levels as well [86].

Two major theories have been proposed to explain the negative correlation between growth and survival rates. The first is the physiological constraints theory, which states that an organ or an individual cannot attain both high productivity and high persistence [87,88]. Such a trade-off may arise because of the selective investment of resources between traits related to high activity and those related to persistence [71,75,89]. For example, an investment of nutrients into the photosynthetic apparatus enhances photosynthetic rates [90] at the expense of investment in defense chemicals that reduce herbivory [75] or into mechanical toughness [91]. Likewise, stems comprising dense wood with narrow conduits are durable but may be less effective at transport [88] (but see [73]). Thus far, the physiological constraints underlying growth–persistence trade-offs have been intensively studied both theoretically [92] and empirically [73,91,93]. The second explanation for the negative correlation between growth and survival rates is the theory of optimal longevity (TOL): an organ should be replaced at an optimal time to maximize whole-plant carbon gain [9,29,31,38–40,94–96]. The key idea is that plant organs should be continuously moved into more productive environments [9,29,31,39,40,71,97] because a plant generally competes with its neighbors for light [25,26,71]. The production and deployment of new leaves and shoots into better-lit positions cause self-shading of the shaded inner part of the plant canopy [1,28,36] resulting in the death of the shaded leaves or shoots [1,3,9,29,30,39,97]. Such organ redeployment would be faster if the plants or shoots were growing faster [3,9,14,31,36,65,97]. TOL predicts that a higher growth rate is one of the reasons for a shorter organ lifespan. The two theories are not mutually exclusive, and the difference between them is a difference between proximate and ultimate or evolutionary factors [98].

Previous studies of shoots, however, have focused primarily on testing physiological constraints (i.e., causes of short organ lifespan), whereas few studies [13,40] have directly investigated the consequences of a shortened organ lifespan. Specifically, although the concept of the economics spectrum may apply to any organ [71], quantitative assessment of the TOL is limited to leaves [39,99] or fine roots [100]. It has been suggested that TOL can also be applied to shoots [40,101]; however, those studies did not quantitatively evaluate the effects of shoot shedding on light environments within crowns. Our results quantitatively demonstrated a consequence of rapid shoot shedding; i.e., redeployment of shoots into better-lit positions within a tree crown. The present results are consistent with the predictions of TOL, which, to date, have been thoroughly tested with leaves but not with shoots.

Based on the wood economics spectrum theory [73], we analyzed the wood density of four of the five species from the published literature and reconfirmed the predicted relationship [73] that the shoot survival rate is higher for species with higher wood density (Figure 6). A recent study [102] further proposed that a variation in crown structure (e.g., those maximizing light capture vs. those maximizing vertical growth) can be considered a new trait trade-off continuum: the structural economics spectrum [102]. This spectrum can be integrated with the previously described concepts of the leaf and wood economics spectra [102], in which the turnover rate of organs (i.e., leaf lifespan) plays a central role as an indicator of plant strategy [71–73]. Quantitative results regarding the consequence of organ turnover on the crown structure and light environment, such as those presented in this study, would be useful for investigating the suggested linkage among those economics spectra in future studies.

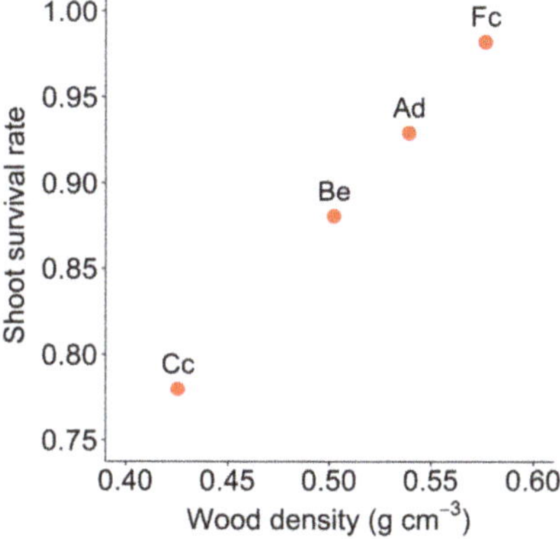

Figure 6. Shoot survival rates in relation to wood density. Fc: *Fagus crenata* (data from [24]); the original publication contained a typographical error (the same numbers were expressed in the wrong units [kg m^{-3}], and therefore the correct units [g cm^{-3}] have been used here, based on personal communication with the first author of [24]. Ad: *Acer distylum* (calculated from values in [103]), Be: *Betula ermanii* (data from [104]), and Cc: *Castanea crenata* (data from [24]). The wood density of *A. sieboldiana* was not found in our literature survey.

A simulation that was based on a dynamic functional-structural model [1] suggests that the death of shade shoots among shaded understory saplings is merely a passive response to the negative carbon balance of shaded shoots. Later, Laurans and Vincent [6] conducted an empirical study to test that suggestion. Their results showed that shoot shedding was not accelerated, whereas shoot production was suppressed for saplings in more heavily shaded understories than the conspecific individuals in less-shaded gaps. Laurans and Vincent [6] intentionally excluded pioneers that grew only in open habitats from their experimental design because their objective was to clarify how the suppression of shoot production in the shade and the long lifespan of the branches of shade-tolerant species determined crown morphological variation as a mechanism of shade tolerance. However, neither of those previous studies [1,6] clarified the consequences of shoot shedding on the within-crown light environment of vigorously growing trees in open, well-lit places. In the present study, we focused on plants in high-light environments, in which the continuous production and growth of shoots are expected. Our results, therefore, cannot be directly compared with those of these previous studies. In contrast to the cases of trees, there have been many results from experiments on herbaceous species grown in well-lit places that clarified the consequences of leaf shedding on the within-canopy light environment (e.g., [9,28–30,39]). These studies reported the function of the redeployment of leaves from shaded into well-lit places. Thus, our results of the shoot redeployment process observed in open, well-lit experimental gardens are more similar to the results of herbaceous species in open places than to the results of woody species in shaded understories.

Our study had several limitations. First, we monitored the trees for only a short time. The shoot mortality rate of the slow-growing trees was very low during the single growth period investigated in this study. It has been reported that late-successional old trees maintain their within-canopy structure by continuously replacing branches [105]. The implication may therefore be that fast- and slow-growing species maintain their within-crown light environments by replacing shoots rapidly and slowly, respectively. Further studies that monitor trees for more extended periods of time are therefore needed. Second, competition with neighbors was excluded in our experiment, which involved growing each sapling in an isolated condition. Nonetheless, the development of a tree crown in a real stand is also affected by competition with neighboring trees [102,106,107] and shading by a surrounding forest canopy [1,4,6,8]. Further studies are therefore needed before the present results can be generalized to real forest conditions.

5. Conclusions

We demonstrated the first empirical evidence that production and shedding of shoots for the saplings of fast-growing woody plants in high-light environments are part of a process of the redeployment of shoots into better-lit parts of the crown, similar to the redeployment of leaves. By contrast, the slow-growing trees displayed slow and determinate growth, with a low mortality rate of the shoots at all positions in the crown. These results indicate that fast- and slow-growing species maintain their within-crown light environments by replacing shoots rapidly and slowly, respectively.

Supplementary Materials: The following are available online at http://www.mdpi.com/1999-4907/11/12/1301/s1.

Author Contributions: Conceptualization, K.K. (Kohei Koyama), H.S., and K.K. (Kihachiro Kikuzawa); formal analysis, K.K. (Kohei Koyama) and H.S.; investigation, H.S.; methodology, H.S.; writing—original draft, K.K. (Kohei Koyama) and H.S.; writing—review and editing, K.K. (Kohei Koyama), H.S., and K.K. (Kihachiro Kikuzawa). All authors have read and agreed to the published version of the manuscript.

Funding: This work was funded by the Japan Society for the Promotion of Science (KAKENHI Grant Number 18K06406).

Acknowledgments: We thank Maki Suzuki, Adriano Dal Bosco, Martin Lechowicz, Atsushi Takayanagi, and Michimasa Yamasaki for their useful comments. We also thank Iwao Kojima for assistance in the field experiments. The present study was conducted with Joint Usage of the Center for Ecological Research (2018jurc-cer27), Kyoto University. The data may be used with proper citation of this article without contacting the authors.

Conflicts of Interest: The authors declare no conflict of interest.

Data Availability Statement: The data presented in this study are available in the Supplementary Materials.

References

1. Sterck, F.J.; Schieving, F.; Lemmens, A.; Pons, T.L. Performance of trees in forest canopies: Explorations with a bottom-up functional-structural plant growth model. *New Phytol.* **2005**, *166*, 827–843. [CrossRef]
2. Bazzaz, F.A.; Harper, J.L. Demographic analysis of the growth of *Linum usitatissimum*. *New Phytol.* **1977**, *78*, 193–208. [CrossRef]
3. Niinemets, U.; Lukjanova, A. Total foliar area and average leaf age may be more strongly associated with branching frequency than with leaf longevity in temperate conifers. *New Phytol.* **2003**, *158*, 75–89. [CrossRef]
4. Sterck, F.J.; Schieving, F. 3-D growth patterns of trees: Effects of carbon economy, meristem activity, and selection. *Ecol. Monogr.* **2007**, *77*, 405–420. [CrossRef]
5. Beyer, R.; Letort, V.; Cournede, P.H. Modeling tree crown dynamics with 3D partial differential equations. *Front. Plant Sci.* **2014**, *5*, 329. [CrossRef] [PubMed]
6. Laurans, M.; Vincent, G. Are inter- and intraspecific variations of sapling crown traits consistent with a strategy promoting light capture in tropical moist forest? *Ann. Bot.* **2016**, *118*, 983–996. [CrossRef] [PubMed]
7. Koyama, K.; Yamamoto, K.; Ushio, M. A lognormal distribution of the lengths of terminal twigs on self-similar branches of elm trees. *Proc. R. Soc. B Biol. Sci.* **2017**, *284*, 20162395. [CrossRef]
8. Rani, R.; Abramowicz, K.; Falster, D.S.; Sterck, F.; Brannstrom, A. Effects of bud-flushing strategies on tree growth. *Tree Physiol.* **2018**, *38*, 1384–1393. [CrossRef]
9. Hikosaka, K. A model of dynamics of leaves and nitrogen in a plant canopy: An integration of canopy photosynthesis, leaf life span, and nitrogen use efficiency. *Am. Nat.* **2003**, *162*, 149–164. [CrossRef]
10. Hikosaka, K. Leaf canopy as a dynamic system: Ecophysiology and optimality in leaf turnover. *Ann. Bot.* **2005**, *95*, 521–533. [CrossRef]
11. Kaitaniemi, P.; Lintunen, A.; Sievänen, R. Power-law estimation of branch growth. *Ecol. Model.* **2020**, *416*, 108900. [CrossRef]
12. Wang, F.; Kang, M.; Lu, Q.; Letort, V.; Han, H.; Guo, Y.; de Reffye, P.; Li, B. A stochastic model of tree architecture and biomass partitioning: Application to Mongolian Scots pines. *Ann. Bot.* **2011**, *107*, 781–792. [CrossRef] [PubMed]
13. Shirakawa, H.; Kikuzawa, K. Crown hollowing as a consequence of early shedding of leaves and shoots. *Ecol. Res.* **2009**, *24*, 839–845. [CrossRef]
14. Koyama, K.; Kikuzawa, K. Is whole-plant photosynthetic rate proportional to leaf area? A test of scalings and a logistic equation by leaf demography census. *Am. Nat.* **2009**, *173*, 640–649. [CrossRef] [PubMed]
15. Koyama, K.; Hidaka, Y.; Ushio, M. Dynamic scaling in the growth of a non-branching plant, *Cardiocrinum cordatum*. *PLoS ONE* **2012**, *7*, e45317. [CrossRef]
16. Ishii, H.; Asano, S. The role of crown architecture, leaf phenology and photosynthetic activity in promoting complementary use of light among coexisting species in temperate forests. *Ecol. Res.* **2010**, *25*, 715–722. [CrossRef]
17. Bentley, L.P.; Stegen, J.C.; Savage, V.M.; Smith, D.D.; von Allmen, E.I.; Sperry, J.S.; Reich, P.B.; Enquist, B.J. An empirical assessment of tree branching networks and implications for plant allometric scaling models. *Ecol. Lett.* **2013**, *16*, 1069–1078. [CrossRef]
18. Smith, D.D.; Sperry, J.S.; Enquist, B.J.; Savage, V.M.; McCulloh, K.A.; Bentley, L.P. Deviation from symmetrically self-similar branching in trees predicts altered hydraulics, mechanics, light interception and metabolic scaling. *New Phytol.* **2014**, *201*, 217–229. [CrossRef]
19. Savage, V.M.; Bentley, L.P.; Enquist, B.J.; Sperry, J.S.; Smith, D.D.; Reich, P.B.; von Allmen, E.I. Hydraulic trade-offs and space filling enable better predictions of vascular structure and function in plants. *Proc. Natl. Acad. Sci. USA* **2010**, *107*, 22722–22727. [CrossRef]
20. Enquist, B.J.; Bentley, L.P.; Shenkin, A.; Maitner, B.; Savage, V.; Michaletz, S.; Blonder, B.; Buzzard, V.; Espinoza, T.E.B.; Farfan-Rios, W.; et al. Assessing trait-based scaling theory in tropical forests spanning a broad temperature gradient. *Glob. Ecol. Biogeogr.* **2017**, *26*, 1357–1373. [CrossRef]
21. Kurosawa, Y.; Mori, S.; Wang, M.; Ferrio, J.P.; Yamaji, K.; Koyama, K.; Haruma, T.; Doyama, K. Initial burst of root development with decreasing respiratory carbon cost in *Fagus crenata* Blume seedlings. *Plant Spec. Biol.* **2020**. [CrossRef]

22. Martin-Ducup, O.; Ploton, P.; Barbier, N.; Momo Takoudjou, S.; Mofack, G., II; Kamdem, N.G.; Fourcaud, T.; Sonké, B.; Couteron, P.; Pélissier, R. Terrestrial laser scanning reveals convergence of tree architecture with increasingly dominant crown canopy position. *Funct. Ecol.* **2020**. [CrossRef]

23. Lau, A.; Martius, C.; Bartholomeus, H.; Shenkin, A.; Jackson, T.; Malhi, Y.; Herold, M.; Bentley, L.P. Estimating architecture-based metabolic scaling exponents of tropical trees using terrestrial LiDAR and 3D modelling. *For. Ecol. Manag.* **2019**, *439*, 132–145. [CrossRef]

24. Komiyama, A.; Nakagawa, M.; Kato, S. Common allometric relationships for estimating tree biomasses in cool temperate forests of Japan. *J. Jpn. For. Soc.* **2011**, *93*, 220–225. [CrossRef]

25. Givnish, T.J. On the adaptive significance of leaf height in forest herbs. *Am. Nat.* **1982**, *120*, 353–381. [CrossRef]

26. Anten, N.P.R.; During, H.J. Is analysing the nitrogen use at the plant canopy level a matter of choosing the right optimization criterion? *Oecologia* **2011**, *167*, 293–303. [CrossRef]

27. Boonman, A.; Anten, N.P.R.; Dueck, T.A.; Jordi, W.; van der Werf, A.; Voesenek, L.; Pons, T.L. Functional significance of shade-induced leaf senescence in dense canopies: An experimental test using transgenic tobacco. *Am. Nat.* **2006**, *168*, 597–607. [CrossRef]

28. Field, C. Allocating leaf nitrogen for the maximization of carbon gain—Leaf age as a control on the allocation program. *Oecologia* **1983**, *56*, 341–347. [CrossRef]

29. Oikawa, S.; Hikosaka, K.; Hirose, T. Leaf lifespan and lifetime carbon balance of individual leaves in a stand of an annual herb, *Xanthium canadense*. *New Phytol.* **2006**, *172*, 104–116. [CrossRef]

30. Oikawa, S.; Hikosaka, K.; Hirose, T. Dynamics of leaf area and nitrogen in the canopy of an annual herb, *Xanthium canadense*. *Oecologia* **2005**, *143*, 517–526. [CrossRef]

31. Kikuzawa, K. A cost-benefit analysis of leaf habit and leaf longevity of trees and their geographical pattern. *Am. Nat.* **1991**, *138*, 1250–1263. [CrossRef]

32. Hikosaka, K.; Terashima, I.; Katoh, S. Effects of leaf age, nitrogen nutrition and photon flux-density on the distribution of nitrogen among leaves of a vine (*Ipomoea tricolor* Cav.) grown horizontally to avoid mutual shading of leaves. *Oecologia* **1994**, *97*, 451–457. [CrossRef] [PubMed]

33. Hikosaka, K. Effects of leaf age, nitrogen nutrition and photon flux density on the organization of the photosynthetic apparatus in leaves of a vine (*Ipomoea tricolor* Cav.) grown horizontally to avoid mutual shading of leaves. *Planta* **1996**, *198*, 144–150. [CrossRef]

34. Franklin, O.; Ågren, G.I. Leaf senescence and resorption as mechanisms of maximizing photosynthetic production during canopy development at N limitation. *Funct. Ecol.* **2002**, *16*, 727–733. [CrossRef]

35. Kikuzawa, K.; Lechowicz, M.J. Theories of Leaf Longevity. In *Ecology of Leaf Longevity*; Kikuzawa, K., Lechowicz, M.J., Eds.; Springer: Tokyo, Japan, 2011; pp. 41–56.

36. Osada, N.; Takeda, H.; Kitajima, K.; Pearcy, R.W. Functional correlates of leaf demographic response to gap release in saplings of a shade-tolerant tree, *Elateriospermum tapos*. *Oecologia* **2003**, *137*, 181–187. [CrossRef]

37. Escudero, A.; Mediavilla, S. Decline in photosynthetic nitrogen use efficiency with leaf age and nitrogen resorption as determinants of leaf life span. *J. Ecol.* **2003**, *91*, 880–889. [CrossRef]

38. Tanaka, T.; Oikawa, S.; Kurokawa, C. Leaf shedding increases the photosynthetic rate of the canopy in N_2-fixing and non-N_2-fixing woody species. *Tree Physiol.* **2018**, *38*, 1903–1911. [CrossRef]

39. Oikawa, S.; Hikosaka, K.; Hirose, T. Does leaf shedding increase the whole-plant carbon gain despite some nitrogen being lost with shedding? *New Phytol.* **2008**, *178*, 617–624. [CrossRef]

40. Seiwa, K.; Kikuzawa, K.; Kadowaki, T.; Akasaka, S.; Ueno, N. Shoot life span in relation to successional status in deciduous broad-leaved tree species in a temperate forest. *New Phytol.* **2006**, *169*, 537–548. [CrossRef]

41. Stoll, P.; Schmid, B. Plant foraging and dynamic competition between branches of *Pinus sylvestris* in contrasting light environments. *J. Ecol.* **1998**, *86*, 934–945. [CrossRef]

42. Umeki, K.; Kikuzawa, K.; Sterck, F.J. Influence of foliar phenology and shoot inclination on annual photosynthetic gain in individual beech saplings: A functional–structural modeling approach. *For. Ecol. Manag.* **2010**, *259*, 2141–2150. [CrossRef]

43. Umeki, K.; Seino, T.; Lim, E.-M.; Honjo, T. Patterns of shoot mortality in *Betula platyphylla* in northern Japan. *Tree Physiol.* **2006**, *26*, 623–632. [CrossRef] [PubMed]

44. Takenaka, A. Shoot growth responses to light microenvironment and correlative inhibition in tree seedlings under a forest canopy. *Tree Physiol.* **2000**, *20*, 987–991. [CrossRef] [PubMed]

45. Sugiura, D.; Tateno, M. Concentrative nitrogen allocation to sun-lit branches and the effects on whole-plant growth under heterogeneous light environments. *Oecologia* **2013**, *172*, 949–960. [CrossRef]

46. Chen, L.; Sumida, A. Effects of light on branch growth and death vary at different organization levels of branching units in Sakhalin spruce. *Trees* **2018**, *32*, 1123–1134. [CrossRef]

47. Sprugel, D.G. When branch autonomy fails: Milton's Law of resource availability and allocation. *Tree Physiol.* **2002**, *22*, 1119–1124. [CrossRef]

48. Yamanaka, T.; Okabe, H.; Kawai, S. Growth and nodulation in *Alnus sieboldiana* in response to *Frankia* inoculation and nitrogen treatments. *Trees* **2016**, *30*, 539–544. [CrossRef]

49. Kamijo, T.; Kitayama, K.; Sugawara, A.; Urushimichi, S.; Sasai, K. Primary succession of the warm-temperate broad-leaved forest on a volcanic island, Miyake-jima, Japan. *Folia Geobot.* **2002**, *37*, 71–91. [CrossRef]

50. Kikuzawa, K. Phenological and morphological adaptations to the light environment in two woody and two herbaceous plant species. *Funct. Ecol.* **2003**, *17*, 29–38. [CrossRef]

51. Poorter, L.; Bongers, F. Leaf traits are good predictors of plant performance across 53 rain forest species. *Ecology* **2006**, *87*, 1733–1743. [CrossRef]

52. Gray, E.F.; Wright, I.J.; Falster, D.S.; Eller, A.S.D.; Lehmann, C.E.R.; Bradford, M.G.; Cernusak, L.A. Leaf:wood allometry and functional traits together explain substantial growth rate variation in rainforest trees. *AoB Plants* **2019**, *11*, plz024. [CrossRef] [PubMed]

53. Wright, I.J.; Cooke, J.; Cernusak, L.A.; Hutley, L.B.; Scalon, M.C.; Tozer, W.C.; Lehmann, C.E.R. Stem diameter growth rates in a fire-prone savanna correlate with photosynthetic rate and branch-scale biomass allocation, but not specific leaf area. *Austral Ecol.* **2019**, *44*, 339–350. [CrossRef]

54. Seiwa, K.; Watanabe, A.; Irie, K.; Kanno, H.; Saitoh, T.; Akasaka, S. Impact of site-induced mouse caching and transport behaviour on regeneration in *Castanea crenata*. *J. Veg. Sci.* **2002**, *13*, 517–526. [CrossRef]

55. Koike, T. Leaf structure and photosynthetic performance as related to the forest succession of deciduous broad-leaved trees. *Plant Spec. Biol.* **1988**, *3*, 77–87. [CrossRef]

56. Cao, K.F.; Ohkubo, T. Allometry, root/shoot ratio and root architecture in understory saplings of deciduous dicotyledonous trees in central Japan. *Ecol. Res.* **1998**, *13*, 217–227. [CrossRef]

57. Ohkubo, T.; Kaji, M.; Hamaya, T. Structure of primary Japanese beech (*Fagus japonica* maxim.) forests in the Chichibu Mountains, central Japan, with special reference to regeneration processes. *Ecol. Res.* **1988**, *3*, 101–116. [CrossRef]

58. Sakai, S. Patterns of branching and extension growth of vigorous saplings of Japanese *Acer* species in relation to their regeneration strategies. *Can. J. Bot.* **1987**, *65*, 1578–1585. [CrossRef]

59. Miyashita, A.; Tateno, M. A novel index of leaf RGR predicts tree shade tolerance. *Funct. Ecol.* **2014**, *28*, 1321–1329. [CrossRef]

60. Okaura, T.; Harada, K. Phylogeographical structure revealed by chloroplast DNA variation in Japanese beech (*Fagus crenata* Blume). *Heredity* **2002**, *88*, 322. [CrossRef]

61. Koyama, K.; Kikuzawa, K. Reduction of photosynthesis before midday depression occurred: Leaf photosynthesis of *Fagus crenata* in a temperate forest in relation to canopy position and a number of days after rainfall. *Ecol. Res.* **2011**, *26*, 999–1006. [CrossRef]

62. Koyama, K.; Kikuzawa, K. Can we estimate forest gross primary production from leaf lifespan? A test in a young *Fagus crenata* forest. *J. Ecol. Field Biol.* **2010**, *33*, 253–260. [CrossRef]

63. Koyama, K.; Kikuzawa, K. Geometrical similarity analysis of photosynthetic light response curves, light saturation and light use efficiency. *Oecologia* **2010**, *164*, 53–63. [CrossRef] [PubMed]

64. Watanabe, M.; Hiroshima, H.; Kinose, Y.; Okabe, S.; Izuta, T. Nitrogen use efficiency for growth of *Fagus crenata* seedlings under elevated ozone and different soil nutrient conditions. *Forests* **2020**, *11*, 371. [CrossRef]

65. Koyama, K.; Kikuzawa, K. Intraspecific variation in leaf life span for the semi-evergreen liana *Akebia trifoliata* is caused by both seasonal and aseasonal factors in a temperate forest. *J. Ecol. Field Biol.* **2008**, *31*, 207–211. [CrossRef]

66. Hollinger, D.Y. Optimality and nitrogen allocation in a tree canopy. *Tree Physiol.* **1996**, *16*, 627–634. [CrossRef]

67. Muraoka, H.; Hirota, H.; Matsumoto, J.; Nishimura, S.; Tang, Y.; Koizumi, H.; Washitani, I. On the convertibility of different microsite light availability indices, relative illuminance and relative photon flux density. *Funct. Ecol.* **2001**, *15*, 798–803. [CrossRef]

68. Muraoka, H.; Tang, Y.; Koizumi, H.; Washitani, I. Combined effects of light and water availability on photosynthesis and growth of *Arisaema heterophyllum* in the forest understory and an open site. *Oecologia* **1997**, *112*, 26–34. [CrossRef]

69. Bates, D.; Mächler, M.; Bolker, B.; Walker, S. Fitting linear mixed-effects models using lme4. *J. Stat. Softw.* **2015**, *67*, 48. [CrossRef]

70. R Core Team. *R: A Language and Environment for Statistical Computing*; R Foundation for Statistical Computing: Vienna, Austria, 2020.

71. Reich, P.B. The world-wide 'fast–slow' plant economics spectrum: A traits manifesto. *J. Ecol.* **2014**, *102*, 275–301. [CrossRef]

72. Wright, I.J.; Reich, P.B.; Westoby, M.; Ackerly, D.D.; Baruch, Z.; Bongers, F.; Cavender-Bares, J.; Chapin, T.; Cornelissen, J.H.; Diemer, M.; et al. The worldwide leaf economics spectrum. *Nature* **2004**, *428*, 821–827. [CrossRef]

73. Chave, J.; Coomes, D.; Jansen, S.; Lewis, S.L.; Swenson, N.G.; Zanne, A.E. Towards a worldwide wood economics spectrum. *Ecol. Lett.* **2009**, *12*, 351–366. [CrossRef] [PubMed]

74. Roumet, C.; Birouste, M.; Picon-Cochard, C.; Ghestem, M.; Osman, N.; Vrignon-Brenas, S.; Cao, K.-f.; Stokes, A. Root structure–function relationships in 74 species: Evidence of a root economics spectrum related to carbon economy. *New Phytol.* **2016**, *210*, 815–826. [CrossRef] [PubMed]

75. Coley, P.D.; Bryant, J.P.; Chapin, F.S. Resource availability and plant antiherbivore defense. *Science* **1985**, *230*, 895–899. [CrossRef] [PubMed]

76. Lambers, H.; Poorter, H. Inherent variation in growth rate between higher plants: A search for physiological causes and ecological consequences. *Adv. Ecol. Res.* **1992**, *23*, 187–261.

77. Freschet, G.T.; Cornelissen, J.H.C.; van Logtestijn, R.S.P.; Aerts, R. Evidence of the 'plant economics spectrum' in a subarctic flora. *J. Ecol.* **2010**, *98*, 362–373. [CrossRef]

78. Iida, Y.; Sun, I.F.; Price, C.A.; Chen, C.T.; Chen, Z.S.; Chiang, J.M.; Huang, C.L.; Swenson, N.G. Linking leaf veins to growth and mortality rates: An example from a subtropical tree community. *Ecol. Evol.* **2016**, *6*, 6085–6096. [CrossRef]

79. Wright, S.J.; Kitajima, K.; Kraft, N.J.B.; Reich, P.B.; Wright, I.J.; Bunker, D.E.; Condit, R.; Dalling, J.W.; Davies, S.J.; Díaz, S.; et al. Functional traits and the growth-mortality trade-off in tropical trees. *Ecology* **2010**, *91*, 3664–3674. [CrossRef]

80. Karst, A.L.; Lechowicz, M.J. Are correlations among foliar traits in ferns consistent with those in the seed plants? *New Phytol.* **2007**, *173*, 306–312. [CrossRef]

81. Waite, M.; Sack, L. How does moss photosynthesis relate to leaf and canopy structure? Trait relationships for 10 Hawaiian species of contrasting light habitats. *New Phytol.* **2010**, *185*, 156–172. [CrossRef]

82. Reich, P.B.; Walters, M.B.; Ellsworth, D.S. From tropics to tundra: Global convergence in plant functioning. *Proc. Natl. Acad. Sci. USA* **1997**, *94*, 13730–13734. [CrossRef]

83. Reich, P.B.; Walters, M.B.; Ellsworth, D.S. Leaf age and season influence the relationships between leaf nitrogen, leaf mass per area and photosynthesis in maple and oak trees. *Plant Cell Environ.* **1991**, *14*, 251–259. [CrossRef]

84. Shiodera, S.; Rahajoe, J.S.; Kohyama, T. Variation in longevity and traits of leaves among co-occurring understorey plants in a tropical montane forest. *J. Trop. Ecol.* **2008**, *24*, 121–133. [CrossRef]

85. Saura-Mas, S.; Shipley, B.; Lloret, F. Relationship between post-fire regeneration and leaf economics spectrum in Mediterranean woody species. *Funct. Ecol.* **2009**, *23*, 103–110. [CrossRef]

86. Vincent, G. Leaf life span plasticity in tropical seedlings grown under contrasting light regimes. *Ann. Bot.* **2006**, *97*, 245–255. [CrossRef] [PubMed]

87. Reich, P.; Walters, M.; Ellsworth, D. Leaf life-span in relation to leaf, plant, and stand characteristics among diverse ecosystems. *Ecol. Monogr.* **1992**, *62*, 365–392. [CrossRef]

88. Baas, P.; Ewers, F.W.; Davis, S.D.; Wheeler, E.A. Evolution of xylem physiology. In *The Evolution of Plant Physiology*; Hemsley, A.R., Poole, I., Eds.; Elsevier: Amsterdam, The Netherlands, 2004; pp. 273–295.

89. Rosas, T.; Mencuccini, M.; Barba, J.; Cochard, H.; Saura-Mas, S.; Martínez-Vilalta, J. Adjustments and coordination of hydraulic, leaf and stem traits along a water availability gradient. *New Phytol.* **2019**, *223*, 632–646. [CrossRef]

90. Mooney, H.A.; Gulmon, S.L. Environmental and evolutionary constraints on the photosynthetic characteristics of higher plants. In *Topics in Plant Population Biology*; Solbrig, O.T., Jain, S., Johnson, G.B., Raven, P.H., Eds.; Macmillan Education: London, UK, 1979; pp. 316–337.

91. Onoda, Y.; Westoby, M.; Adler, P.B.; Choong, A.M.; Clissold, F.J.; Cornelissen, J.H.; Diaz, S.; Dominy, N.J.; Elgart, A.; Enrico, L.; et al. Global patterns of leaf mechanical properties. *Ecol. Lett.* **2011**, *14*, 301–312. [CrossRef]

92. Shipley, B.; Lechowicz, M.J.; Wright, I.; Reich, P.B. Fundamental trade-offs generating the worldwide leaf economics spectrum. *Ecology* **2006**, *87*, 535–541. [CrossRef]

93. Niinemets, Ü.; Díaz-Espejo, A.; Flexas, J.; Galmés, J.; Warren, C.R. Role of mesophyll diffusion conductance in constraining potential photosynthetic productivity in the field. *J. Exp. Bot.* **2009**, *60*, 2249–2270. [CrossRef]

94. Chabot, B.F.; Hicks, D.J. The ecology of leaf life spans. *Ann. Rev. Ecol. Syst.* **1982**, *13*, 229–259. [CrossRef]

95. Osada, N.; Oikawa, S.; Kitajima, K. Implications of life span variation within a leaf cohort for evaluation of the optimal timing of leaf shedding. *Funct. Ecol.* **2015**, *29*, 308–314. [CrossRef]

96. Oikawa, S.; Suno, K.; Osada, N. Inconsistent intraspecific pattern in leaf life span along nitrogen-supply gradient. *Am. J. Bot.* **2017**, *104*, 342–346. [CrossRef] [PubMed]

97. Osada, N.; Takeda, H.; Furukawa, A.; Awang, M. Leaf dynamics and maintenance of tree crowns in a Malaysian rain forest stand. *J. Ecol.* **2001**, *89*, 774–782. [CrossRef]

98. Reich, P.B.; Wright, I.J.; Cavender-Bares, J.; Craine, J.M.; Oleksyn, J.; Westoby, M.; Walters, M.B. The evolution of plant functional variation: Traits, spectra, and strategies. *Int. J. Plant Sci.* **2003**, *164*, S143–S164. [CrossRef]

99. Kikuzawa, K.; Ackerly, D. Significance of leaf longevity in plants. *Plant Spec. Biol.* **1999**, *14*, 39–45. [CrossRef]

100. Eissenstat, D.; Wells, C.; Yanai, R.; Whitbeck, J. Building roots in a changing environment: Implications for root longevity. *New Phytol.* **2000**, *147*, 33–42. [CrossRef]

101. Fujiki, D.; Kikuzawa, K. Stem turnover strategy of multiple-stemmed woody plants. *Ecol. Res.* **2006**, *21*, 380–386. [CrossRef]

102. Verbeeck, H.; Bauters, M.; Jackson, T.; Shenkin, A.; Disney, M.; Calders, K. Time for a plant structural economics spectrum. *Front. For. Glob. Chang.* **2019**, *2*, 43. [CrossRef]

103. Ohtsuka, T.; Akiyama, T.; Hashimoto, Y.; Inatomi, M.; Sakai, T.; Jia, S.; Mo, W.; Tsuda, S.; Koizumi, H. Biometric based estimates of net primary production (NPP) in a cool-temperate deciduous forest stand beneath a flux tower. *Agric. For. Meteorol.* **2005**, *134*, 27–38. [CrossRef]

104. Ameztegui, A.; Paquette, A.; Shipley, B.; Heym, M.; Messier, C.; Gravel, D. Shade tolerance and the functional trait: Demography relationship in temperate and boreal forests. *Funct. Ecol.* **2017**, *31*, 821–830. [CrossRef]

105. Ishii, H.; Ford, E.D. The role of epicormic shoot production in maintaining foliage in old *Pseudotsuga menziesii* (Douglas-fir) trees. *Can. J. Bot.* **2001**, *79*, 251–264.

106. Sumida, A.; Terazawa, I.; Togashi, A.; Komiyama, A. Spatial arrangement of branches in relation to slope and neighbourhood competition. *Ann. Bot.* **2002**, *89*, 301–310. [CrossRef] [PubMed]

107. Guillemot, J.; Kunz, M.; Schnabel, F.; Fichtner, A.; Madsen, C.P.; Gebauer, T.; Härdtle, W.; von Oheimb, G.; Potvin, C. Neighbourhood-mediated shifts in tree biomass allocation drive overyielding in tropical species mixtures. *New Phytol.* **2020**, *228*, 1256–1268. [CrossRef] [PubMed]

Publisher's Note: MDPI stays neutral with regard to jurisdictional claims in published maps and institutional affiliations.

 *forests*

Article

Environmental Effects on Carbon Isotope Discrimination from Assimilation to Respiration in a Coniferous and Broad-Leaved Mixed Forest of Northeast China

Haoyu Diao [1,2], **Anzhi Wang** [1], **Fenghui Yuan** [1], **Dexin Guan** [1], **Guanhua Dai** [1,3] **and Jiabing Wu** [1,*]

1   CAS Key Laboratory of Forest Ecology and Management, Institute of Applied Ecology,
Chinese Academy of Sciences, Shenyang 110016, China; diaohaoyu17@mails.ucas.ac.cn (H.D.);
waz@iae.ac.cn (A.W.); fhyuan@iae.ac.cn (F.Y.); dxguan@iae.ac.cn (D.G.); daiguanhua@126.com (G.D.)

2   College of Resources and Environment, University of Chinese Academy of Sciences, Beijing 100049, China

3   Research Station of Changbai Mountain Forest Ecosystems, Chinese Academy of Sciences,
Erdaobaihe 133600, China

*   Correspondence: wujb@iae.ac.cn; Tel.: +86-24-83-970-336

Received: 1 September 2020; Accepted: 26 October 2020; Published: 30 October 2020

Abstract: Carbon (C) isotope discrimination during photosynthetic CO_2 assimilation has been extensively studied, but the whole process of fractionation from leaf to soil has been less well investigated. In the present study, we investigated the $\delta^{13}C$ signature along the C transfer pathway from air to soil in a coniferous and broad-leaved mixed forest in northeast China and examined the relationship between $\delta^{13}C$ of respiratory fluxes (leaf, trunk, soil, and the entire ecosystem) and environmental factors over a full growing season. This study found that the $\delta^{13}C$ signal of CO_2 from canopy air was strongly imprinted in the organic and respiratory pools throughout C transfer due to the effects of discrimination and isotopic mixing on C assimilation, allocation, and respiration processes. A significant difference in isotopic patterns was found between conifer and broadleaf species in terms of seasonal variations in leaf organic matter. This study also found that $\delta^{13}C$ in trunk respiration, compared with that in leaf and soil respiration, was more sensitive to seasonal variations of environmental factors, especially soil temperature and soil moisture. Variation in the $\delta^{13}C$ of ecosystem respiration was correlated with air temperature with no time lag and correlated with soil temperature and vapor pressure deficit with a lag time of 10 days, but this correlation was relatively weak, indicating a delayed linkage between above- and belowground processes. The isotopic linkage might be confounded by variations in atmospheric aerodynamic and soil diffusion conditions. These results will help with understanding species differences in isotopic patterns and promoting the incorporation of more influencing factors related to isotopic variation into process-based ecosystem models.

Keywords: carbon isotopes; climate change; respiration; discrimination; mixed forest; keeling plot

---

1. Introduction

Stable carbon (C) isotope measurement is a useful tool for disentangling C cycling processes in forest ecosystems, as a unique C isotopic signature is imprinted along the C transfer pathway from CO_2 in the atmosphere to C in different ecosystem pools, which is affected by isotopic discrimination (Δ). During photosynthesis, C_3 plants discriminate heavily against ^{13}C through photosynthetic discrimination; the resulting C isotopic signature ($\delta^{13}C$) in the photosynthate is more negative compared to the $\delta^{13}C$ of atmospheric CO_2 ($\delta^{13}C_{air}$) and is linearly dependent on the ratio of the intercellular and atmospheric CO_2 concentration (c_i/c_a) [1,2]. After photosynthesis, plants rapidly release photo-assimilated C into the soil, and the post-photosynthetic discrimination effects become involved in a number of biochemical processes alongside mixing and exchange with other C storages, which leads to different $\delta^{13}C$ levels among woody plant parts and soil [3,4]. Subsequently, respiration-associated discrimination and the consumption of substrates with different $\delta^{13}C$ values further modify the $\delta^{13}C$ of respiratory fluxes, resulting in apparent differences between the $\delta^{13}C$ of respired CO_2 ($\delta^{13}C_R$) and the respiration substrates [5–8]. The accurate measurement of $\delta^{13}C$ in different transfer pathways is important to explore the complexity of the C cycle under the changing climate.

Moreover, the discrimination processes encode biochemical and physiological information into $\delta^{13}C$, and this information is transferred through different stages of the C cycle. Thus, thoroughly investigating $\delta^{13}C$ in each C pool and respiratory flux can help evaluate the coupling between each of the processes in the pathways of C cycling. For example, strong evidence of the linkage between plant photosynthesis and soil respiration was found by Kuzyakov and Gavrichkova [9], indicating that the $\delta^{13}C$ of soil-respired CO_2 may also be related to photosynthetic discrimination via root respiration or exudation. Ecosystem respiration combines the CO_2 emitted from leaves, trunks, and soil within the ecosystem, and it also integrates their C isotopic signals. Thus, the $\delta^{13}C$ in ecosystem respiration ($\delta^{13}C_{eco}$) can vary substantially with changes in the relative contribution of the respiration of each component.

Discrimination also encodes information about the dynamics of environmental drivers into $\delta^{13}C$. Seasonal variation in environmental factors, such as vapor pressure deficit (VPD) and soil moisture, influence stomatal conductance, which then alters photosynthetic discrimination and further leads to seasonal dynamics in the $\delta^{13}C$ of CO_2 respired from leaves [7]. Due to the linkages between the photosynthetic metabolites and the substrates of different respiratory fluxes, several studies also found that changes in environmental drivers may be responsible for seasonal dynamics in the $\delta^{13}C_R$ of trunks, roots, and soil [10–12]. Accordingly, the $\delta^{13}C$ of ecosystem respiration, which integrates all respiratory components of the ecosystem, was also found to have significant short-term and seasonal variations [13–15], which have been directly or implicitly linked to environmental factors with a time lag. This indicates either a fast linkage between assimilates and ecosystem respiration or a linkage delayed by the time in which the assimilates are transported to the sites of respiration [10,16].

However, a significant response of $\delta^{13}C_{eco}$ to environmental factors does not always exist. For example, McDowell et al. [13] found only a weak relationship between canopy conductance and $\delta^{13}C_{eco}$ over a 2-week period, which they attributed to noncanopy controls, such as belowground respiration on $\delta^{13}C_{eco}$. Similarly, the results of long-term measurements with high time resolution showed that $\delta^{13}C_{eco}$ was not tightly coupled with environmental factors in some of the study periods [17]. These studies collectively indicate that the response of $\delta^{13}C_{eco}$ to environmental changes is highly variable and that the influence of complicated environmental conditions, combined with other factors, may weaken the linkage between assimilates and ecosystem respiration. Thus, there is a practical need to explore what factors control $\delta^{13}C_{eco}$ and to what extent $\delta^{13}C_{eco}$ reveals the links between C cycling processes.

Photosynthetic discrimination can be calculated at the leaf level by determining the difference between the $\delta^{13}C$ of leaf organic matter and $\delta^{13}C_{air}$ [1], or it can be measured online using branch chambers [17]. Alternatively, photosynthetic discrimination can be estimated using the comprehensive Farquhar's model at a canopy scale by combining $\delta^{13}C$ and CO_2 flux measurements [18,19]. On the other hand, respiration-associated discrimination can be evaluated by chamber measurements [20,21] using the Keeling plot approach [22,23]. This approach is based on the theory of the two-component gas-mixing model, where $\delta^{13}C_R$ emerges as the y-intercept of the mixing relationship between $\delta^{13}C$ and the inverse of the CO_2 concentration (CO_2) [24,25]. The Keeling plot approach can also determine the $\delta^{13}C_{eco}$ using nighttime CO_2 and $\delta^{13}C$ across the vertical profile of the forest [14,17]. Recently, optical laser spectroscopy techniques have allowed in situ continuous monitoring of CO_2 isotopes within open forest canopies or closed gas exchange chambers, thus tracing the high-resolved dynamics in $\delta^{13}C_{air}$, $\delta^{13}C_R$, and $\delta^{13}C_{eco}$ over more weather conditions [12,17]. However, few studies have simultaneously measured $\delta^{13}C_{air}$, $\delta^{13}C_R$, and $\delta^{13}C_{eco}$, which represents a limitation of our knowledge on discrimination in the process of C exchange.

Coniferous and broad-leaved mixed forests are widely distributed in northeast China. These forests are well-protected old-growth virgin forests characterized by rich species diversity as well as high biomass. Previous studies have shown that coniferous and broad-leaved mixed forests are strong C sinks and have experienced climate warming over the past 50 years [26]. To our best knowledge, variations of $\delta^{13}C$ in forest storage C pools and respiratory fluxes have mainly been studied in pure forests [17,27–29], while mixed forests with multiple tree species have rarely been covered by previous studies [12]. In the present study, we continuously investigated in situ seasonal variations in the $\delta^{13}C_{air}$ and $\delta^{13}C_R$ of leaves, trunks, and soil as well as $\delta^{13}C_{eco}$ in a 200-year-old coniferous and broad-leaved mixed forest over a full growing season. We also explored the Δ variation of the dominant species in this mixed forest. Firstly, we aimed to identify species-specific differences in photosynthetic Δ and variations in the Δ subsequent photosynthesis in this mixed forest. Secondly, we assessed the environmental effects on seasonal variations in $\delta^{13}C$ for CO_2 respired by leaves, trunks, and soil as well as the ecosystem. Finally, we tested whether $\delta^{13}C_{eco}$ reveals links between aboveground and belowground processes.

2. Materials and Methods

2.1. Site Description

The study was carried out in a coniferous and broad-leaved mixed forest site (42° 24.149′ N, 128° 05.768′ E, 732 m elevation) in the Changbai mountain natural reserve, Jilin, northeast China. The site is pristine and undisturbed by anthropic activities and characterized by richness of species in both its overstorey and understorey. The density of the trees is about 560 trees ha^{-1} (stem diameter > 8 cm). Dominant species include one evergreen needle species, Korean pine (*Pinus koraiensis*), and four deciduous broadleaf species, including Manchurian ash (*Fraxinus mandshurica*), Mono maple (*Acer mono*), Mongolian oak (*Quercus mongolica*), and Tuan linden (*Tilia amurensis*). *P. koraiensis*, *F. mandshurica*, and *T. amurensis* account for about 40%, 30%, and 20% of the total tree number, respectively, while *Q. mongolica* and *A. mono* account for the remaining 10%. The average age of the dominant species has been estimated to be over 200 years, and the mean canopy height is about 26 m. The growing season of the vegetation extends from May to September. The site's topography is nearly flat. The soil is classified as dark brown forest soil with a depth of 60–100 cm. The climate is temperate, with a mean annual temperature of 3.6 °C and annual precipitation of 695.3 mm measured between 1982 and 2003; over 80% of the rainfall occurs within the growing season.

2.2. Eddy Covariance and Meteorological Measurements

A 62 m high meteorological tower is located at the study site. Temperature and relative humidity probes (HMP155A, Vaisala, Helsinki, Finland) and anemometers (WindSonic1, Campbell Scientific, Logan, UT, USA) were installed at seven heights (2.5, 8, 22, 26, 32, 50, and 60 m) on the tower. A pyranometer (CMP22, Kipp&Zonen, Delft, Netherlands) was installed at 40 m on the tower. Rainfall was measured at 62 m using a rain gauge (TE525MM, Texas Electronics, Dallas, TX, USA). Soil temperature and moisture were measured at a 5 cm depth using a thermocouple burial probe (105E-L, Campbell Scientific, Logan, UT, USA) and a water content reflectometer (CS616, Campbell Scientific, Logan, UT, USA), respectively. All of the meteorological data were recorded via a CR1000X data logger (Campbell Scientific, Logan, UT, USA) at 30 min intervals.

An eddy covariance system was installed on the tower at 40 m above the ground. The system consisted of a three-axis anemometer (WindMaster Pro, Gill Instruments, Lymington, UK) and an open-path CO_2/H_2O infrared gas analyzer (Li-7500DS, LI-COR, Lincoln, NE, USA). Raw data were collected by a SmartFlux 3 System (LI-COR, Lincoln, NE, USA).

2.3. Canopy CO_2 Concentration and Carbon Isotope Composition Profile

In this study, the canopy CO_2 and $\delta^{13}C_{air}$ were measured using a wavelength-scanned cavity ring down spectroscopy (WS-CRDS) analyzer (G2201-i, Picarro Inc., Santa Clara, CA, USA). This instrument performed well in measuring the $\delta^{13}C$ of CO_2 with a 5 min precision of 0.11‰ and a maximum drift of 0.25‰ day^{-1}. Data were collected at about 1 Hz and expressed on the VPDB (Vienna Pee Dee Belemnite) scale.

A profile control system (PRI-8500, PRI-ECO Inc., Beijing, China) was used as a peripheral of the WS-CRDS analyzer to automatically change ports in the $\delta^{13}C$ profile measurements (Figure 1a). Air inlets were installed at the seven heights on the tower. Different levels of air were simultaneously pumped from the inlets by a vacuum pump (R410, Rocker Scientific Co., Ltd., Taiwan, China) through 1/4-inch Teflon tubes followed by flow control valves (ASC200-08, AirTAC, Taiwan, China). Each of the inlets was selected and then scanned for 5 min one after another from lower to higher levels through a 16-port rotation valve (Valco Instruments Co. Inc., Houston, TX, USA). The air was pumped into the WS-CRDS analyzer at a flow rate of 30 mL/min using a Picarro external pump. The first minute of each measurement was discarded to remove the data of the response time after the port was switched.

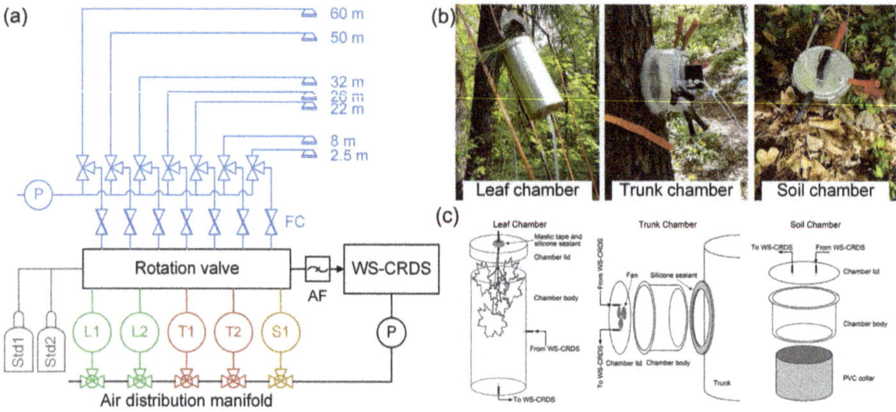

Figure 1. Schematic of the atmospheric profile and respiration chamber system (**a**) and photographs (**b**) and diagrammatic representations (**c**) of the custom-made chambers. L1 and L2, leaf chambers; T1 and T2, trunk chambers; S1, soil chamber; Std1 and Std2, standard gas tanks; FC, flow controller; AF, air filter; P, pump.

Two standard gases (Std1: 370.7 ppm for CO_2 and −20.8709‰ for $\delta^{13}C$; Std2: 799.0 ppm for CO_2 and −21.5685‰ for $\delta^{13}C$) were scanned for 15 min once a day to conduct two-point gain and offset calibrations for CO_2 measurements and single-point delta value offset calibrations for $\delta^{13}C$ measurements [30]. The dependence of $\delta^{13}C$ on CO_2 for G2201-i was 0.09‰ per 100 ppm in the range of 368.1–550.1 ppm, as tested by Pang et al. [31]. Thus, it can be presumed that the CO_2 dependence of $\delta^{13}C$ is linear within the small CO_2 range (about 400–495 ppm for the air profile measurement and 400–600 ppm for the respiration measurements; see Section 2.4) in the present study. The first 5 min and last 1 min of each standard gas measurement were discarded to remove biased data introduced by port switching.

2.4. Carbon Isotope Composition of CO_2 Respired from the Leaves, Trunks, Soil, and Ecosystem

Five ports of the profile control system along with the respiration chambers were used to build a closed-circuit system under non-steady-state conditions [21,32] to conduct measurements of the $\delta^{13}C$ in the leaf-, trunk-, and soil-respired CO_2 ($\delta^{13}C_R$) (Figure 1a). The outlet of each chamber was directly connected to the rotation valve in the profile control system and then connected to the inlet of the WS-CRDS analyzer. The outlet of the WS-CRDS analyzer, which is the outlet of the Picarro external pump, was divided into five sub-outlets by a 5-way air distribution manifold connected to the inlet of each chamber. All tubes were 1/8-inch Teflon tubes, thereby reducing the system volume and response time.

The leaf chambers were made from cylindrical PET (polyethylene terephthalate) jars (10 cm in diameter and 20 cm in height) (Figure 1b,c). The lid of each jar was fixed on the branch by letting the branch pass through the center of the lid. Mastic tape and silicone sealant were used to fill the space between the branch and the lid to make the leaf chamber airtight. The inlet and the outlet were on the sides and the bottoms of the jar bodies. No fans were installed inside the jars because of limited space.

The trunk chambers were made from cylindrical PP (polypropylene) airtight containers (18 cm in diameter and 9.5 cm in height) (Figure 1b,c). The bottoms of each container were removed, and the rest of the body was fixed onto the trunk using silicone sealant after gently sanding the bark. An inlet and an outlet were located on the lid of the container. A mini fan was installed inside the lid for air mixing.

The soil chamber was made from the same material and of the same size as the trunk chambers (Figure 1b,c). The body of the soil chamber was fixed on a soil collar (16 cm in diameter and 10 cm in height) using hot-melt adhesive. The soil collar was inserted into the soil to a depth of 5 cm. Here, only one soil chamber was used due to the limited number of sampling ports. The leaf chambers and the soil chamber were permanently covered by aluminum foil tape to isolate sunlight and maintain a steady temperature.

Two leaf chambers were installed on a branch of a *P. koraiensis* and an *A. mono*. The branches were mature and healthy with a diameter of about 1.5 cm and located at about 5 m in height on the sunny side of the trees. Two trunk chambers were installed on the sunny side of the trunks on a *P. koraiensis* and a *F. mandshurica* at breast height (1.3 m). The selected trees and the soil collar were about 5 m apart from each other, thereby sharing similar meteorological and environmental conditions. The chambers were installed one month before the first sampling. Each of the chambers included a silicon sealing ring on the inside of the lid to ensure airtightness. The airtightness of the respiration chamber was tested by connecting a CO_2/H_2O analyzer (LI-850, LI-COR, Lincoln, NE, USA) to a closed chamber (screwing the body back on the lid for leaf chambers or locking the lid with clamps on the body for trunk and soil chambers). If there was no obvious change in the measured values of CO_2 concentration in the chamber when blowing along the joints of the chamber, then the sealing performance was considered good.

The $\delta^{13}C_R$ measurements were conducted from early May to late September of 2019. To begin measuring, the chamber was first closed after selecting an inlet and an outlet corresponding to the chamber by manually operating the rotation valve and the air distribution manifold. The measurement continued until the internal CO_2 concentration increased by about 200 ppm (about 20–40 min for the leaf and trunk chambers and about 5–10 min for the soil chamber), as suggested in [21], and then we switched to the next chamber. The chambers were closed only during measuring. We periodically closed the chambers in the following sequence: the *P. koraiensis* leaf chamber (L1), the *A. mono* leaf chamber (L2), the *P. koraiensis* trunk chamber (T1), the *F. mandshurica* trunk chamber (T2), and the soil chamber (S1). Four rounds of measurements were conducted from around 7:00 to 18:00 each day. The $\delta^{13}C_R$ of each measurement was estimated based on the widely used Keeling plot approach [22,23]. The daily $\delta^{13}C_R$ of the leaf, trunk, and soil was averaged from the four measurements each day.

The $\delta^{13}C_{eco}$ was determined by the Keeling plot approach [22,23] using the nighttime (21:00–03:00) CO_2 and $\delta^{13}C_{air}$ profile measurements. Considering that small a CO_2 range would cause uncertainties in fitting the Keeling plot [24], we set a restriction that only nights with a CO_2 range greater than 60 ppm were used to conduct the regression, and the ordinary least squares regression (OLS, model I) was used to obtain the intercept of regression, as suggested by Chen et al. [25].

2.5. Sampling and Carbon Isotope Analysis of the Ecosystem Compartments

Leaves of the five dominant species were collected in the mornings of sunny days once a month from May to September 2019 around the flux tower. Specifically, nine mature trees of each species were randomly selected on each sampling day; then, mature and healthy leaves were collected at three canopy heights (lower, middle, and upper). Since the trees are different in height and the canopy structure is different among species, the three canopy heights of each tree were determined relative to their canopy height range rather than using absolute heights. We collected bark and xylem at breast height and coarse and fine roots of the five species, as well as litters in the undecomposed layer (Oi) and decomposed layer (Oe + Oa) and the soil at four depths (0–5, 5–10, 10–20, and 20–40 cm) with nine replicates around the flux tower at the end of September 2019.

All samples were immediately transported to the lab and dried at 65 °C for 4 days to a constant weight and then ground with a ball mill (MM 400, Retsch, Haan, Germany). The C isotope compositions of nine sample replicates were determined separately using an elemental analyzer (Flash EA1112, Thermo Finnigan, Milan, Italy) coupled with a mass spectrometer (Finnigan MAT 253, Bremen, Germany). The overall precision of the $\delta^{13}C$ measurement was < ±0.2‰. All $\delta^{13}C$ values are reported on the VPDB (Vienna Pee Dee Belemnite) scale.

2.6. Calculation of Photosynthetic Carbon Isotope Discrimination

C isotope discrimination of the leaf (Δ_{leaf}) was calculated as

$$\Delta_{leaf} = \frac{\delta_{air} - \delta_{leaf}}{\delta_{leaf}/1000 + 1} \tag{1}$$

where δ_{air} is the averaged C isotope composition of the canopy CO_2 (‰) measured at 26 m within the canopy during the sampling day, and δ_{leaf} is the averaged C isotope composition of the leaves (‰) for all dominant species and canopy heights.

Canopy-scale photosynthetic discrimination (Δ_{canopy}) was calculated using Farquhar's classical model ($\Delta_{classical}$) [1] and compared with Δ_{leaf} to provide robust evidence for variations in C isotope discrimination over time. $\Delta_{classical}$ describes the fractionation in CO_2 diffusion, carboxylation fractionation, and respiratory fractionation as

$$\Delta_{classical} = \bar{a} + (b - \bar{a})\frac{c_c}{c_a} - f\frac{\Gamma^*}{c_a} - e\frac{R_d}{kc_a} \tag{2}$$

where $\bar{a}$ is the overall diffusional fractionation for CO_2 as calculated from the diffusional fractionation factors (fractionation across the boundary layer (2.9‰), fractionation across the stoma (4.4‰), and fractionation across the mesophyll, including dissolution (1.1‰ at 25 °C) and diffusion (0.7‰) in water) and the conductance factors of CO_2 across the foliar boundary layer, stoma, and mesophyll during photosynthesis based on the big-leaf model according to Wehr and Saleska [33]; b is the Rubisco fractionation in C_3 plants and assumed to be 27.5‰ [34]; f is the respiratory fractionation for photorespiration (11‰) [33]; e is the respiratory fractionation for daytime dark respiration (−5‰) [33]; Γ^* is the CO_2 compensation point (μmol mol^{-1}), which was calculated from the leaf temperature according to Brooks and Farquhar [35]; R_d is the daytime respiration rate (μmol m^{-2} s^{-1}), which is assumed to be 3 μmol m^{-2} s^{-1} [36]; k is the carboxylation efficiency (μmol m^{-2} s^{-1}), which is assumed to be 0.1 μmol m^{-2} s^{-1} [36]; c_c is the CO_2 concentration in the chloroplast (ppm); and c_a is the CO_2 concentration in the canopy air (ppm) measured at 26 m above the ground (see Section 2.3). The c_c/c_a was obtained by the numerical solution of the equation derived from the theory of isoflux-based isotopic flux partitioning (IFP) [19], which is expressed as

$$\frac{c_c}{c_a} = \frac{-(\delta_a + b - \delta_{NR} - 2\bar{a} + \frac{f\Gamma^*}{c_a} + \frac{eR_d}{kc_a}) \pm \sqrt{(\delta_a + b - \delta_{NR} - 2\bar{a} + \frac{f\Gamma^*}{c_a} + \frac{eR_d}{kc_a})^2 - 4(\bar{a} - b)(\delta_{NR} + \bar{a} - \delta_a - \frac{f\Gamma^*}{c_a} - \frac{eR_d}{kc_a} - \frac{isoflux - \delta_{NR}NEE}{\bar{g}c_a})}}{2(\bar{a} - b)} \tag{3}$$

where δ_a is the $\delta^{13}C$ of canopy CO_2 measured at 26 m above the ground (‰) (see Section 2.3); δ_{NR} is $\delta^{13}C_{eco}$ (see Section 2.4) but was substituted by the weekly smoothed value to capture the seasonal variation; isoflux is the C isotopic flux calculated from the eddy covariance measurement (see Section 2.2), the canopy CO_2 $\delta^{13}C$ profile measurement (see Section 2.3), and the intercept of the daytime (7:00–17:00) Keeling plot according to [18,37]; and NEE is the net ecosystem exchange flux (μmol CO_2 m^{-2} s^{-1}). Note that Δ_{canopy} was calculated using data from the daytime and expressed as the daily average.

2.7. Statistical Analyses

Student's *t*-test was used to assess the differences in mean $\delta^{13}C$ (air, organic matter, and respiration) among sampling positions, C pools, or species. A one-way analysis of variance (ANOVA) was performed to assess the significance of the seasonal variations of $\delta^{13}C_{air}$, leaf $\delta^{13}C$, Δ_{leaf}, $\delta^{13}C_R$, and $\delta^{13}C_{eco}$. To evaluate how the $\delta^{13}C_R$ of the leaf, trunk, and soil respond to environmental factors, we conducted correlation analyses and linear regression analyses using the $\delta^{13}C_R$ of each measurement and the environmental data of the corresponding timespans. Correlation analyses were also conducted to assess the environmental effects on $\delta^{13}C_{eco}$ and the potential lagged responses of $\delta^{13}C_{eco}$ to environmental factors using daily $\delta^{13}C_{eco}$ and daily average meteorological data. The environmental factors include air temperature, vapor pressure deficit (VPD), global radiation, soil temperature, and soil moisture. These factors were taken into consideration because they are expected to influence C isotope discriminations in above- and belowground processes. The lagged correlation was tested with shifted time periods from zero to 10 days.

3. Results

3.1. Seasonal Variation of $\delta^{13}C_{air}$

In the year 2019, the mean annual temperature was 5.4 °C and annual rainfall was 627 mm, making 2019 warmer and drier than the long-term average (see Section 2.1). During the growing season of 2019, the meteorological conditions showed significant seasonal fluctuations (Figure 2). The vapor pressure deficit (VPD) was higher in May and September, and air temperature was also higher during these periods, which resulted in remarkable soil water deficits. Soil temperature and air temperature reached their highest points at the end of July. In July and August, precipitation and soil moisture were the highest, especially at the end of July and in the middle of August, when the accumulated rainfall accounted for 15.7% and 19.6% of the annual total, respectively.

We found a significant seasonal variation in $\delta^{13}C_{air}$ during the growing season ($p < 0.01$, ANOVA; Figure 3a). The average $\delta^{13}C_{air}$ of the growing season was −9.25‰. At the beginning of the growing season, $\delta^{13}C_{air}$ showed a slight decrease and then reached a maximum of about −8.5‰ in early July and then started to decrease until it reached a minimum of below −10‰ in the middle of August. In September, $\delta^{13}C_{air}$ first showed an increase and then started to decrease in the middle of September. Moreover, $\delta^{13}C_{air}$ showed remarkable day-to-day variation. The largest difference in $\delta^{13}C_{air}$ between two consecutive days was up to about 2‰. Short-term rapid depletions of ^{13}C in the forest occasionally occurred during the growing season, such as the relative depletion of ^{13}C at the beginning of September. The difference of $\delta^{13}C_{air}$ between the above-canopy (−9.13‰) and within-canopy (−9.18‰) was small but statistically significant ($p < 0.01$, *t*-test). $\delta^{13}C_{air}$ within the canopy was significantly more positive ($p < 0.001$, *t*-test) than the $\delta^{13}C_{air}$ below the canopy (−9.44‰), and $\delta^{13}C_{air}$ below the canopy was always the lowest throughout the growing season.

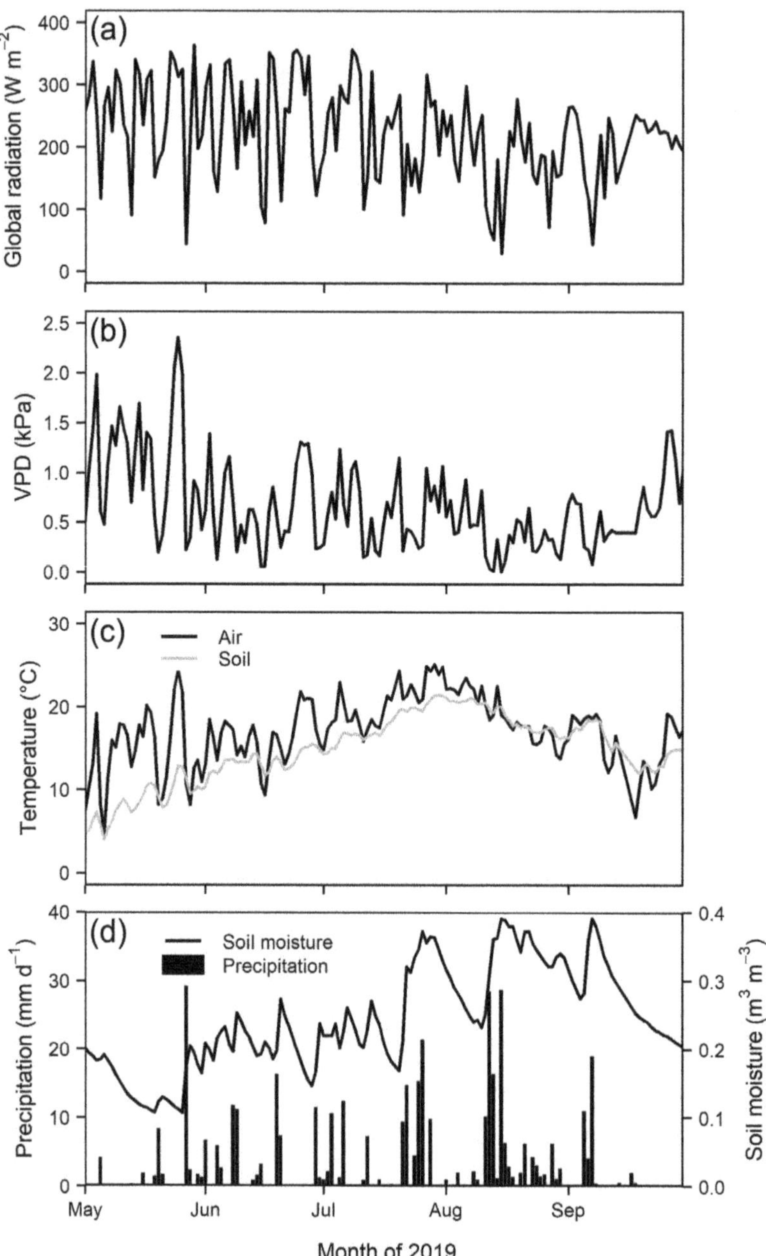

Figure 2. Seasonal variations in global radiation (**a**), vapor pressure deficit (VPD) (**b**), air and soil temperature (**c**), and soil moisture and precipitation (**d**) during the 2019 growing season. Global radiation, VPD, and air temperature are the averages of the seven heights in the forest. Soil temperature and soil moisture were measured at a 5 cm depth.

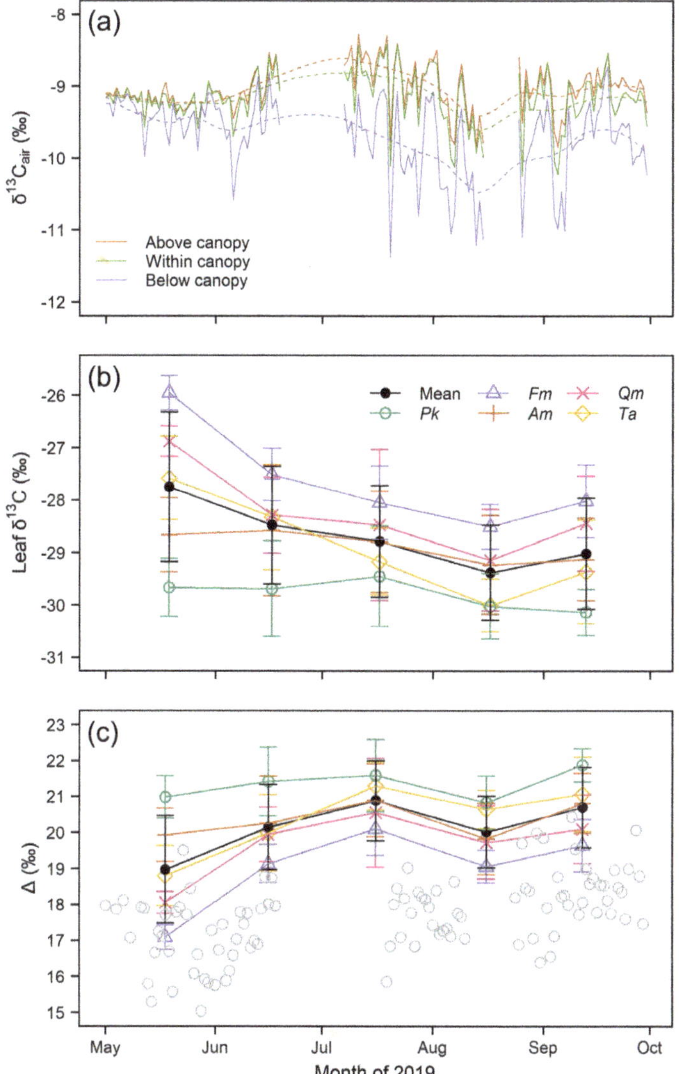

Figure 3. Seasonal variation in the $\delta^{13}C$ of atmospheric CO_2 ($\delta^{13}C_{air}$) (**a**), seasonal variation in the $\delta^{13}C$ of leaf bulk organic matter (**b**), and C isotope discrimination at the leaf and canopy levels (**c**) during the growing season of 2019. $\delta^{13}C_{air}$ values are given as daily averages. The $\delta^{13}C_{air}$ values below the canopy were averaged from data collected at heights of 2.5 and 8 m; the $\delta^{13}C_{air}$ values within the canopy were averaged by data collected at heights of 22, 26, and 32 m; the $\delta^{13}C_{air}$ values above the canopy were averaged by data collected at heights of 50 and 60 m. The dashed line is the locally weighted scatterplot smoothing (LOESS) with span = 0.3. The two periods of missing data in June and August were caused by power failure. Leaf $\delta^{13}C$ values (mean ± SE, n = 9) were aggregated from three canopy heights (lower, middle, and upper). Δ_{leaf} was calculated from leaf $\delta^{13}C$ and $\delta^{13}C_{air}$ and separated by species; gray circles represent Δ_{canopy} derived from Farqhar's $\Delta_{classical}$ model. *Pk*, Korean pine (*P. koraiensis*); *Fm*, Manchurian ash (*F. mandshurica*); *Am*, Mono maple (*A. mono*); *Qm*, Mongolian oak (*Q. mongolica*); *Ta*, Tuan linden (*T. amurensis*). Error bars indicate standard deviation.

3.2. Carbon Isotopic Compositions of Leaves and Photosynthetic Carbon Isotope Discrimination

Considering the complexity of the mixed forest, we determined the leaf $\delta^{13}C$ for all five dominant species and along the vertical position of their canopies. In terms of temporal variation, leaf $\delta^{13}C$ in broadleaved species showed significant seasonal variations ($p < 0.001$, ANOVA) except for *A. mono* ($p = 0.516$, ANOVA), with a gradual decrease from May to August by a magnitude of 2‰, followed by an increase after August (Figure 3b). In contrast to broadleaves, the $\delta^{13}C$ in the needles of *P. koraiensis* did not vary significantly over time ($p = 0.275$, ANOVA) and showed a slight decrease throughout the growing season, except for a relatively small increase in July (Figure 3b). $\delta^{13}C$ in the needles of *P. koraiensis* (−29.79‰) was also the lowest among the species (Figure 3b). Since broadleaved trees are the main components of the mixed forest in this study, variation in the mean leaf $\delta^{13}C$ values of the five dominant species followed the same seasonal patterns as the four broadleaved species. The temporal variation of leaf $\delta^{13}C$ corresponded to that of $\delta^{13}C_{air}$, except in May, when $\delta^{13}C_{air}$ was relatively low, while leaf $\delta^{13}C$ was the highest compared to the rest of the growing season (Figure 3a,b), indicating that leaf $\delta^{13}C$ depends not only on the C isotopic signal carried by CO_2 in the air.

C isotope discrimination at both the leaf and canopy level is shown in Figure 3c. Δ_{leaf} was calculated directly from the leaf $\delta^{13}C$ and $\delta^{13}C_{air}$ of the corresponding sampling days (Equation (1)). Species-specific differences in Δ_{leaf} were found over the growing season. Except for *A. mono* ($p = 0.084$, ANOVA), Δ_{leaf} varied significantly ($p < 0.05$ for *P. koraiensis* and $p < 0.001$ for the other species, ANOVA) over time. On average, Δ_{leaf} showed an increasing trend from May (18.97‰) to July (20.70‰), while the Δ_{leaf} of both species showed a decrease in August compared to July and September (Figure 3c). This variation pattern of leaf level discrimination was strongly supported by the Δ_{canopy} estimated by the classical Farquhar's model (Equation (2)), although Δ_{canopy} was generally lower than Δ_{leaf} (Figure 3c). Δ_{canopy} also showed a general increasing trend over time, with the lowest value observed in May. Since there are two periods of missing measurements, we can only deduce that the Δ_{canopy} reached its maximum in early to middle July, followed by relatively lower values in the middle of August, according to the increasing trend in June and the decreasing trend at the beginning of August (Figure 3c).

3.3. Carbon Isotopic Compositions of Ecosystem Organic Pools and Respired CO_2

Seasonal variation in the $\delta^{13}C_R$ of different ecosystem pools (leaf, trunk, and soil), as well as $\delta^{13}C_{eco}$, is shown in Figure 4. Over the entire growing season, the $\delta^{13}C_R$ of leaves and trunks varied within a range of 3.2‰ and 4.5‰, respectively. A statistically significant general increase of $\delta^{13}C_R$ was found in the needles of *P. koraiensis* ($p < 0.001$, ANOVA), while no such phenomenon was found in the leaves of *A. mono* ($p = 0.422$, ANOVA; Figure 4a). The trunk $\delta^{13}C_R$ of both *P. koraiensis* and *F. mandshurica* varied significantly ($p < 0.001$, ANOVA) throughout the growing season. Trunk $\delta^{13}C_R$ showed a decreasing trend from May to August and a slight increase in September (Figure 4b). This variation trend was similar to the general seasonal pattern of leaf $\delta^{13}C$ (Figures 3a and 4b). Soil $\delta^{13}C_R$ did not vary significantly over time ($p = 0.084$, ANOVA) and was also relatively stable (with a range of 2.3‰) compared to leaf and trunk $\delta^{13}C_R$ (Figure 4c). Furthermore, a decrease of $\delta^{13}C_R$ was found from May to June in leaf, trunk, and soil (Figure 4a–c).

$\delta^{13}C_R$ exhibited a slight depletion from leaves (−26.33‰) to trunks (−25.67‰) and soil (−25.48‰), as did $\delta^{13}C$ in the corresponding organic matter (leaf: −28.68‰, trunk: −26.55‰, soil: −25.72‰; Figure 5). A distinct spatial pattern of leaf $\delta^{13}C$ was found throughout the growing season (Figure 5). Similar to $\delta^{13}C_{air}$, leaf $\delta^{13}C$ values increased significantly with an increase of height (−29.16‰, −28.73‰, and −28.16‰ from the lower to middle and upper canopy; $p < 0.001$, ANOVA). Leaf $\delta^{13}C_R$ was generally more enriched compared to the $\delta^{13}C$ in leaf organic matter but closer to the $\delta^{13}C$ in the upper leaf (−28.16‰), which indicates that a substantial quantity of photosynthate from the upper canopy was used as respiratory substrate. Trunk $\delta^{13}C_R$ was also more enriched than the $\delta^{13}C$ in trunk organic matter but showed a larger range of variation (Figure 5). Soil $\delta^{13}C_R$ was more enriched than the $\delta^{13}C$ of organic matter on the surface (0–5 cm) soil (−26.59‰), in the roots (−28.02‰), and especially in litters ($-28.40‰$). However, the soil $\delta^{13}C_R$ was similar ($p = 0.527$, *t*-test) to the $\delta^{13}C$ of soil in deeper layers (5–40 cm) ($-25.43‰$; Figure 5).

Nights with a CO_2 range lower than 60 ppm were excluded from $\delta^{13}C_{eco}$ calculations because a small CO_2 range could cause uncertainties in fitting the Keeling plot. After applying this exclusion, all of the values for R^2 in the regression were greater than 0.85 and yielded a sufficient quantity of reliable $\delta^{13}C_{eco}$ data (Figure 4d). The variation in $\delta^{13}C_{eco}$ was significant over the growing season ($p < 0.001$, ANOVA), with an average of $-25.79‰$. $\delta^{13}C_{eco}$ steadily increased as the growing season progressed (Figure 4d), and the $\delta^{13}C_{eco}$ values were within the range of the $\delta^{13}C_R$ in the leaves, trunks, and soil (Figure 5). Some short-term $\delta^{13}C_{eco}$ variations were observed, such as two gradual decreases in May and July, which may be attributed to the gradual ^{13}C depletion of CO_2 in the air and respired from C pools during these periods (Figure 3; Figure 4).

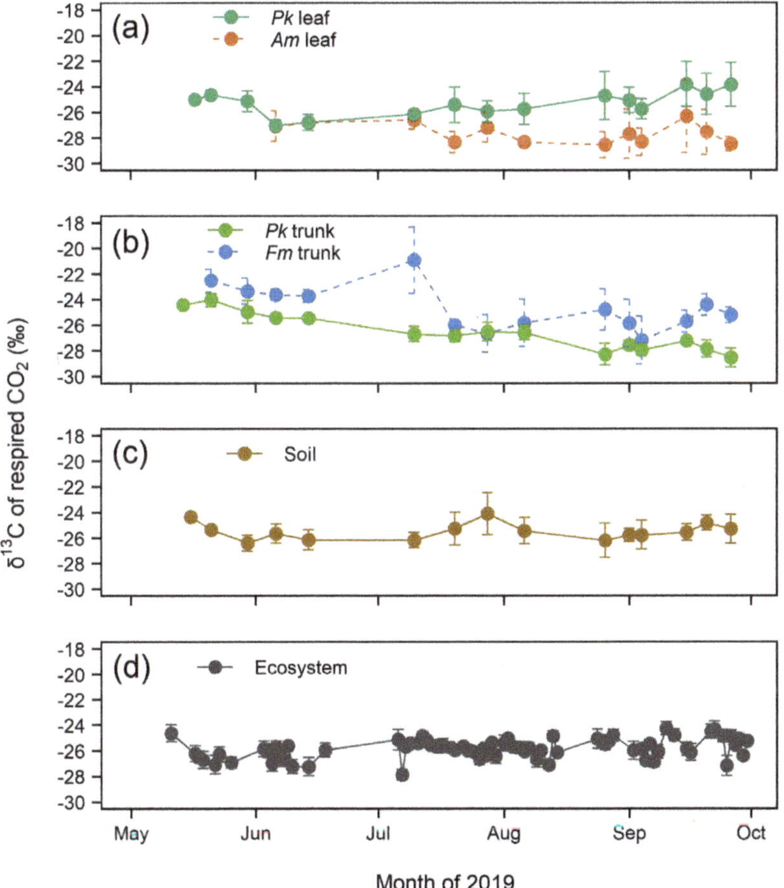

Figure 4. Seasonal variations in the $\delta^{13}C_R$ of leaves (**a**), trunks (**b**), and soil (**c**), and in $\delta^{13}C_{eco}$ (**d**). $\delta^{13}C_R$ and $\delta^{13}C_{eco}$ were both derived using the Keeling plot method. The $\delta^{13}C_R$ of leaves, trunks, and soil were averaged from four measurements between 7:00 and 18:00, and the error bars indicate standard deviation. $\delta^{13}C_{eco}$ was determined from the nighttime (21:00–03:00) CO_2 and $\delta^{13}C_{air}$ profiles, and the error bars indicate standard errors of the intercept of the Keeling plot. *Pk*, Korean pine (*P. koraiensis*); *Fm*, Manchurian ash (*F. mandshurica*); *Am*, Mono maple (*A. mono*); *Qm*, Mongolian oak (*Q. mongolica*).

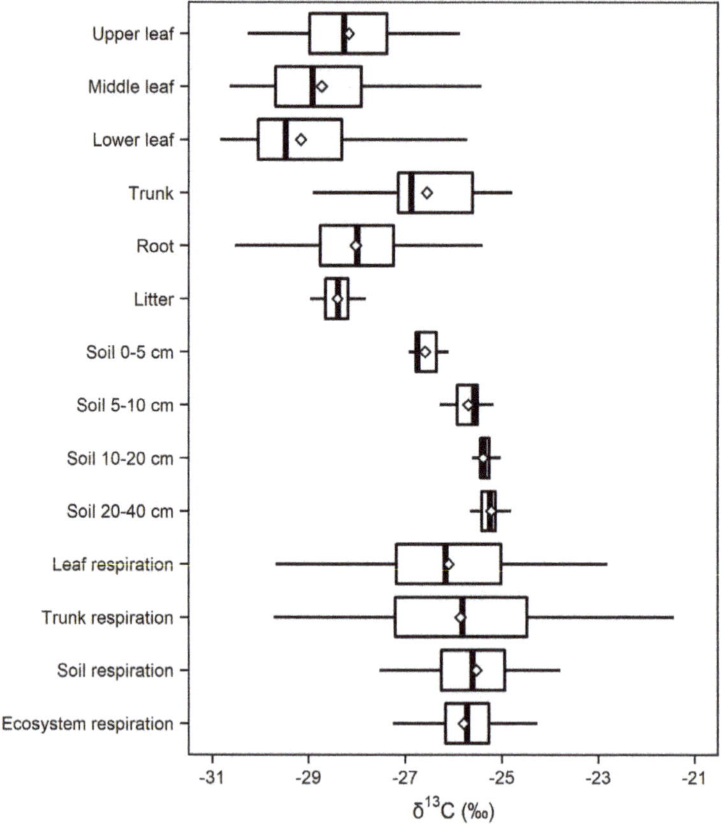

Figure 5. A comparison of the C isotopic composition of ecosystem organic pools and respiratory fluxes. Diamonds represent the means. The box represents the median and the 25% upper/lower quartiles. The tails represent the 10% and 90% limits of the data. Leaf samples were taken once a month from May to September 2019. The trunk values are the average of the bark and xylem, the root values are the average of the coarse and fine roots, and the litter values are the average of the undecomposed layer (Oi) and the decomposed layer (Oe + Oa). These samples, along with the soil samples, were sampled at the end of September 2019. The leaf, trunk, and soil $\delta^{13}C_R$ were measured 15 times from May to September 2019. $\delta^{13}C_{eco}$ was determined at a daily scale from May to September 2019.

3.4. Relationship between $\delta^{13}C_R$ and Environmental Factors

The correlation analyses showed that the $\delta^{13}C_R$ in the trunks of both *P. koraiensis* and *F. mandshurica* had a significant negative correlation with air temperature, soil temperature, and soil moisture ($p < 0.01$; Figure 6 and Table 1). Soil $\delta^{13}C_R$ only showed a significant positive correlation with global radiation ($r = 0.312$, $p < 0.05$; Figure 6 and Table 1). No significant correlation was found between leaf $\delta^{13}C_R$ and any environmental factors (Figure 6 and Table 1).

We conducted a correlation analysis between $\delta^{13}C_{eco}$ and environmental factors with a time window from zero to 10 days. It was found that $\delta^{13}C_{eco}$ was negatively correlated with air temperature ($p < 0.05$; Table 2) measured on the same day as $\delta^{13}C_{eco}$. And $\delta^{13}C_{eco}$ was negatively correlated with VPD ($p < 0.05$; Table 2) and positively correlated with soil moisture ($p < 0.01$; Table 2) with a lag time of 10 days. Further, $\delta^{13}C_{eco}$ was negatively correlated with global radiation and positively correlated with soil temperature, but this correlation was not statistically significant (Table 2).

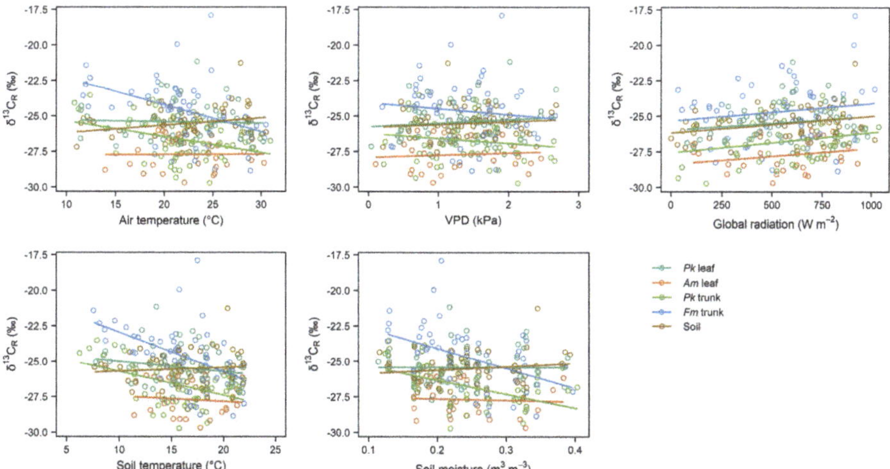

Figure 6. Relationship between the $\delta^{13}C_R$ of different ecosystem organic pools and environmental factors. The data shown are the $\delta^{13}C_R$ of each measurement and the environmental factors of the corresponding half hour when the $\delta^{13}C_R$ was measured. VPD, vapor pressure deficit; *Pk*, Korean pine (*P. koraiensis*); *Fm*, Manchurian ash (*F. mandshurica*); *Am*, Mono maple (*A. mono*); *Qm*, Mongolian oak (*Q. mongolica*).

Table 1. Pearson correlation coefficients for the $\delta^{13}C_R$ from the C pools with environmental factors in the growing season. The $\delta^{13}C_R$ of each measurement and the corresponding half-hour meteorological data were used to conduct this analysis.

Pearson's Correlation Coefficient (r)	$\delta^{13}C_R$ (‰)				
	Pk **Leaf**	*Am* **Leaf**	*Pk* **Trunk**	*Fm* **Trunk**	**Soil**
Air temperature (°C)	−0.035	0.009	−0.363 **	−0.443 **	0.234
VPD (kPa)	0.110	0.086	−0.146	−0.120	0.101
Global radiation (W m^{-2})	0.166	0.242	0.244	0.156	0.312 *
Soil temperature (°C)	−0.199	−0.101	−0.462 **	−0.513 **	0.092
Soil moisture (m^3 m^{-3})	−0.000	−0.056	−0.484 **	−0.469 **	0.153

Note: * $p < 0.05$, ** $p < 0.01$.

Table 2. Pearson correlation coefficients for the $\delta^{13}C_{eco}$ with environmental factors in the growing season. $\delta^{13}C_{eco}$ and daily average meteorological data were used to conduct this analysis. The lag time in days and the Pearson's correlation coefficient (r) are presented, and only the correlation that provided the best r value is presented.

	Lag	**r**
Air temperature (°C)	0	−0.256 *
VPD (kPa)	10	−0.280 *
Global radiation (W m^{-2})	0	−0.217
Soil temperature (°C)	10	0.085
Soil moisture (m^3 m^{-3})	10	0.295 **

Note: * $p < 0.05$, ** $p < 0.01$.

4. Discussion

4.1. Variations in Carbon Isotope Discrimination from Assimilation to Respiration

Photosynthetic C isotope discrimination and $\delta^{13}C_{air}$ both influenced leaf $\delta^{13}C$. In this study, the influence of photosynthetic discrimination on leaf $\delta^{13}C$ was greater in the growing season most of the time. For example, from May to July, although $\delta^{13}C_{air}$ showed a moderate increase, both Δ_{leaf} and Δ_{canopy} showed a strong increase, which directly led to a continuous decrease in leaf $\delta^{13}C$ during this period (Figure 3). This increase in Δ_{leaf} and Δ_{canopy} may have resulted from an increase in stomatal conductance under the gradually warmer and wetter weather conditions from May to July (Figure 2). However, the influence of photosynthetic discrimination may have become weak for determining the leaf $\delta^{13}C$ in August, at which point we also found that leaf $\delta^{13}C$ reached its minimum, and Δ_{leaf} and Δ_{canopy} were not the highest (Figure 3). On the other hand, $\delta^{13}C_{air}$ influenced leaf $\delta^{13}C$ more profoundly in terms of height. Since ^{13}C depleted CO_2 relative to the atmospheric CO_2 respired by the soil and understorey and due to the reduction in mixing between the air inside the forest and the atmosphere with a decrease in height, canopy $\delta^{13}C_{air}$ was usually lower, as it was closer to the ground [38,39], which resulted in a similar spatial pattern between leaf $\delta^{13}C$ and $\delta^{13}C_{air}$ (Figure 5).

The seasonal pattern of photosynthetic C isotope discrimination was found to be species-specific. Compared with deciduous broadleaf species, the coniferous species (*P. koraiensis*) was found to have the highest Δ_{leaf} in the growing season (Figure 3), which is consistent with a previous study conducted at the same site as our study [40]. Low photosynthetic activity and higher stomatal conductance could result in a higher c_i/c_a and, consequently, higher Δ_{leaf} [41]. Therefore, *P. koraiensis* in this temperate mixed forest may have a lower photosynthetic rate and higher stomatal conductance compared to deciduous species. Moreover, $\delta^{13}C$ in the needles of *P. koraiensis* showed smooth depletion during the growing season, resulting in less varied Δ_{leaf} over time compared to the deciduous broadleaf species (Figure 3). This difference is most likely due to the large proportion of slow-turnover compounds (e.g., starch and pinitol) in needles compared to broadleaves. These compounds generally have more stable seasonal $\delta^{13}C$ variation than other carbohydrates [11,42,43] and can strongly dampen the overall $\delta^{13}C$ signals of needle bulk organic matter [43,44].

Our study also found apparent post-photosynthetic C isotope discrimination among the C pools from leaf to trunk and root (Figure 5). More importantly, our results show that the leaf (leaves of *P. koraiensis* and *A. mono*) and trunk (trunks of *P. koraiensis* and *F. mandshurica*) respired CO_2 that was ^{13}C-enriched in comparison to the leaf and trunk organic matter of the domain species (Figure 5). Due to the long duration of a single measurement (see Section 2.4), it was unfeasible to establish more chambers to cover more species and at different heights or to study more replicates. Despite this weakness in the $\delta^{13}C_R$ measurements, the results obtained by our study are more abundant than those in many previous studies [20,45,46] and indicate the effects of post-photosynthetic C isotope discrimination in the process of dark respiration. In the process of dark respiration, ^{13}C-enriched C-3 and C-4 positions in glucose are preferably used to produce pyruvate during glycolysis; then, pyruvate is decarboxylated by pyruvate dehydrogenase (PDH) to release ^{13}C-enriched CO_2. A greater contribution of CO_2 decarboxylated in the PDH will result in more positive $\delta^{13}C_R$ [47].

Soil $\delta^{13}C_R$ was more positive than $\delta^{13}C$ in the potential respiratory substrates, including roots, litter, and soil on the surface; however, it was similar to $\delta^{13}C$ in the deeper soil (Figure 5). Soil respiration is mainly performed by autotrophic components (roots and rhizosphere microbes) and heterotrophic components (the decomposition of litter and soil organic matter). Among the autotrophic components, root respiration contributes significantly to soil respiration with a generally more negative $\delta^{13}C_R$ signal compared to $\delta^{13}C$ in root organic matter, as reviewed in [4,46,48]. Along with the relatively negative root $\delta^{13}C_R$, if we further assume that litter $\delta^{13}C_R$ was mildly more positive than $\delta^{13}C$ in litter organic matter during the process of decomposition, as shown in previous studies [49,50], more positive $\delta^{13}C_R$ signals would still be needed from other belowground processes to produce the exact soil $\delta^{13}C_R$ values in our forest, considering that litter respiration generally contributes less to soil respiration.

Many studies have shown that $\delta^{13}C$ in microbial biomass is more positive (1–2‰) relative to that in total soil C [4,51,52], and that microbial CO_2 is also more ^{13}C-enriched compared to soil organic matter [53]. Furthermore, the $\delta^{13}C$ in soil CO_2 seems to behave similarly to the typical patterns of soil $\delta^{13}C$ isotopic enrichment with soil depth. For example, Goffin et al. [54] found that soil air $\delta^{13}C$ in the deeper layer (10–20 cm) was more positive than that on the surface (0–10 cm) in a Scots pine stand. Using incubated soil, Formánek and Ambus [55] also found more positive $\delta^{13}C$ in the CO_2 respired from the deeper layer (15–38 cm) than that from the surface (3–10 cm). Thus, the soil $\delta^{13}C_R$ values observed in this study may have resulted from mixing between the $\delta^{13}C$ signal from autotrophic respiration and a relatively large portion of the $\delta^{13}C$ signal from microbe decomposition of soil organic matter, especially soil organic matter in the deeper layers, which indicates that the soil CO_2 flux in our forest was mainly from heterotrophic components.

4.2. Seasonal Variations in the $\delta^{13}C_R$ and its Linkage to Environmental Factors

The lack of replication of in situ online $\delta^{13}C_R$ measurements using non-automated systems is a common problem in previous studies [27,56], especially for studies measuring the $\delta^{13}C_R$ of different C pools simultaneously [12,17], including the present study. Nonetheless, the continuous $\delta^{13}C_R$ measurements in the present study accurately captured the seasonal variation of $\delta^{13}C_R$ in the target components. No consistent pattern was found in the $\delta^{13}C_R$ between the needles of *P. koraiensis* and the leaves of *A. mono* throughout the growing season (Figure 4). One possible reason is that the $\delta^{13}C$ in the leaf respiratory substrates of *P. koraiensis* and *A. mono* experienced relatively weak seasonal variations (Figure 3), because leaf $\delta^{13}C_R$ found to be linked with the $\delta^{13}C$ of leaf metabolites [5,57]. By contrast, both of the $\delta^{13}C_R$ values in the trunks of *P. koraiensis* and *F. mandshurica* showed a decreasing trend over the growing season (Figure 4). This result is consistent with a previous study of *Quercus petraea* in a temperate forest [8]; they ascribed that to the use of ^{13}C-enriched respiratory substrates (i.e., starch) in trunks at the beginning of the growth period. The seasonal changes in leaf $\delta^{13}C_R$ differed from the $\delta^{13}C$ of leaf organic matter (Figures 3 and 4), which may indicate changes in the proportion of different respiratory substrates used in leaf respiration during the growing season [42]. However, trunk $\delta^{13}C_R$ shared a similar seasonal variation pattern to the $\delta^{13}C$ in leaf organic matter (Figures 3 and 4). This may be because a large amount of assimilated photosynthate was transported through the phloem to the trunk, and these were subsequently used as substrates for trunk respiration over the entire growing season.

Furthermore, due to the weak seasonal variation of leaf $\delta^{13}C_R$ in *P. koraiensis* and *A. mono*, we found no significant correlation between leaf $\delta^{13}C_R$ and the environmental variables (Table 1). It is still unclear how the leaf $\delta^{13}C_R$ of other dominant species in our forest are linked to environmental variables due to our limited chamber measurements. However, as the $\delta^{13}C$ in the leaf organic matter of other dominant species showed larger seasonal variations (Figure 3), it is expected that the leaf $\delta^{13}C_R$ of other dominant species could have more significant seasonal variations and could be more sensitive to environmental conditions. The trunk $\delta^{13}C_R$ in our forest was found to be negatively correlated with air temperature (Table 1), which is consistent with the results of Maunoury et al. [8] on sessile oak in France. Interestingly, our study found that the $\delta^{13}C$ in trunk respiration was more sensitive to environmental changes at a seasonal scale than that in other flux components, regardless of species. Comparably, the successful utilization of $\delta^{13}C$ in tree rings to reconstruct climate demonstrates that $\delta^{13}C$ in the structural carbohydrates in wood is closely related to environmental changes [58]. This suggests that the specific compounds responsible for tree ring $\delta^{13}C$ may also be responsible for the $\delta^{13}C$ in trunk respiration, resulting in the sensitivity of trunk $\delta^{13}C_R$ to environmental factors.

Soil $\delta^{13}C_R$ showed a relatively stable variation over the growing season (Figure 4). Variability in soil $\delta^{13}C_R$ has been linked to environmental factors such as air temperature, soil moisture, VPD, and photosynthetically active radiation [6,13,59,60]. Our results found that soil $\delta^{13}C_R$ had no significant correlation with environmental factors except for global radiation (Figure 6 and Table 1). This result is partly supported by Bowling et al. [56], who measured soil $\delta^{13}C_R$ in a US subalpine forest using chambers. The authors' results showed that environmental forcing does not induce temporal variation in soil $\delta^{13}C_R$. The lack of a linkage between soil $\delta^{13}C_R$ and environmental factors at our site could be due to the lack of in soil $\delta^{13}C_R$ variation during the growing season. The large range in potential soil respiratory substrates [3] and the correspondingly large range of $\delta^{13}C$ values could also have dampened the isotopic linkage between soil $\delta^{13}C_R$ and current environmental conditions. Moreover, a recent study suggested that variations in soil $\delta^{13}C_R$ are mainly derived from diffusive fractionation rather than biological causes [32], which provides another possible explanation for the weak relationship between soil $\delta^{13}C_R$ and the environmental factors we found.

4.3. $\delta^{13}C_{eco}$ Reveals Short-Term Links between Aboveground and Belowground Processes

The $\delta^{13}C_{eco}$ calculated by the Keeling plot method using nighttime $\delta^{13}C$ and CO_2 profile data represents the integrated $\delta^{13}C$ of the CO_2 produced by all above- and belowground respiring components in the forest. The results showed that the average $\delta^{13}C_{eco}$ in the growing season is between the average $\delta^{13}C_R$ of the trunk and the soil (Figure 5), which is consistent with the results reported by Wingate et al. [17].

From a short-term perspective, aboveground vegetation could determine the biogeochemical processes belowground through, for example, newly fixed photoassimilates transport to roots, which then contribute to soil respiration. This linkage between assimilates and respiration could be revealed by relating the $\delta^{13}C$ of ecosystem respiration to environmental factors but with a time lag from hours to days [9,10,16,61]. For example, Fessenden and Ehleringer [59] found that soil moisture measured at a 40 cm depth was negatively correlated with $\delta^{13}C_{eco}$ on the day of the measurements in an old-growth coniferous forest. Scartazza et al. [62] also found a negative relationship between soil moisture measured at 70–88 cm depth and $\delta^{13}C_{eco}$ measured on the same day in a beech forest, whereas Mortazavi et al. [6] found that $\delta^{13}C_{eco}$ was negatively correlated to soil moisture for 7 days before $\delta^{13}C_{eco}$ measurements. The negative correlation between soil moisture and $\delta^{13}C_{eco}$ was also found to be significant in a Douglas-fir forest but not in a ponderosa pine forest [63]. However, our results suggest that the correlation between soil moisture and $\delta^{13}C_{eco}$ is positive but weak and that a lag of 10 days provides the best correlation (Table 2).

The VPD was found to be positively related to $\delta^{13}C_{eco}$ after a lag of 3 days [63], 3–4 days [6], 4–5 days [14,61], or 5–10 days [16]. Again, our results show an opposite correlation pattern between VPD and $\delta^{13}C_{eco}$ compared to previous studies, and a lag of 10 days gave the best (but still weak) correlation (Table 2). In previous studies, Bowling et al. [64] and Schaeffer et al. [65] found no correlation between $\delta^{13}C_{eco}$ and soil moisture or VPD. We also found that $\delta^{13}C_{eco}$ has no significant correlation with the other environmental factors we measured, regardless of the time lag (Table 2). The weak impact of air temperature, VPD, and soil moisture on $\delta^{13}C_{eco}$ in our study reflects the weak link between assimilation and respiration at a seasonal scale. Additionally, stronger correlations were reported by Schaeffer et al. [65] after they separately calculated the $\delta^{13}C_{eco}$ using data within or below the canopy by selectively removing periods when the air within and below the canopy was well mixed. These findings suggest that the environmental controls for $\delta^{13}C_{eco}$ may also be influenced by other factors, such as the canopy's aerodynamic properties.

Although we found different seasonal isotopic patterns in different respiratory C pools and different environmental factors can contribute differently to the $\delta^{13}C_R$ of those pools over the entire growing season, some strong correlations were still observed between the $\delta^{13}C_R$ of the leaf, trunk, and soil and the $\delta^{13}C_{eco}$. The most obvious occurred in May, when the $\delta^{13}C$ from respiratory C pools as well as the ecosystem showed decreasing trends (Figure 4). These distinct decreases may have resulted from unique extreme weather events in May, when the ecosystem received a large amount of global radiation but less rainfall, which led to an overall increase in VPD, a decrease in soil moisture, and a large temperature difference between the air and the soil (Figure 2). The concurrent decrease in $\delta^{13}C_R$ and $\delta^{13}C_{eco}$ suggests that although the linkage between assimilates and respiration was weak at a seasonal scale in our study, recently fixed assimilates could be allocated to the location of respiration in a short period of time, indicating a quick link between different respiration pools and assimilation at the beginning of the growing season in the coniferous and broad-leaved mixed forest.

5. Conclusions

Overall, the observed difference in seasonal variation in photosynthetic discrimination between the deciduous broadleaved and evergreen needle species in our forest indicated a remarkable difference in C assimilation between them. As expected, post-photosynthetic discrimination explained the substantial isotopic differences among C pools and their respiratory fluxes during the process of photosynthate allocation and respiration. Compared to leaf and soil $\delta^{13}C_R$, trunk $\delta^{13}C_R$ was more sensitive to environmental changes and related to the seasonal patterns of leaf $\delta^{13}C$, which suggests that trunk $\delta^{13}C_R$ has great potential as a way to record discrimination-encoded biochemical and physiological information related to C cycling. In addition, the linkages between environmental factors and $\delta^{13}C_{eco}$ were not constant, which complicates the use of $\delta^{13}C_{eco}$ to determine the linkage between assimilation and respiration. We conclude that the effects of environmental changes on C isotope discrimination can be distracted through a number of processes during C transfer.

We suggest that the combined effect of multiple environmental factors and the potential time lags should be addressed carefully when evaluating the long-term temporal changes of $\delta^{13}C_R$ and $\delta^{13}C_{eco}$. Apart from the biological impacts initiated by controlling the photosynthetic rate and stomatal openness, the physical parameters, such as aerodynamic conditions (e.g., atmospheric stability) and the conditions of the soil–atmosphere diffusion systems may also regulate $\delta^{13}C_{eco}$, although they seem to act as distracters, impairing the relationship between environmental factors and $\delta^{13}C_{eco}$. Furthermore, soil $\delta^{13}C_R$ and $\delta^{13}C_{eco}$ integrate the $\delta^{13}C$ signals of multiple tree species with different C assimilation and transfer patterns in mixed forests. In these respects, the present study expands our knowledge of C isotope discrimination in mixed forests with multiple tree species and effectively promotes a comprehensive consideration of the influencing factors related to isotopic variation within process-based ecosystem models.

Author Contributions: J.W. designed the study; H.D. performed the research; A.W., F.Y. and G.D. provided the resources; H.D., D.G. and J.W. contributed to the data analysis; D.G. and J.W. contributed to the funding acquisition; H.D. wrote the first draft and all authors contributed to the final manuscript. All authors have read and agreed to the published version of the manuscript.

Funding: This research was funded by the National Natural Science Foundation of China (grant numbers 31870625, 41975150).

Conflicts of Interest: The authors declare no conflict of interest.

References

1. Farquhar, G.D.; O'Leary, M.H.; Berry, J.A. On the Relationship between Carbon Isotope Discrimination and the Intercellular Carbon Dioxide Concentration in Leaves. *Funct. Plant Biol.* **1982**, *9*, 121–137. [CrossRef]
2. Cernusak, L.A.; Ubierna, N.; Winter, K.; Holtum, J.A.M.; Marshall, J.D.; Farquhar, G.D. Environmental and physiological determinants of carbon isotope discrimination in terrestrial plants. *New Phytol.* **2013**, *200*, 950–965. [CrossRef] [PubMed]
3. Ehleringer, J.R.; Buchmann, N.; Flanagan, L.B. Carbon isotope ratios in belowground carbon cycle processes. *Ecol. Appl.* **2000**, *10*, 412–422. [CrossRef]
4. Bowling, D.R.; Pataki, D.E.; Randerson, J.T. Carbon isotopes in terrestrial ecosystem pools and CO_2 fluxes. *New Phytol.* **2008**, *178*, 24–40. [CrossRef] [PubMed]
5. Ghashghaie, J.; Duranceau, M.; Badeck, F.-W.; Cornic, G.; Adeline, M.-T.; Deleens, E. $\delta^{13}C$ of CO_2 respired in the dark in relation to $\delta^{13}C$ of leaf metabolites: Comparison between *Nicotiana sylvestris* and *Helianthus annuus* under drought. *Plant Cell Environ.* **2001**, *24*, 505–515. [CrossRef]
6. Mortazavi, B.; Chanton, J.P.; Prater, J.L.; Oishi, A.C.; Oren, R.; Katul, G. Temporal variability in ^{13}C of respired CO_2 in a pine and a hardwood forest subject to similar climatic conditions. *Oecologia* **2005**, *142*, 57–69. [CrossRef]

7. Prater, J.L.; Mortazavi, B.; Chanton, J.P. Diurnal variation of the $\delta^{13}C$ of pine needle respired CO_2 evolved in darkness. *Plant Cell Environ.* **2006**, *29*, 202–211. [CrossRef]

8. Maunoury, F.; Berveiller, D.; Lelarge, C.; Pontailler, J.-Y.; Vanbostal, L.; Damesin, C. Seasonal, daily and diurnal variations in the stable carbon isotope composition of carbon dioxide respired by tree trunks in a deciduous oak forest. *Oecologia* **2007**, *151*, 268–279. [CrossRef]

9. Kuzyakov, Y.; Gavrichkova, O. Time lag between photosynthesis and carbon dioxide efflux from soil: A review of mechanisms and controls. *Glob. Chang. Biol.* **2010**, *16*, 3386–3406. [CrossRef]

10. Ekblad, A.; Högberg, P. Natural abundance of ^{13}C in CO_2 respired from forest soils reveals speed of link between tree photosynthesis and root respiration. *Oecologia* **2001**, *127*, 305–308. [CrossRef]

11. Damesin, C.; Lelarge, C. Carbon isotope composition of current-year shoots from *Fagus sylvatica* in relation to growth, respiration and use of reserves. *Plant Cell Environ.* **2003**, *26*, 207–219. [CrossRef]

12. Kuptz, D.; Matyssek, R.; Grams, T.E.E. Seasonal dynamics in the stable carbon isotope composition ($\delta^{13}C$) from non-leafy branch, trunk and coarse root CO_2 efflux of adult deciduous (*Fagus sylvatica*) and evergreen (*Picea abies*) trees. *Plant Cell Environ.* **2011**, *34*, 363–373. [CrossRef] [PubMed]

13. McDowell, N.G.; Bowling, D.R.; Bond, B.J.; Irvine, J.; Law, B.E.; Anthoni, P.; Ehleringer, J.R. Response of the carbon isotopic content of ecosystem, leaf, and soil respiration to meteorological and physiological driving factors in a *Pinus ponderosa* ecosystem. *Glob. Biogeochem. Cycles* **2004**, *18*. [CrossRef]

14. Werner, C.; Unger, S.; Pereira, J.S.; Maia, R.; David, T.S.; Kurz-Besson, C.; David, J.S.; Máguas, C. Importance of short-term dynamics in carbon isotope ratios of ecosystem respiration ($\delta^{13}C_R$) in a Mediterranean oak woodland and linkage to environmental factors. *New Phytol.* **2006**, *172*, 330–346. [CrossRef]

15. Barbour, M.M.; Hunt, J.E.; Kodama, N.; Laubach, J.; McSeveny, T.M.; Rogers, G.N.D.; Tcherkez, G.; Wingate, L. Rapid changes in $\delta^{13}C$ of ecosystem-respired CO_2 after sunset are consistent with transient ^{13}C enrichment of leaf respired CO_2. *New Phytol.* **2011**, *190*, 990–1002. [CrossRef] [PubMed]

16. Bowling, D.R.; McDowell, N.G.; Bond, B.J.; Law, B.E.; Ehleringer, J.R. ^{13}C content of ecosystem respiration is linked to precipitation and vapor pressure deficit. *Oecologia* **2002**, *131*, 113–124. [CrossRef] [PubMed]

17. Wingate, L.; Ogée, J.; Burlett, R.; Bosc, A.; Devaux, M.; Grace, J.; Loustau, D.; Gessler, A. Photosynthetic carbon isotope discrimination and its relationship to the carbon isotope signals of stem, soil and ecosystem respiration. *New Phytol.* **2010**, *188*, 576–589. [CrossRef]

18. Zobitz, J.M.; Burns, S.P.; Reichstein, M.; Bowling, D.R. Partitioning net ecosystem carbon exchange and the carbon isotopic disequilibrium in a subalpine forest. *Glob. Chang. Biol.* **2008**, *14*, 1785–1800. [CrossRef]

19. Chen, C.; Wei, J.; Wen, X.; Sun, X.; Guo, Q. Photosynthetic carbon isotope discrimination and effects on daytime NEE partitioning in a subtropical mixed conifer plantation. *Agric. For. Meteorol.* **2019**, *272–273*, 143–155. [CrossRef]

20. Damesin, C.; Barbaroux, C.; Berveiller, D.; Lelarge, C.; Chaves, M.; Maguas, C.; Maia, R.; Pontailler, J.-Y. The carbon isotope composition of CO_2 respired by trunks: Comparison of four sampling methods. *Rapid Commun. Mass Spectrom.* **2005**, *19*, 369–374. [CrossRef]

21. Kammer, A.; Tuzson, B.; Emmenegger, L.; Knohl, A.; Mohn, J.; Hagedorn, F. Application of a quantum cascade laser-based spectrometer in a closed chamber system for real-time $\delta^{13}C$ and $\delta^{18}O$ measurements of soil-respired CO_2. *Agric. For. Meteorol.* **2011**, *151*, 39–48. [CrossRef]

22. Keeling, C.D. The concentration and isotopic abundances of atmospheric carbon dioxide in rural areas. *Geochim. Cosmochim. Acta* **1958**, *13*, 322–334. [CrossRef]

23. Keeling, C.D. The concentration and isotopic abundances of carbon dioxide in rural and marine air. *Geochim. Cosmochim. Acta* **1961**, *24*, 277–298. [CrossRef]

24. Zobitz, J.M.; Keener, J.P.; Schnyder, H.; Bowling, D.R. Sensitivity analysis and quantification of uncertainty for isotopic mixing relationships in carbon cycle research. *Agric. For. Meteorol.* **2006**, *136*, 56–75. [CrossRef]

25. Chen, C.; Pang, J.; Wei, J.; Wen, X.; Sun, X. Inter-comparison of three models for $\delta^{13}C$ of respiration with four regression approaches. *Agric. For. Meteorol.* **2017**, *247*, 229–239. [CrossRef]

26. Guan, D.; Wu, J.; Zhao, X.; Han, S.; Yu, G.; Sun, X.; Jin, C. CO_2 fluxes over an old, temperate mixed forest in northeastern China. *Agric. For. Meteorol.* **2006**, *137*, 138–149. [CrossRef]

27. Kodama, N.; Barnard, R.L.; Salmon, Y.; Weston, C.; Ferrio, J.P.; Holst, J.; Werner, R.A.; Saurer, M.; Rennenberg, H.; Buchmann, N.; et al. Temporal dynamics of the carbon isotope composition in a *Pinus sylvestris* stand: From newly assimilated organic carbon to respired carbon dioxide. *Oecologia* **2008**, *156*, 737. [CrossRef]

28. Brandes, E.; Kodama, N.; Whittaker, K.; Weston, C.; Rennenberg, H.; Keitel, C.; Adams, M.A.; Gessler, A. Short-term variation in the isotopic composition of organic matter allocated from the leaves to the stem of *Pinus sylvestris*: Effects of photosynthetic and postphotosynthetic carbon isotope fractionation. *Glob. Chang. Biol.* **2006**, *12*, 1922–1939. [CrossRef]

29. Gessler, A.; Keitel, C.; Kodama, N.; Weston, C.; Winters, A.J.; Keith, H.; Grice, K.; Leuning, R.; Farquhar, G.D. δ^{13}C of organic matter transported from the leaves to the roots in *Eucalyptus delegatensis*: Short-term variations and relation to respired CO_2. *Funct. Plant Biol.* **2007**, *34*, 692–706. [CrossRef]

30. Wen, X.F.; Meng, Y.; Zhang, X.Y.; Sun, X.M.; Lee, X. Evaluating calibration strategies for isotope ratio infrared spectroscopy for atmospheric $^{13}CO_2/^{12}CO_2$ measurement. *Atmos. Meas. Tech.* **2013**, *6*, 1491–1501. [CrossRef]

31. Pang, J.; Wen, X.; Sun, X.; Huang, K. Intercomparison of two cavity ring-down spectroscopy analyzers for atmospheric $^{13}CO_2/^{12}CO_2$ measurement. *Atmos. Meas. Tech.* **2016**, *9*, 3879–3891. [CrossRef]

32. Zhou, J.; Yang, Z.; Wu, G.; Yang, Y.; Lin, G. The relationship between soil CO_2 efflux and its carbon isotopic composition under non-steady-state conditions. *Agric. For. Meteorol.* **2018**, *256-257*, 492–500. [CrossRef]

33. Wehr, R.; Saleska, S.R. An improved isotopic method for partitioning net ecosystem–atmosphere CO_2 exchange. *Agric. For. Meteorol.* **2015**, *214–215*, 515–531. [CrossRef]

34. Brugnoli, E.; Farquhar, G.D. Photosynthetic fractionation of carbon isotopes. In *Photosynthesis: Physiology and Metabolism*; Leegood, R.C., Sharkey, T.D., von Caemmerer, S., Eds.; Springer: Dordrecht, The Netherlands, 2000; pp. 399–434. [CrossRef]

35. Brooks, A.; Farquhar, G.D. Effect of temperature on the CO_2/O_2 specificity of ribulose-1,5-bisphosphate carboxylase/oxygenase and the rate of respiration in the light. *Planta* **1985**, *165*, 397–406. [CrossRef] [PubMed]

36. Tcherkez, G.; Schäufele, R.; Nogués, S.; Piel, C.; Boom, A.; Lanigan, G.; Barbaroux, C.; Mata, C.; Elhani, S.; Hemming, D.; et al. On the $^{13}C/^{12}C$ isotopic signal of day and night respiration at the mesocosm level. *Plant Cell Environ.* **2010**, *33*, 900–913. [CrossRef] [PubMed]

37. Bowling, D.R.; Tans, P.P.; Monson, R.K. Partitioning net ecosystem carbon exchange with isotopic fluxes of CO_2. *Glob. Chang. Biol.* **2001**, *7*, 127–145. [CrossRef]

38. Ehleringer, J.R.; Field, C.B.; Lin, Z.-f.; Kuo, C.-y. Leaf carbon isotope and mineral composition in subtropical plants along an irradiance cline. *Oecologia* **1986**, *70*, 520–526. [CrossRef]

39. Brooks, J.R.; Flanagan, L.B.; Varney, G.T.; Ehleringer, J.R. Vertical gradients in photosynthetic gas exchange characteristics and refixation of respired CO_2 within boreal forest canopies. *Tree Physiol.* **1997**, *17*, 1–12. [CrossRef]

40. Tan, W.; Wang, G.; Han, J.; Liu, M.; Zhou, L.; Luo, T.; Cao, Z.; Cheng, S. δ^{13}C and water-use efficiency indicated by δ^{13}C of different plant functional groups on Changbai Mountains, Northeast China. *Chin. Sci. Bull.* **2009**, *54*, 1759–1764. [CrossRef]

41. Farquhar, G.D.; Ehleringer, J.R.; Hubick, K.T. Carbon isotope discrimination and photosynthesis. *Annu. Rev. Plant Physiol. Plant Mol. Biol.* **1989**, *40*, 503–537. [CrossRef]

42. Scartazza, A.; Moscatello, S.; Matteucci, G.; Battistelli, A.; Brugnoli, E. Seasonal and inter-annual dynamics of growth, non-structural carbohydrates and C stable isotopes in a Mediterranean beech forest. *Tree Physiol.* **2013**, *33*, 730–742. [CrossRef] [PubMed]

43. Rinne, K.T.; Saurer, M.; Kirdyanov, A.V.; Bryukhanova, M.V.; Prokushkin, A.S.; Churakova, O.V.; Siegwolf, R.T.W. Examining the response of needle carbohydrates from Siberian larch trees to climate using compound-specific δ^{13}C and concentration analyses. *Plant Cell Environ.* **2015**, *38*, 2340–2352. [CrossRef]

44. Jäggi, M.; Saurer, M.; Fuhrer, J.; Siegwolf, R. The relationship between the stable carbon isotope composition of needle bulk material, starch, and tree rings in *Picea abies*. *Oecologia* **2002**, *131*, 325–332. [CrossRef] [PubMed]

45. Cernusak, L.A.; Marshall, J.D.; Comstock, J.P.; Balster, N.J. Carbon isotope discrimination in photosynthetic bark. *Oecologia* **2001**, *128*, 24–35. [CrossRef]

46. Badeck, F.-W.; Tcherkez, G.; Nogués, S.; Piel, C.; Ghashghaie, J. Post-photosynthetic fractionation of stable carbon isotopes between plant organs—A widespread phenomenon. *Rapid Commun. Mass Spectrom.* **2005**, *19*, 1381–1391. [CrossRef] [PubMed]

47. Werner, C.; Gessler, A. Diel variations in the carbon isotope composition of respired CO_2 and associated carbon sources: A review of dynamics and mechanisms. *Biogeosciences* **2011**, *8*, 2437–2459. [CrossRef]

48. Ghashghaie, J.; Badeck, F.W. Opposite carbon isotope discrimination during dark respiration in leaves versus roots—A review. *New Phytol.* **2014**, *201*, 751–769. [CrossRef]

49. Millard, P.; Midwood, A.J.; Hunt, J.E.; Barbour, M.M.; Whitehead, D. Quantifying the contribution of soil organic matter turnover to forest soil respiration, using natural abundance δ^{13}C. *Soil Biol. Biochem.* **2010**, *42*, 935–943. [CrossRef]

50. Ngao, J.; Cotrufo, M.F. Carbon isotope discrimination during litter decomposition can be explained by selective use of substrate with differing δ^{13}C. *Biogeosci. Discuss.* **2011**, *8*, 51–82. [CrossRef]

51. ŠantRůČková, H.; Bird, M.I.; Lloyd, J. Microbial processes and carbon-isotope fractionation in tropical and temperate grassland soils. *Funct. Ecol.* **2000**, *14*, 108–114. [CrossRef]

52. Dijkstra, P.; Ishizu, A.; Doucett, R.; Hart, S.C.; Schwartz, E.; Menyailo, O.V.; Hungate, B.A. ^{13}C and ^{15}N natural abundance of the soil microbial biomass. *Soil Biol. Biochem.* **2006**, *38*, 3257–3266. [CrossRef]

53. Werth, M.; Kuzyakov, Y. ^{13}C fractionation at the root–microorganisms–soil interface: A review and outlook for partitioning studies. *Soil Biol. Biochem.* **2010**, *42*, 1372–1384. [CrossRef]

54. Goffin, S.; Aubinet, M.; Maier, M.; Plain, C.; Schack-Kirchner, H.; Longdoz, B. Characterization of the soil CO_2 production and its carbon isotope composition in forest soil layers using the flux-gradient approach. *Agric. For. Meteorol.* **2014**, *188*, 45–57. [CrossRef]

55. Formánek, P.; Ambus, P. Assessing the use of δ^{13}C natural abundance in separation of root and microbial respiration in a Danish beech (*Fagus sylvatica L.*) forest. *Rapid Commun. Mass Spectrom.* **2004**, *18*, 897–902. [CrossRef] [PubMed]

56. Bowling, D.R.; Egan, J.E.; Hall, S.J.; Risk, D.A. Environmental forcing does not induce diel or synoptic variation in the carbon isotope content of forest soil respiration. *Biogeosciences* **2015**, *12*, 5143–5160. [CrossRef]

57. Duranceau, M.; Ghashghaie, J.; Badeck, F.; Deleens, E.; Cornic, G. δ^{13}C of CO_2 respired in the dark in relation to δ^{13}C of leaf carbohydrates in *Phaseolus vulgaris L.* under progressive drought. *Plant Cell Environ.* **1999**, *22*, 515–523. [CrossRef]

58. Loader, N.J.; McCarroll, D.; Gagen, M.; Robertson, I.; Jalkanen, R. Extracting climatic information from stable isotopes in tree rings. In *Terrestrial Ecology*; Elsevier: Amsterdam, The Netherlands, 2007; Volume 1, pp. 25–48.

59. Fessenden, J.E.; Ehleringer, J.R. Temporal variation in δ^{13}C of ecosystem respiration in the Pacific Northwest: Links to moisture stress. *Oecologia* **2003**, *136*, 129–136. [CrossRef]

60. Ekblad, A.; Boström, B.; Holm, A.; Comstedt, D. Forest soil respiration rate and δ^{13}C is regulated by recent above ground weather conditions. *Oecologia* **2005**, *143*, 136–142. [CrossRef]

61. Knohl, A.; Werner, R.A.; Brand, W.A.; Buchmann, N. Short-term variations in δ^{13}C of ecosystem respiration reveals link between assimilation and respiration in a deciduous forest. *Oecologia* **2005**, *142*, 70–82. [CrossRef]

62. Scartazza, A.; Mata, C.; Matteucci, G.; Yakir, D.; Moscatello, S.; Brugnoli, E. Comparisons of δ^{13}C of photosynthetic products and ecosystem respiratory CO_2 and their responses to seasonal climate variability. *Oecologia* **2004**, *140*, 340–351. [CrossRef]

63. McDowell, N.G.; Bowling, D.R.; Schauer, A.; Irvine, J.; Bond, B.J.; Law, B.E.; Ehleringer, J.R. Associations between carbon isotope ratios of ecosystem respiration, water availability and canopy conductance. *Glob. Chang. Biol.* **2004**, *10*, 1767–1784. [CrossRef]

64. Bowling, D.R.; Burns, S.P.; Conway, T.J.; Monson, R.K.; White, J.W.C. Extensive observations of CO_2 carbon isotope content in and above a high-elevation subalpine forest. *Glob. Biogeochem. Cycles* **2005**, *19*. [CrossRef]

65. Schaeffer, S.M.; Anderson, D.E.; Burns, S.P.; Monson, R.K.; Sun, J.; Bowling, D.R. Canopy structure and atmospheric flows in relation to the δ^{13}C of respired CO_2 in a subalpine coniferous forest. *Agric. For. Meteorol.* **2008**, *148*, 592–605. [CrossRef]

Publisher's Note: MDPI stays neutral with regard to jurisdictional claims in published maps and institutional affiliations.

Article

Detecting Growth Phase Shifts Based on Leaf Trait Variation of a Canopy Dipterocarp Tree Species (*Parashorea chinensis*)

Yun Deng [1,2,3,4], Xiaobao Deng [1,2,4], Jinlong Dong [1,2,4], Wenfu Zhang [1,2,4], Tao Hu [1,2,3], Akihiro Nakamura [1,2], Xiaoyang Song [1,2], Peili Fu [1,2] and Min Cao [1,2,*]

1   CAS Key Laboratory of Tropical Forest Ecology, Xishuangbanna Tropical Botanical Garden, Chinese Academy of Sciences, Mengla, Menglun 666303, China; dy@xtbg.org.cn (Y.D.); dxb@xtbg.ac.cn (X.D.); dongjinlong@xtbg.org.cn (J.D.); zwf@xtbg.org.cn (W.Z.); hutao@xtbg.ac.cn (T.H.); a.nakamura@xtbg.ac.cn (A.N.); songxiaoyang@xtbg.ac.cn (X.S.); fpl@xtbg.org.cn (P.F.)
2   Center for Plant Ecology, Core Botanical Gardens, Chinese Academy of Sciences, Mengla, Menglun 666303, China
3   Department of Life Sciences, University of Chinese Academy of Sciences, Beijing 100049, China
4   National Forest Ecosystem Research Station at Xishuangbanna, Xishuangbanna Tropical Botanical Garden, Chinese Academy of Sciences, Mengla, Menglun 666303, China
*   Correspondence: caom@xtbg.ac.cn; Tel.: +86-137-0066-9659

Received: 25 September 2020; Accepted: 26 October 2020; Published: 29 October 2020

Abstract: Canopy species need to shift their adaptive strategy to acclimate to very different light environments as they grow from seedlings in the understory to adult trees in the canopy. However, research on how to quantitively detect ecological strategy shifts in plant ontogeny is scarce. In this study, we hypothesize that changes in light and tree height levels induce transitions in ecological strategies, and growth phases representing different adaptive strategies can be classified by leaf trait variation. We examined variations in leaf morphological and physiological traits across a vertical ambient light (represented by the transmittance of diffuse light, %TRANS) and tree height gradient in *Parashorea chinensis*, a large canopy tree species in tropical seasonal rainforest in Southwestern China. Multivariate regression trees (MRTs) were used to detect the split points in light and height gradients and classify ontogenetic phases. Linear piecewise regression and quadratic regression were used to detect the transition point in leaf trait responses to environmental variation and explain the shifts in growth phases and adaptive strategies. Five growth phases of *P. chinensis* were identified based on MRT results: (i) the vulnerable phase, with tree height at less than 8.3 m; (ii) the suppressed phase, with tree height between 8.3 and 14.9 m; (iii) the growth release phase, with tree height between 14.9 and 24.3 m; (iv) the canopy phase, with tree height between 24.3 and 60.9 m; and (v) the emergent phase, with tree height above 60.9 m. The suppressed phase and canopy phase represent "stress-tolerant" and "competitive" strategies, respectively. Light conditions drive the shift from the "stress-tolerant" to the "competitive" strategy. These findings help us to better understand the regeneration mechanisms of canopy species in forests.

Keywords: ontogenetic phases; adaptive strategies; leaf functional traits; light environment; canopy tree species

---

1. Introduction

Ecological strategies represent adaptations to the intensity of environmental filters involving competition, stress and disturbance [1]. When considered as an operating process for resource allocation, plants' ecological strategies can be expected to shift from seedling to mature stages [1]. Differences in light requirements among tree species underlie the major changes in vegetation composition during succession [2,3]. Compared with pioneers and understory species, canopy tree species exhibit a greater change in tree stature, and very different light environments occur as they grow up from seedlings and saplings in the understory to adult trees in the canopy. Several forest tree species have different light requirements throughout their developmental stages [4], and ontogenetic variation in traits has received research attention in evaluating the adaptive strategies of specific species [5]. The adaptive strategies of canopy species throughout their ontogeny cannot be characterized by a single style.

Grime [6] proposed that there are three primary adaptive strategies in plants (CSR theory): (1) C-selected strategy, usually with long life span and large organ size, which maximizes resource acquisition to improve competitiveness in stable and productive habitats; (2) R-selected strategy, in which more resources are invested in rapid growth to avoid frequent disturbance events; and (3) S-selected strategy, which favors resource conservation for survival in resource-poor environments. These ecological strategies are expected to be habitat- and stage-specific [7] and are widely accepted in classifying inter-species functional types [8]. However, although CSR theory could also be used to evaluate adaptive strategies in ontogeny [1], it is rarely applied to explain the ontogenetic shifts of adaptive strategies in specific species, especially for tall species in forest canopies.

Leaf morphophysiological traits are the most sensitive indicators of in situ microenvironmental changes from the understory to the canopy life stages of individual species [9], and this leaf trait-based method can be used to classify adaptive strategies in the different growth phases of canopy species. Pierce et al. [10] summarized three main extreme trait syndromes concerned with adaptive strategies: (i) tall plants with large leaves, intermediate leaf economics and intermediate flowering start and flowering period (C-selected strategy); (ii) short plants with small leaves, conservative leaf economic trait values (e.g., low specific leaf area and leaf nitrogen concentrations) and a brief reproductive phase (S-selected strategy); and (iii) short plants with small leaves, highly acquisitive leaf economics and early, prolonged reproductive development (R-selected strategy). This triangle suggests that leaf traits reflect key components of the leaf economics spectrum, and leaf size could provide "a dependable common reference frame for the quantitative comparison of the wider primary adaptive strategies of plants from highly contrasting habitats" [11].

For trees in closed-canopy forests, light levels associated with canopy position and local tree crowding are a major axis of ontogenetic environmental change [12]. As a result, forest tree species have different light requirements throughout their developmental stages [4]. For canopy species at the community level, the leaf mass per area (LMA), photosynthetic capacity at the light saturation point (LSP), light compensation point (LCP) and maximal net assimilation rate (A_{max}) increase with light availability and tree height [13]. These effects are associated with positive responses of Rubisco, total nitrogen and light-harvesting complex (LHC) polypeptide contents to the vertical light gradient [4]. Not only the light conditions in a habitat but also inherent ontogenetic traits such as size or height growth could also lead to variations in traits [5]. An increasing canopy height can increase hydraulic limitations of the leaf structure, which can increase the leaf tissue density and cell wall thickness and reduce the mesophyll air space, potentially restricting mesophyll conductance to CO_2 and photosynthesis [14]. As a result, leaf structure or defense-related traits such as LMA, leaf thickness and leaf carbon content increased monotonically with tree height [15]. Moreover, age- and size-dependent declines in leaf nitrogen concentration and A_{max} in tall trees were reported [16,17], and reproductive onset [15,18] over the canopy layer could lead plants to invest more resources in competition and affect leaf morphological and physiological adaptations. All these variations in leaf traits throughout the life history of canopy species indicated the existence of strategic shifts in tree ontogeny.

Parashorea chinensis (formerly *Shorea chinensis* or *Shorea wantianshuea*), in the family Dipterocarpaceae, is a dominant canopy tree species in the Xishuangbanna tropical seasonal rainforest. Adult individuals of *P. chinensis* occur in the emergent layer at a height of approximately 60 m [19–21]. *P. chinensis* begins reproducing at 30–40 cm diameter at breast height (DBH) and approximately 60–80 years old [22]; the largest individuals can reach 1.5 m DBH and over 180 years old [23]. *P. chinensis* is a typical canopy species with shade-tolerant seedlings; previous studies suggested that irradiance is not a limiting factor for seed germination of *P. chinensis*, and partial shading favored seedling growth [24]. He et al. [25] reported that leaf morphological traits such as LMA have a positive relation with tree height of *P. chinensis* in the canopy layer, but we still need more ontogenetic evidence to explain the variations in strategy from understory to canopy. In this study, a canopy crane was used to measure morphology and photosynthesis-related leaf traits of sample trees of *P. chinensis* in the forest at different life stages. The following hypotheses were tested: (1) insufficient ambient light in the understory leads to a stress-tolerant strategy in *P. chinensis*, and with the growth in tree height, the increased light conditions induce a shift from stress-tolerant strategy to release growth; (2) with the continuous height growth above the main canopy, hydraulic limitation limits plant growth and changes the adaptive strategy to relatively competitive; (3) all these shifts in adaptive strategy can be detected by variation in leaf traits, and based on the threshold level of light or height, the ontogenetic growth of *P. chinensis* can be classified into several phases. To test these hypotheses, variations in both morphological traits (LMA, leaf area and leaf thickness) and physiological traits (leaf dark respiration rate (R_d), A_{max}, LCP, LSP, leaf nitrogen content and carbon-nitrogen ratio (C:N ratio)) across a vertical ambient light and tree height (sapling to adult) gradient in *P. chinensis* were examined, and different growth phases of *P. chinensis* were also detected by quantitative evaluation of leaf traits in response to height and light variance.

2. Materials and Methods

2.1. Study Area

This study was performed in a tropical seasonal rainforest in Xishuangbanna, Southwestern China (101°34′ E, 21°36′ N). This region is predominated by a typical monsoon climate that alternates between a dry season (November–April) and a rainy season (May–October). The mean air temperature is 21.8 °C, and the mean annual precipitation is 1493 mm [20].

The tropical seasonal rainforest in this area is distributed in lowlands, valleys and hills with sufficient water supply [19]. The emergent layer of the forest in the present study was dominated by *P. chinensis*; its canopy can reach an average height > 45 m and approaches approximately 60 m for individuals in the emergent layer. *P. chinensis* also showed the largest basal area among all the tree species in the 20 ha Xishuangbanna tropical seasonal rainforest dynamics plot (XTRDP) [20]. The main canopy layer (approximately 30 m high) is dominated by *Sloanea tomentosa*, *Pometia pinnata* (formerly *Pometia tomentosa*) and *Semecarpus reticulatus* [20].

The canopy crane in the same tropical seasonal rainforest (TCT7015-10E, Zoomlion Heavy Industry, Changsha, China) was built in December 2014, approximately 300 m away from the XTRDP. The maximum operational height of this crane is 80 m (the maximum canopy height around the crane is approximately 60 m), and the jib length is 60 m (Figure 1). Centered at the base of the crane, a 120 m × 120 m ancillary plot was established for plant species inventory in 2014. A total of 6928 individuals of 217 woody tree species with a DBH ≥ 1 cm were recorded in the plot of 1.44 ha, and *P. chinensis* had the highest relative importance values (21.73%) (Table S1). The forest structure and species composition in the crane plot are similar to those in the XTRDP [20].

A total of 21 *P. chinensis* individuals from 0.6 to 66.5 m tall (with diffuse transmittance from 0.30% to 100%) in the plot were sampled and measured from October to November in 2016 and September to October in 2017, at the end of the rainy season, which is also the main growing season for most local plants. The weather had enough sunny days so we could operate the crane and conduct measurements in the field. In total, 34 workdays were actually taken for field measurements. The tree height of each selected individual was measured with a tape measure from the crane.

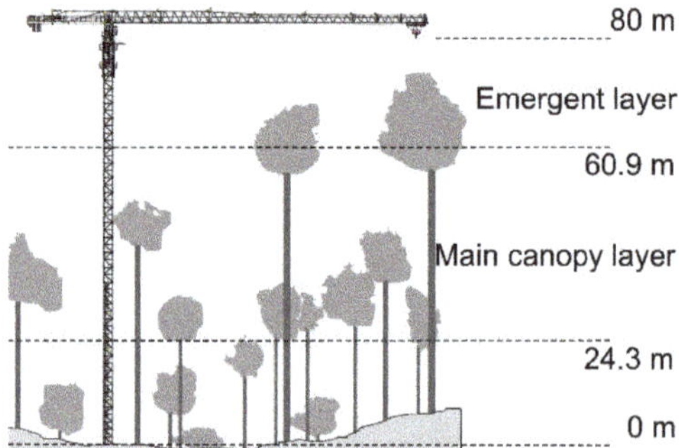

Figure 1. Vertical profile of the Xishuangbanna tropical seasonal rainforest with the canopy crane.

2.2. Light Conditions

The spatial and temporal distribution of light in the understory is quite heterogeneous, and the measurement of direct light becomes difficult for long time spans or with too many samples [26]. Indirect measurements, such as the transmittance of diffuse light (%TRANS, sometimes presented as the diffuse non-interceptance, DIFN), are also acceptable to estimate light conditions. A methodological comparison previously showed that %TRANS measured with LAI-2000 has a good fit with directly measured long-term PPFD in the understory of a tropical forest [26]. In the present study, we measured %TRANS above each sampled tree with LAI-2200 (LI-COR, Inc., Lincoln, NE, USA), an upgraded vision of LAI-2000 (LI-COR, Inc., Lincoln, NE, USA). Open sky measurements were collected on top of the crane, and each individual was measured at the crown top with five repeats. A 270° view cap was used to prevent the crane jib from obstructing the view. The repeated measurements of each individual were used to adjust the open sky measurements with FV2200 software (LI-COR, Inc., Lincoln, NE, USA) and to estimate diffuse transmittance.

2.3. Leaf Dark Respiration Rate (R_d)

Five leaves on sun-exposed terminal shoots were selected from the upper- and outermost crown of each individual, and all measurements (including R_d, light-response curve, leaf area, thickness, LMA, leaf nitrogen and carbon) were performed in same leaf. Prior to sunset (approximately 6.00 p.m.) on the day before the respiration measurements, the leaves were covered with thin aluminum foil so that they were not exposed to sunlight until the measurements. The abaxial side was not completely covered to allow free gas exchange. Because the leaves were covered overnight, the effects on respiration of light-enhanced dark respiration [27] and light-induced metabolites and respiratory gene expression [28] were avoided in the measurements.

The R_d was measured using a portable infrared gas analyzer (LI-6400, LI-COR Inc., Lincoln, NE, USA) at ambient humidity of 60%–90%, while the CO_2 concentration was maintained at 400 ppm with the built-in CO_2 mixer. The block temperature of the LI-6400 instrument was set to the ambient temperature in this study, approximately 26.9 ± 3.5 °C (mean value ± standard deviation), and the vapor pressure deficit (VPD) in the cuvette was maintained at 1.1 ± 0.3 kPa. The system stability was assessed by "stableF" in LI-6400 output, which indicated the comprehensive stability of CO_2, H_2O and flow in the cuvette, and the value varied from 0 to 1 (1 = stable, 0 = not) (LI-COR 2011). In this measurement, the stability is 0.9 ± 0.1.

Each leaf was measured for 1 min (recorded every 3 s) the following morning (between 8:00 a.m. and 11.00 a.m.) at ambient temperature, and a mean value of 1 min was used as the R_d for each leaf.

After the R_d measurement, leaves with their shoots were cut, and the bases were immersed in deaerated water and transported to the laboratory for additional trait measurements.

2.4. Light-Response Curve

The swinging of the crane basket made it difficult to measure a light-response curve in situ, because this measurement requires clamping a leaf for more than one hour and the photosynthesis of *P. chinensis* decreased at midday [29]; the use of detached shoots could eliminate these immediate effects of path length and gravity on gas exchange [30]. In this study, we used detached shoots to isolate the influence of any height-related trends in foliar structure on the various gas exchange parameters, a method applied in previous studies on *P. chinensis* [31]. Although excised branches lead to a reduction in photosynthetic rate and respiration rate [32], the purpose of this study was to detect the tendency of variation in leaf traits between ontogenetic phases rather than measuring the absolute value of traits, so the systematic bias from excising branches was acceptable. The R_{d_area} in situ and the respiration rate under 0 $\mu mol \cdot m^{-2} \cdot s^{-1}$ PPFD in the light-response curve (R_{d_area}') were consistent, with little bias due to excision (Figure S1).

The light-response curve (A-Q curve) was measured with a LI-6400 instrument equipped with a 6400–02B LED source and CO_2 injector. The cuvette CO_2 was set to 400 ppm, the relative humidity was 40%–70%, the air flow was 500 $\mu mol \cdot s^{-1}$, the leaf temperature was approximately 30 °C, the VPD was 1.3 ± 0.4 kPa, and the system stability was 0.9 ± 0.1. After light adaptation under 1000 $\mu mol \cdot m^{-2} \, s^{-1}$ (the adaptation time depended on the stomatal situation and was often more than 30 min), the PPFD gradient was set as follows: 1800, 1500, 1200, 900, 600, 300, 150, 70, 30, 15, and 0 $\mu mol \cdot m^{-2} \, s^{-1}$. The leaves were allowed to equilibrate for at least 3 min under each PPFD gradient based on the stability of stomatal conductance (g_s) (the fluctuation of g_s in same leaf in same gradient during 1 min was 1.9 ± 1.7 mmol $m^{-2} \cdot s^{-1}$). The instrument was matched before the measurement in each PPFD gradient. The branch samples were measured within 8 h of excision.

The following model was used to analyze the A-Q curve [33]:

$$A_n = A_{max}\left(1 - C_0 \cdot e^{-\alpha \, \cdot \, PPFD \, \cdot \, A_{max}^{-1}}\right) \tag{1}$$

where A_{max} is the area-based maximal net assimilation rate, α is the efficiency of the quantum yield in weak light, and C_0 is a constant.

The LSP and LCP calculations were performed using the following equations [33]:

$$LCP = A_{max} \times \ln (C_0) \times \alpha^{-1} \tag{2}$$

$$LSP = A_{max} \times \ln (100 \times C_0) \times \alpha^{-1} \tag{3}$$

2.5. Leaf Mass per Area (LMA), Leaf Nitrogen, and Carbon

After the gas exchange measurements, the leaf lamina thickness of fresh leaves was determined using a digital caliper (accuracy: 0.01 mm, No. 111–101–40, Guanglu Co., Guilin, China) based on the mean value of three repetitive measurements across the lamina while avoiding major veins. The areas of the leaves were measured with a leaf area meter (LI-3000C, LI-COR Inc., Lincoln, NE, USA). After then, all leaves were dried at 60 °C in an oven for 3 days and the leaf dry mass was measured. The petiole of *P. chinensis* is short (1–3 cm), and we included it in the measurement of leaf area and leaf mass.

The dried leaves were then pulverized, and the leaf nitrogen and carbon contents were determined with an elemental analyzer (Vario ISOTOPE Cube, Elementar Analysensysteme GmbH, Langenselbold, Germany) at the Central Laboratory of Xishuangbanna Tropical Botanical Garden, Chinese Academy of Sciences. Photosynthetic nitrogen use efficiency (PNUE) was calculated by the ratio of the area-based maximal net assimilation rate to leaf nitrogen content per area (N_{area}) [34].

2.6. Statistical Analysis

The value of light intensity was log-transformed in this study, for light often shows a curved, asymptotic relationship with height [13] and LMA [5,35].

We used multivariate regression trees (MRTs) to cluster growth phases. This method is applied to analyze complex ecological data that may include imbalanced or missing values, nonlinear relationships between variables, and high-order interactions [36]. Height and ln(%TRANS) were explanatory variables, and leaf traits including LCP, LSP, PNUE, C:N ratio, leaf thickness, LMA, R_{d_area}, A_{max_area}, N_{area}, R_{d_mass}, A_{max_mass}, N_{mass}, and leaf area were response variables in MRTs. All variables were standardized in this analysis.

Regression trees are summarized by their size (the number of leaves) and overall fit. Fit is defined by relative error (RE): the total impurity of the leaves divided by the impurity of the root node (the undivided data). RE gives an over-optimistic estimate of how accurately an MRT will predict new data, and predictive accuracy is better estimated from the cross-validated relative error (CVRE). CVRE varies from zero for a perfect predictor to close to one for a poor predictor [36].

The size of the MRT was selected by cross-validation that generates a series of MRTs with only marginally worse predictive ability than the best predictive MRT. From this series, the smallest MRT within one standard error of the best is often selected (the 1-SE rule; [37]).

Because of the discontinuous variation of light availability with increasing tree height, traits such as LMA and leaf area may be highly correlated with light in the lower crown, but the relationships become weaker in the upper crown [5,38]. Hence, linear approaches such as multiple regression could not be used to distinguish the effects of height versus light on these variables [38]. Therefore, linear piecewise regression was used to model traits vs. height (or ln(%TRANS)) to determine the stage at which the trait value may not respond to increasing height (or ln(%TRANS)) levels. Model parameters are presented in Tables S2 and S3.

To better predict the "hump-shaped" variation in traits such as area-based maximal net assimilation rate (A_{max_area}), leaf area, and the C:N ratio with increasing tree height, quadratic regressions (traits = β_0 + β_1 × height + β_2 × height^2) were also used, and the model parameters are presented in Table S4. R program v3.6.0 [39] was used for this data analysis, and MRTs were constructed using the "mvpart" package in R.

3. Results

3.1. Tree Size and Light Availability

Quadratic regressions described the positive relationship between DBH and tree height of *P. chinensis* very well (height = 2.8598 + 0.8863 × DBH − 0.0030 × DBH^2, r^2 = 0.973, F = 362.1, P < 0.01; Figure 2). Considering that the tree height could better reflect the effect of hydrostatic limitation, we selected tree height instead of DBH to represent tree size in this study.

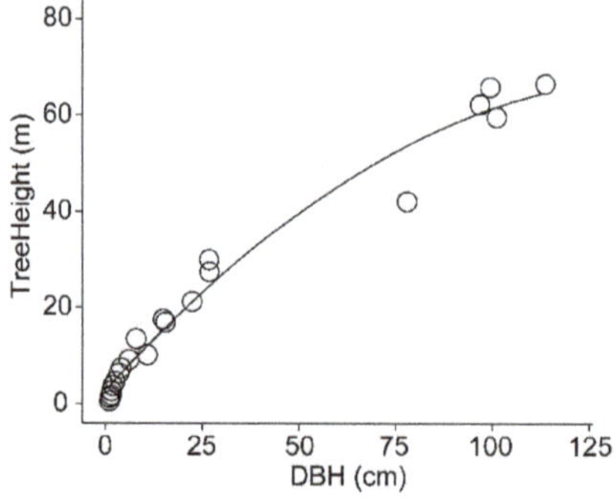

Figure 2. Relationship between tree height and diameter at breast height (DBH) of *P. chinensis*. The curve indicates the quadratic regression result.

Light availability was promoted as a result of increasing tree height up to 27.4 m ($r^2 = 0.804$, $F = 54.14$, $P < 0.01$; Figure 3; Table S2), and the transmittance of diffuse light varied from 0.3% at 3.7 m to 100% at $\geq$ 21.2 m. When a tree grew higher than 27.4 m, the linear model became nonsignificant ($r^2 = 0.262$, $F = 3.12$, $P = 0.13$).

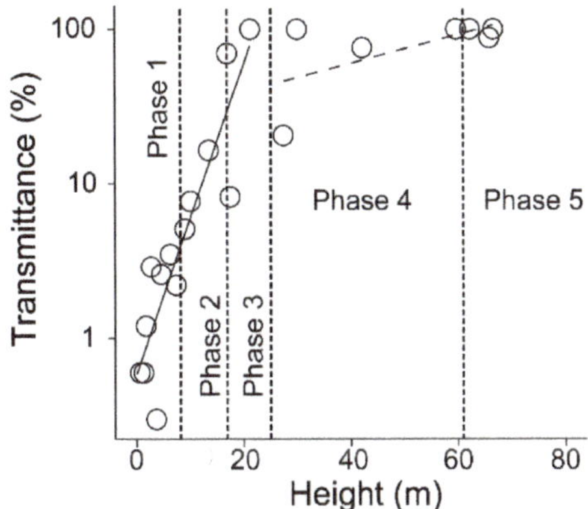

Figure 3. Relationship between height (x) and ln(%TRANS) (y). Lines indicate linear piecewise regression results; solid lines denote significant relationships ($P < 0.05$), and broken lines denote nonsignificant relationships. Phase partitions were determined based on the results of multivariate regression tree (MRT). Note that a natural log scale is used on the Y-axis.

3.2. Growth Phases

Based on the 1-SE rule, the CVRE result generated an MRT size of five leaves (when size of tree = 5, CVRE ± SE = 0.544 ± 0.086, RE = 0.204; Figure 4). Four splits were presented in MRT (Figure 5): %TRANS determined the first split (18.33%), and other splits were mainly driven by tree height (8.3, 24.3, 60.9 m; Figure 5).

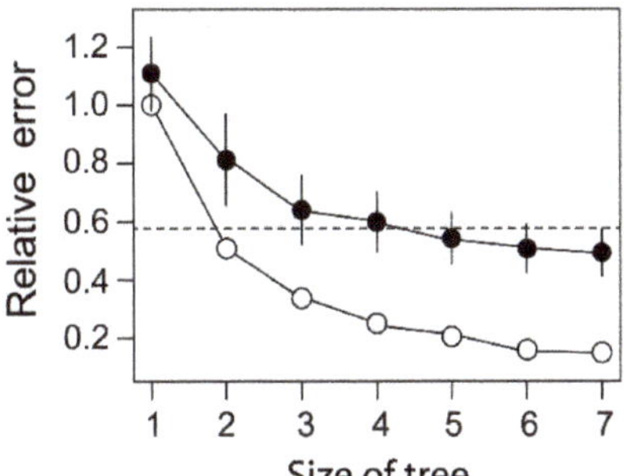

Figure 4. Variation in relative error with the size of the multivariate regression tree. Open circles indicate the relative error, and filled circles indicate the cross-validated relative error (CVRE). The vertical bars indicate one standard error for the CVRE, and the dashed line indicates one standard error above the minimum cross-validated relative error.

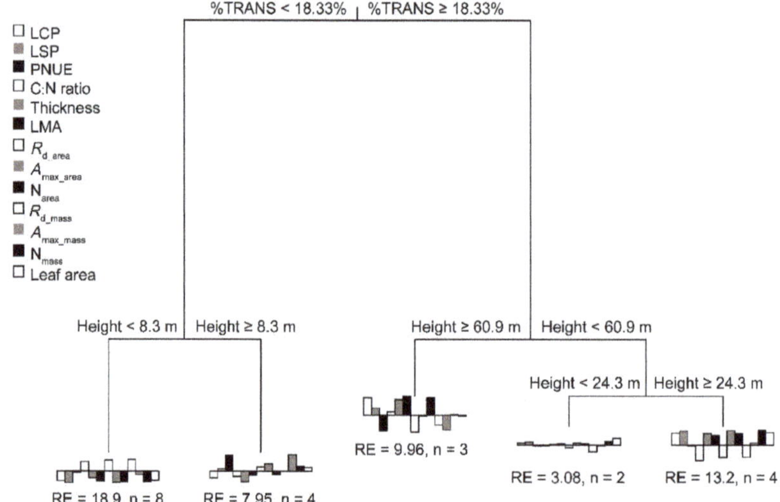

Figure 5. Multivariate regression tree for the leaf traits data. All data include explanatory variables (height and ln(%TRANS)), and response variables (leaf traits) were standardized in MRT analysis, but the explanatory variables were restored to raw values in this figure for clarity. Bar plots show the multivariate traits' mean at each node. The cyclical shadings (white, gray, and black) indicate the various traits and run from left to right across the bar plots.

3.3. Leaf Morphological Trait Transition Point of P. chinensis

LMA had a positive relationship with both tree height (Figure 6a) and light gradient (Figure 6b). The piecewise regression indicated that the LMA of *P. chinensis* increased significantly with tree height growth up to 21.2 m, although the slope of the linear model decreased at tree heights above 21.2 m (with height < 21.2 m, slope = 2.0735, r^2 = 0.853, F = 70.81, P < 0.001; with height ≥ 21.2 m, slope = 0.8900, r^2 = 0.490, F = 7.72, P = 0.03; Figure 6a). Increasing light also significantly promoted LMA in the understory (with ln(%TRANS) < −0.278 or diffuse transmittance < 75.73%, r^2 = 0.721, F = 34.7, P < 0.001; Figure 6b).

Although the negative relationship between leaf area and tree height above the main canopy layer appeared to be nonsignificant in the linear piecewise regression (with height < 9.1 m, r^2 = 0.864, F = 45.27, P < 0.01; with height ≥ 9.1 m, P > 0.05; Figure 6c), quadratic regression of leaf area vs. tree height showed a significant effect (r^2 = 0.671, F = 21.35, P < 0.001) and indicated that the highest leaf area appeared at a height of 37.1 m (Table S4).

Leaf area had a significant linear relationship with increasing diffuse light in the understory (with ln(%TRANS) < −0.132 or diffuse transmittance < 87.63%, slope = 10.766, r^2 = 0.601, F = 22.06, P < 0.001). After the tree crown was exposed to high light conditions with diffuse transmittance ≥ 87.63%, the leaf area decreased with increasing ln(%TRANS) (Figure 6d). The quadratic regression detected that leaf area vs. ln(%TRANS) had a lower r^2 (r^2 = 0.449, F = 9.13, P < 0.01) than leaf area vs. tree height (Table S4).

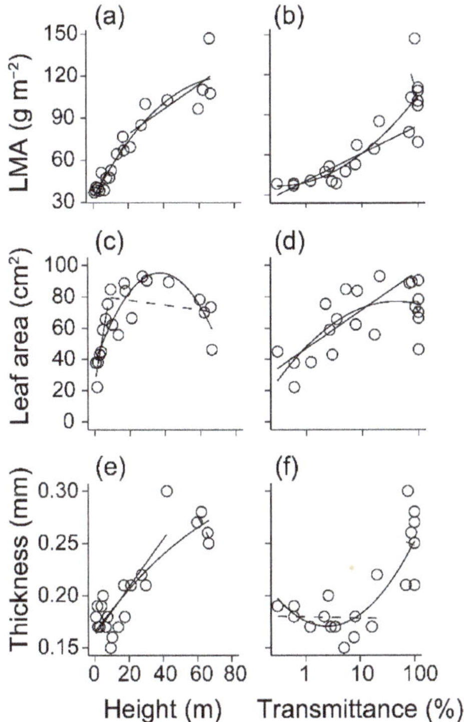

Figure 6. Leaf morphological traits such as LMA, leaf area, and thickness plotted against tree height (**a,c,e**) and light (represented as ln(%TRANS)) (**b,d,f**). Lines indicate the linear piecewise regression results; solid lines denote significant relationships ($P < 0.05$), and broken lines denote nonsignificant relationships. The curves indicate the quadratic regression results. Note that a natural log scale is used on the X-axis in panels (**b,d,f**).

Leaf thickness could increase linearly with increasing tree height (with height < 59.6 m, $r^2 = 0.595$, $F = 24.54$, $P < 0.001$ (Figure 6e)), but the linear relationship between leaf thickness and diffuse transmittance was not significant (Figure 6f). Quadratic regression also indicated that leaf thickness and tree height had positive relationship during the whole life history (Figure 6e), and the lowest thickness appeared at 2.50% diffuse transmittance (Figure 6f; Table S4) at the height of around 6.3 m (converted by ln(%TRANS)-height model in Table S2).

3.4. Physiological Trait Transition Point of P. chinensis

R_{d_area} showed the lowest value at 57.3 m in the quadratic model (Figure 7a), and the height-related piecewise linear model of R_{d_area} was divided into two parts at 17.5 m (Figure 7a). R_{d_area} had only one unique significant linear relationship throughout almost all light gradients in piecewise linear regression (Figure 7b).

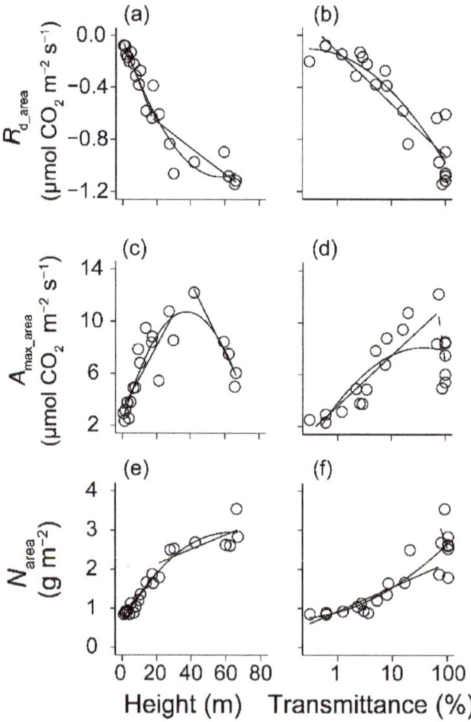

Figure 7. Area-based physiological traits such as R_{d_area}, A_{max_area}, and N_{area} plotted against tree height (**a**,**c**,**e**) and light (represented as ln(%TRANS)) (**b**,**d**,**f**). Lines indicate the linear piecewise regression results; solid lines denote significant relationships ($P < 0.05$), and broken lines denote nonsignificant relationships. The curve indicates the quadratic regression result. Note that a natural log scale is used on the x-axis in panels (**b**,**d**,**f**).

A_{max_area} had a "hump-shaped" correlation with tree height growth, a transition point in linear piecewise regression at an intermediate position in the whole height gradient (42.1 m; Figure 7c), and a similar peak value in the quadratic regression of A_{max_area} vs. tree height (highest A_{max_area} = 10.7 μmol CO_2 m^{-2} s^{-1} at 37.7 m high, r^2 = 0.766, F = 33.66, $P < 0.001$). A_{max_area} had a significant linear correlation with light promotion in the understory (with ln(%TRANS) < −0.278 or diffuse transmittance < 75.73%, r^2 = 0.760, F = 42.24, $P < 0.001$; Figure 7d), but the quadratic model of A_{max_area} vs. ln(%TRANS) had a lower r^2 (r^2 = 0.534, F = 12.45, $P < 0.001$) than leaf area vs. tree height (Table S4).

N_{area} had a positive relationship with tree height growth, but the slope of the linear model decreased at tree height ≥ 21.2 m (slope = 0.019, r^2 = 0.459, F = 6.93, $P < 0.05$; Figure 7e). Only one significant linear model was detected in the understory (with ln(%TRANS) < −0.278 or diffuse transmittance < 75.73%, r^2 = 0.660, F = 26.22, $P < 0.001$; Figure 7f).

The increase in tree height significantly reduced R_{d_mass} up to 17.5 m (r^2 = 0.700, F = 26.73, $P < 0.01$), and the quadratic model indicated the lowest R_{d_mass} occurring at 43.5 m (Figure 8a). The quadratic term in quadratic regression of R_{d_mass} vs. ln(%TRANS) was nonsignificant ($P > 0.05$; Table S4), and R_{d_mass} linearly decreased throughout the process of light improvement (r^2 = 0.719, F = 52.08, $P < 0.01$; Figure 8b).

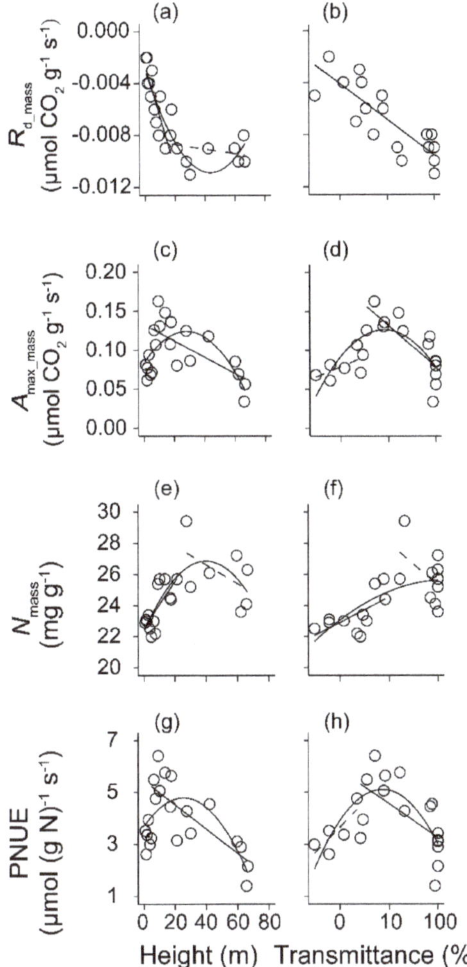

Figure 8. Mass-based physiological traits such as R_{d_mass}, A_{max_mass}, N_{mass}, and photosynthetic nitrogen use efficiency (PNUE) plotted against tree height (**a,c,e,g**) and light (represented as ln(%TRANS)) (**b,d,f,h**). Lines indicate the linear piecewise regression results; solid lines denote significant relationships ($P < 0.05$), and broken lines denote nonsignificant relationships. The curve indicates the quadratic regression result. Note that a natural log scale is used on the X-axis in panels (**b,d,f,h**).

Mass-based maximal net assimilation rate (A_{max_mass}) had a "hump-shaped" curve with both height (Figure 8c) and light (Figure 8d) gradient. A_{max_mass} had a significant negative relationship with tree height growth beginning at a height of 4.6 m (piecewise regression: $r^2 = 0.425$, $F = 12.09$, $P < 0.01$; Figure 8c) or 43.5 m (quadratic regression: $r^2 = 0.778$, $F = 35.98$, $P < 0.001$), and light condition improvement only significantly reduced A_{max_mass} once more than 3.5% of the transmittance consisted of diffuse light in piecewise linear regression (or -3.352 in ln(%TRANS); $r^2 = 0.606$, $F = 20.96$, $P < 0.01$; Figure 8d), or 8.22% of diffuse transmittance in the quadratic model ($r^2 = 0.477$, $F = 10.14$, $P < 0.001$, Table S4).

N_{mass} had the highest value at 39.9 m in the quadratic model ($r^2 = 0.494$, $F = 10.77$, $P < 0.001$; Table S4). Piecewise linear regression showed a significant linear increase in N_{mass} with height growth up to 27.4 m (slope = 0.148, $r^2 = 0.478$, $F = 12.9$, $P < 0.01$). Although the quadratic term in the quadratic model of N_{mass} vs. light was nonsignificant ($P > 0.05$, Table S4), N_{mass} only had a significant linear increase with ln(%TRANS) up to -1.808 (representing 16.4% transmittance; slope = 0.7143, r^2 = 0.337, F = 6.074, $P < 0.05$), and the relationship between N_{mass} and higher height or light gradient was not significant in piecewise regression (Figure 8e,f).

The tendency of PNUE in terms of both height and light was quite similar to that of A_{max_mass} (Figure 8g,h), and the transition point of the linear piecewise regression in PNUE was 4.6 m in the height model and -3.65 ln(%TRANS) (representing 2.60% transmittance) in the light model. In quadratic regression, the transition point in PNUE was 25.3 m in the height model and 7.23% transmittance in the light model (Table S4).

LCP showed a positive relationship with height (Figure 9a) and light (Figure 9b) in either the linear or quadratic model. In piecewise regression, LCP was stable before the tree height reached 13.5 m (Figure 9a). Subsequently, it increased linearly with tree height growth (r^2 = 0.706, F = 24.95, $P < 0.001$). Although LCP increased with improved light conditions and a transition point at 16.40% transmittance of diffuse light (ln(%TRANS) = -1.808) was observed in piecewise regression, no significant piecewise linear relationship was found between LCP and ln(%TRANS) (Figure 9b).

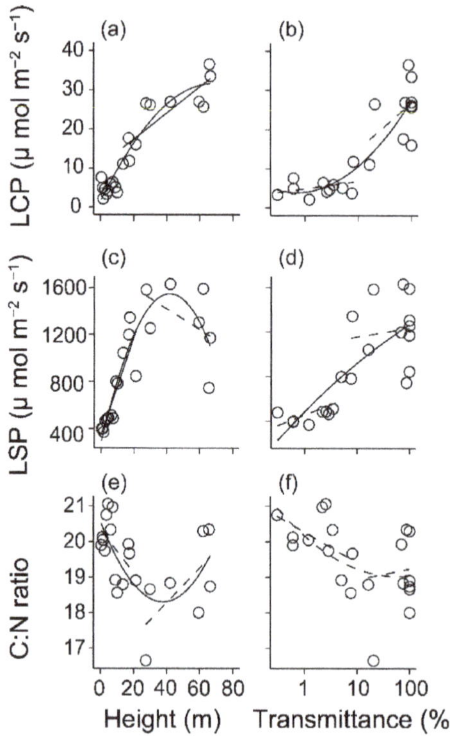

Figure 9. Light compensation point (LCP), light saturation point (LSP), and carbon-nitrogen ratio (C:N ratio) plotted against tree height (**a,c,e**) and light (represented as ln(%TRANS)) (**b,d,f**). Lines indicate the linear piecewise regression results; solid lines denote significant relationships ($P < 0.05$), and broken lines denote nonsignificant relationships. The curve indicates the quadratic regression result. Note that a natural log scale is used on the X-axis in panels (**b,d,f**).

LSP showed a significantly linear increase with tree height up to 27.4 m (r^2 = 0.751, F = 40.28, $P < 0.001$; Figure 9c), after which the trend became nonsignificant (r^2 = 0.077, F = 1.50, $P = 0.275$). The quadratic model of LSP vs. tree height had a peak value at 41.8 m (Table S4). High light could significantly promote LSP in quadratic regression ($r^2 = 0.572$, $F = 14.36$, $P < 0.001$), but the piecewise linear regression was nonsignificant (Figure 9d).

Linear piecewise regression showed a C:N ratio transition point at 27.4 m in the height-related model or 16.40% transmittance (or -1.808 in ln(%TRANS)) in the light-related model, but all the linear models were nonsignificant ($P > 0.05$; Figure 9e,f). By contrast, the relationship of the C:N ratio vs. height could be predicted by a significant quadratic model ($r^2 = 0.333$, $F = 5.99$, $P < 0.05$), and the lowest predicted C:N ratio (18.30) appeared at a height of 37.8 m. In fact, the quadratic model of the C:N ratio vs. height had a higher r^2 than the quadratic model of the C:N ratio vs. ln(%TRANS) ($r^2 = 0.198$, $F = 3.47$, $P > 0.05$).

4. Discussions

4.1. Relationships between Variations in Leaf Traits

Photosynthetic traits such as respiration rate and LCP play an important role in determining a plant's tolerance to low light levels [40]. *P. chinensis* has low LCP and R_{d_area} in the understory, which allows its seedlings and treelets to conserve carbon better than trees in the canopy stage and to gain carbon better, having net photosynthesis at lower light levels [41]. The trade-off between low respiration rates and high maximum photosynthetic rates of *P. chinensis* in habitat shift from understory to canopy reflected its change in adaptive strategy: conservation of carbon in shaded habitats, which indicates a "stress-tolerant" strategy; increased photosynthesis in high light environments, representing a competitive strategy. All the strategies described here were inferred only relative to ontogenetic adaptation in *P. chinensis*, which only reflected the similarity between the leaf traits in the different phases with those of the Grimes' strategies.

Investing more nitrogen in proteins to limit the light saturation rate of photosynthesis can improve photosynthesis under high light conditions [42]; therefore, *P. chinensis* has higher N_{mass} and LSP under high light transmittance (Figure 5f). These increased photosynthetic activities could promote leaf respiration rate in the dark [43], leading to a close negative relationship between R_{d_mass} and the light gradient (Figure 8b).

A_{max_area} and LSP increased with tree height in understory but decreased in very tall trees, while A_{max_mass} and PNUE peaked at low %TRANS (3.50% for A_{max_mass} and 2.60% for PNUE) and then monotonically decreased with height (Figure 8c,g). This difference between area- and mass-based characteristics was similar to the findings in a previous report on other tropical canopy trees [44] and can probably be attributed to the finding that area-based characteristics mainly depend on LMA, while mass-based characteristics mainly depend on chlorophyll content per unit dry mass [42].

Although He et al. [25] reported that both leaf thickness and LMA have positive correlations with tree height of *P. chinensis* in the canopy layer, the increasing tendency with increased tree height differs between leaf thickness and LMA. Increasing height seems to be the main factor affecting leaf area throughout ontogeny (Figure 6e) and is correlated with a linear increase in leaf thickness of *P. chinensis* from seedlings to very tall trees. This pattern corresponds to an increase in palisade and mesophyll cells [42] and results in greater chlorophyll per unit dry mass with tree height growth. However, the leaf area of *P. chinensis* displayed a quadratic relation with height growth (Figure 6c), and the piecewise linear regression of LMA vs. tree height showed different slopes beginning at a height of 27.4 m (Figure 6a), which demonstrates that the effect of light conditions on LMA must be taken into account in the understory. Although hydraulic limitations with the height gradient can potentially limit leaf development as indicated by increasing LMA [14,35], high light availability could also lead to thicker leaves as a result of thicker palisade mesophyll cell layers, which maximize light capture [45]. The light gradient in the understory could enhance the sensitivity of LMA to the change in tree height, but after the tree crown overtopped the main canopy layer, the saturated light conditions could not further increase LMA. Both light and height conditions contributed to the ontogenetic variability of LMA.

4.2. Growth Phases and Adaptive Strategies of P. chinensis

MRT analysis showed that the leaf trait combinations throughout the ontogeny of *P. chinensis* could be divided into five phases. The first split was divided by the diffuse transmittance at 18.33%, which indicated the importance of ambient light conditions for leaf trait acclimation in the ontogeny of *P. chinensis*.

The other three splits were mainly driven by height, at 8.3, 24.3, and 60.9 m. The effects of tree size at 8.3 m may be combined with light acclimation because the height and light are always positively correlated in the understory and within the canopy [5]. The other two splits, at 24.3 and 60.9 m, occurred in trees exposed to a light-saturated environment; thus, other age-specific environmental filters, such as hydraulic limit and reproductive onset, are expected to drive selection towards specific resource-investment strategies for these specific height gradients.

With further combination with the transition points of linear piecewise regression (Tables S2 and S3) (or peak value of the quadratic regression, Table S4) in leaf traits vs. environmental factors, five phases in MRT were redefined as the vulnerable phase, suppressed phase, growth release phase, canopy phase, and emergent phase to better describe the growth strategies in different phases of *P. chinensis*. Split levels of height and corresponding diffuse transmittance are listed in Table 1.

Table 1. Growth phases and changes in the adaptive strategy of leaf traits of *P. chinensis*. The letters in the strategy column indicate the following: C-relatively "competitive" strategy; S-relatively "stress-tolerant" strategy.

Growth Phase	Transmittance of Diffuse Light	Tree Height	DBH	Strategy	Representative Traits (Variation Tendency)
Emergent	100%	60.9 m	$\geq$96.8 cm	–	Leaf area and A_{max_area} (decline)
Canopy	100%	24.3–60.9 m	26.7–96.8 cm	C	Leaf area and A_{max_area} (highest)
Growth release	18.33–100%	14.9–24.3 m	14.8–26.7 cm	–	All traits (rapid change)
Suppressed	3.94–18.33%	8.3–14.9 m	6–14.8 cm	S	LCP (low) PNUE and A_{max_mass} (highest)
Vulnerable	0–3.94%	0–8.3 m	<6 cm	–	LCP (low) PNUE and A_{max_mass} (low)

Tree height less than 8.3 m or diffuse transmittance less than 3.94% (converted by ln(%TRANS)-height model; Table S2) characterize the vulnerable phase. PNUE and A_{max_mass} have similar transition points around the split point of diffuse transmittance (2.60% for PNUE and 3.50% for A_{max_mass} in piecewise linear regression, Table S3; 7.23% for PNUE and 8.22% for A_{max_mass} in quadratic regression, Table S4). Below these values, *P. chinensis* had a lower PNUE (Figure 8d,h) and A_{max_mass} (Figure 8b,f), probably due to the extremely low light intensity in the understory, which led to a larger fraction of investment of leaf nitrogen in light harvesting [42]. PNUE is quite closely related to the core leaf economic traits [34]. Low PNUE in the vulnerable phase might indicate that *P. chinensis* seedlings invested much less photosynthetic product into growth. Some studies have reported that minimum light requirements for species in dipterocarp forests is 1.7–5.2 mol m^{-2} d^{-1} or 6%–19% for sun [46]. We also observed that the height growth of *P. chinensis* seedlings after germination remained unchanged under low transmittance in a greenhouse [24], although high mortality caused by drought, herbivory, and missing occurred within 10 months after germination in the field [47], indicating that *P. chinensis* seedlings were quite vulnerable in this phase.

The highest PNUE and A_{max_mass} occurred in the transitional zone between the vulnerable phase and the suppressed phase and subsequently decreased with increasing tree height growth and light. A high PNUE indicates that treelet leaves partition more nitrogen to photosynthetic enzymes (such as Rubisco) [48] and/or have lower mesophyll resistance because of large air spaces and thin cell walls within the lamina [49]. Treelet leaves in the suppressed phase had the lowest A_{max_area} (Figure 7c,d) and the lowest N_{area} (Figure 7e,f), similar to saplings in the vulnerable phase, suggesting that the A_{max_area} in treelet leaves in the suppressed phase is severely limited by low nitrogen content [50]. In the suppressed phase, PNUE and A_{max_mass} exhibited a continuous and significant linear relationship with light (or height) variation until the canopy phase. That is, *P. chinensis* was physiologically flexible in its adaptation to light environmental change in the understory. Previous studies have reported that treelets of *P. chinensis* can survive in the understory for several years [51], which would be a consequence of this strategy, just as many other tropical canopy species can survive in the understory as seedlings or treelets for decades [52–54]. Tall seedlings of *Shorea leprosula*, *S. parvifolia*, and *S. pauciflora* in a Sabah rainforest also showed distinctly lower mortality than seedlings shorter than 50 cm [54]. Su [55] observed that seedlings of *Pometia pinnata* (formerly, *P. tomentosa*), another canopy tree species in the Xishuangbanna rainforest, had 99.6% mortality before reaching 1.5 m in height; beyond this height, mortality rapidly decreased and remained very low until physiological death, which Silvertown [56] described as Oscar Syndrome. In conclusion, in the suppressed phase, *P. chinensis* has conservative leaf economic trait values such as a smaller leaf area, lower A_{max_area}, and lower area-based leaf nitrogen concentration during ontogenesis, and these characters may indicate that the suppressed phase is an S-selected ("stress-tolerant" strategy) phase [10].

P. chinensis had a lower but relatively stable LCP below 16.40% transmittance or 13.5 m in height (piecewise linear regression; Figure 9a,d), which is close to the split point between the suppressed phase and the growth release phase (18.33% transmittance or 14.9 m in height; Table 1). The lower LCP allowed the treelets and saplings to achieve positive carbon gains under lower light intensities [57]. Based on these findings, diffuse transmittance at 18.33% could be suggested as a threshold value dividing the suppressed phase from the growth release phase.

Most leaf traits, including LMA, N_{area}, N_{mass}, LSP, leaf area, R_{d_area}, and R_{d_mass}, showed significant variations with increasing tree height and light intensity in the growth release phase, in accordance with the hypothesis that tree juveniles exhibit greater intraspecific variability in ecological strategies than adults [1]. The rapid change in leaf traits caused by the improvement of the light environment is the most important feature for the growth release phase of *P. chinensis* and indicates the conversion of the adaptive strategy from the understory to the canopy.

When the tree crown completely entered the main canopy layer at 24.3 m, the growth release phase ended and the canopy phase began. Ambient light in the canopy phase was close to saturation, and hydraulic limitation was achieved because of increasing trunk height, leaf tissue density, and cell wall thickness and decreased mesophyll air space [14]. The decrease in R_{d_mass} (Figure 8a) and increase in N_{mass} (Figure 8e) with height growth were terminated in this phase (transition points: 17.5 m in R_{d_mass} and 27.4 m in N_{mass}, piecewise linear regression; Table 1). Furthermore, leaf area-related traits such as LMA (Figure 6a), R_{d_area} (Figure 7a), and N_{area} (Figure 7e) continued to vary with height growth, although the model slopes were already reduced (transition points: 21.2 m in LMA, 17.5 m in R_{d_area} and 21.2 m in N_{area}, piecewise linear regression; Table 1). The canopy phase had the largest leaf area and the highest A_{max_area} and N_{mass} throughout the life history of *P. chinensis*, implying the highest photosynthetic efficiency in this phase. The leaves of *P. chinensis* in the canopy layer displayed higher A_{max_mass} than other local *Dipterocarp* species [29]; this species does not suffer from irreversible photoinhibition in the uppermost canopy leaves [31]. Although hydraulic limitation becomes a major determinant of photosynthesis in the canopy layer in other tree species [58], Kenzo et al. [59,60] observed that high leaf nitrogen concentration and a well-developed mesophyll structure help to maintain high A_{max} in the upper-canopy leaves of dipterocarp canopy trees, which explains why the peak value of A_{max_area} occurred in the canopy phase. In short, the canopy phase was more similar to a C-selected ("competitive" strategy) phase, and higher LMA and A_{max_area} in this phase implied conservative growth and maximized resource acquisition to improve competitiveness in stable and productive habitats [6,10].

MRT suggests a final phase (emergent phase) at tree height over 60.9 m, which is correlated with the decrease in some leaf traits associated with carbon-gain, including A_{max_area} (Figure 7c), leaf area (Figure 6c), and leaf C:N ratios (Figure 9e); these traits showed "hump-shaped" patterns with increasing tree height. Quadratic regression predicted that these traits had peak values at specific heights (A_{max_area} at 37.7 m, leaf area at 37.1 m, and leaf C:N ratios at 37.8 m; Table S4) that were already emergent over the main canopy layer (approximately 30 m) and then decreased with the further growth in tree height. This strategy is similar to other mid- to late-successional species [15,61]. Age- and size-dependent declines in leaf nitrogen concentration could lead to a reduction in A_{max} in very tall trees [16,17]; this could be a reason for the decreasing A_{max_area} in the emergent phase. Increased reproductive allocation to nonstructural carbohydrates among very large trees [61] could explain the reduction in nitrogen and A_{max}, and the increase in leaf C:N ratios (Figure 9e) provides potential support for this hypothesis. Former research in *P. chinensis* reported that reproduction of this species occurs only after its crown emerges from the main canopy, with DBH approximately 30–40 cm [22], and dipterocarps in Malaysia have a reproductive size threshold of approximately 60 cm in primary forest [62]. This empirical evidence indicated that *P. chinensis* had a reproductive phase only when the individual became huge and tall enough. In addition, increased mechanical perturbation due to herbivory [63] and wind exposure [64] among large canopy trees is also a potential cause for the declines in leaf nitrogen content and photosynthetic capacity with size. Due to the limit of the crane jib (60 m long), we could sample very few individuals of emergent trees in this phase. More experiments are still required to better support our explanations of the emergent phase in the future.

P. chinensis has larger leaves in the canopy phase than in the suppressed phase and vulnerable phase, but in the emergent phase, leaf area again decreases (Figure 6b). Similar phenomena have been reported for canopy tree species in other tropical forests [18]. Energy-demanding and resource-limited old trees could undergo more morphological than biochemical acclimation [65], and a limited leaf area could better control the leaf temperature, rates of carbon dioxide uptake, and water loss for adaptive acclimation to ambient light promotion in the canopy [66]. The decrease in leaf size in the emergent phase may also be a result of the physiological effects of carbon allocation to reproductive structures [18,61].

5. Conclusions

As a canopy tree species, *P. chinensis* adapts to light environments both in the understory and canopy throughout its life history. Based on the different strategies of leaf traits during ontogenetic development in response to light and height, five phases of *P. chinensis* could be recognized: vulnerable phase, suppressed phase, growth release phase, canopy phase, and emergent phase. According to CSR theory [6] and ontogenetic shifts in leaf traits [1,8,10], the suppressed phase and canopy phase could be classified as "stress-tolerant" strategy and "competitive" strategy, respectively, although these strategies were inferred only relative to ontogenetic adaptation in *P. chinensis*. The vulnerable phase and growth release phase may also not fall into any specific strategies because the trait values in these two phases were quite dynamic: the former was the key phase during which the plant achieved the transformation from seeds to seedlings, and the latter served as a spanning period during which light-related traits overcame suppression by low light in the understory and responded rapidly to change within a narrow height range (14.9–24.3 m) with a large change in ambient light (18.33%–100% transmittance) in the canopy. *P. chinensis* had lower carbon assimilation in the emergent phase than in the canopy phase, which may be attributed to reproductive allocation strategy, and this deserves more experiments in the future. Our results suggest that this tropical forest canopy species adjusts its adaptive strategy in leaf functional traits to adapt to variation in the ambient light environment and tree size in different life-history stages. The ecological implications of this pattern of change throughout the lifetime of a tree species for the forest growth cycle deserve further investigation in terms of maintaining forest biodiversity.

Supplementary Materials: The following are available online at http://www.mdpi.com/1999-4907/11/11/1145/s1, Figure S1: Simple linear regression result for R_{d_area}' vs. R_{d_area}. The solid line indicates the simple linear regression result: $R_{d\_area}' = -0.1210 + 1.0337 \times R_{d\_area}$, $r^2 = 0.889$, $F = 161.6$, $P < 0.01$, Table S1: Species list of the 1.44 ha (120 m $\times$ 120 m) canopy crane plot in Xishuangbanna, Southwestern China. This list was based on survey data collected in December 2014 and includes all tree species with DBH $\geq$ 1 cm, Table S2: Linear piecewise regression results for traits (y) vs. height (x). The following regression models were used: Equation (1): $y = a_1 + b_1 \times x$, $x <$ transition point; Equation (2): $y = a_2 + b_2 \times x$, $x \geq$ transition point, Table S3: Linear piecewise regression results for traits (y) vs. light (represented as ln(%TRANS), x). The following regression models were used: Equation (1): $y = a_1 + b_1 \times x$, $x <$ transition point; Equation (2): $y = a_2 + b_2 \times x$, $x \geq$ transition point, Table S4: Quadratic regression results for traits (y) vs. height or ln(%TRANS) (x). The following regression model was used: $y = \beta_0 + \beta_1 \times x + \beta_2 \times x^2$.

Author Contributions: M.C., Y.D., and X.D. conceived, designed, and directed the experiments and coordinated this study. Y.D. wrote the manuscript. M.C. improved and edited the manuscript. J.D., W.Z., and T.H. performed the field work. A.N., X.S., and P.F. revised the manuscript. All authors have read and agreed to the published version of the manuscript.

Funding: This work was supported by the Chinese Academy of Sciences 135 Program (No. 2017XTBG-T01) and Field Station Foundation of Chinese Academy of Sciences.

Acknowledgments: We thank Daowei Wu, Jinhai Mu, Xin Dong, Mingzhong Liu, Tao Jiang, and Pinping Zeng for their assistance with the field work and the Public Technical Service Platform of Xishuangbanna Tropical Botanical Garden, Chinese Academy of Sciences, for the chemical analysis.

References

1. Dayrell, R.L.C.; Arruda, A.J.; Pierce, S.; Negreiros, D.; Meyer, P.B.; Lambers, H.; Silveira, F.A.O. Ontogenetic shifts in plant ecological strategies. *Funct. Ecol.* **2018**, *32*, 2730–2741. [CrossRef]

2. Sack, L.; Grubb, P.J. Why do species of woody seedlings change rank in relative growth rate between low and high irradiance? *Funct. Ecol.* **2001**, *15*, 145–154. [CrossRef]

3. Valladares, F. Light heterogeneity and plants: From ecophysiology to species coexistence and biodiversity. In *Progress in Botany*; Esser, K., Lüttge, U., Beyschlag, W., Hellwig, F., Eds.; Springer: Berlin, Germany, 2003; pp. 439–471.

4. Coopman, R.E.; Briceno, V.F.; Corcuera, L.J.; Reyes-Diaz, M.; Alvarez, D.; Saez, K.; Garcia-Plazaola, J.I.; Alberdi, M.; Bravo, L.A. Tree size and light availability increase photochemical instead of non-photochemical capacities of *Nothofagus nitida* trees growing in an evergreen temperate rain forest. *Tree Physiol.* **2011**, *31*, 1128–1141. [CrossRef] [PubMed]

5. Cavaleri, M.A.; Oberbauer, S.F.; Clark, D.B.; Clark, D.A.; Ryan, M.G. Height is more important than light in determining leaf morphology in a tropical forest. *Ecology* **2010**, *91*, 1730–1739. [CrossRef] [PubMed]

6. Grime, J.P. Evidence for the existence of three primary strategies in plants and its relevance to ecological and evolutionary theory. *Am. Nat.* **1977**, *111*, 1169–1194. [CrossRef]

7. Grubb, P.J. The maintenance of species-richness in plant communities: The importance of the regeneration niche. *Biol. Rev.* **1977**, *52*, 107–145. [CrossRef]

8. Pierce, S.; Negreiros, D.; Cerabolini, B.E.L.; Kattge, J.; Díaz, S.; Kleyer, M.; Shipley, B.; Wright, S.J.; Soudzilovskaia, N.A.; Onipchenko, V.G.; et al. A global method for calculating plant CSR ecological strategies applied across biomes world-wide. *Funct. Ecol.* **2017**, *31*, 444–457. [CrossRef]

9. Lloyd, J.; Patiño, S.; Paiva, R.Q.; Nardoto, G.B.; Quesada, C.A.; Santos, A.J.B.; Baker, T.R.; Brand, W.A.; Hilke, I.; Gielmann, H.; et al. Optimisation of photosynthetic carbon gain and within-canopy gradients of associated foliar traits for Amazon forest trees. *Biogeosciences* **2010**, *7*, 1833–1859. [CrossRef]

10. Pierce, S.; Brusa, G.; Vagge, I.; Cerabolini, B.E.L. Allocating CSR plant functional types: The use of leaf economics and size traits to classify woody and herbaceous vascular plants. *Funct. Ecol.* **2013**, *27*, 1002–1010. [CrossRef]

11. Pierce, S.; Brusa, G.; Sartori, M.; Cerabolini, B.E.L. Combined use of leaf size and economics traits allows direct comparison of hydrophyte and terrestrial herbaceous adaptive strategies. *Ann. Bot.* **2012**, *109*, 1047–1053. [CrossRef]

12. Pacala, S.W.; Canham, C.D.; Saponara, J.; Silander, J.A.; Kobe, R.K.; Ribbens, E. Forest models defined by field measurements: Estimation, error analysis and dynamics. *Ecol. Monogr.* **1996**, *66*, 1–43. [CrossRef]

13. Kenzo, T.; Inoue, Y.; Yoshimura, M.; Yamashita, M.; Tanaka-Oda, A.; Ichie, T. Height-related changes in leaf photosynthetic traits in diverse Bornean tropical rain forest trees. *Oecologia* **2015**, *177*, 191–202. [CrossRef] [PubMed]

14. Koch, G.W.; Sillett, S.C.; Jennings, G.M.; Davis, S.D. The limits to tree height. *Nature* **2004**, *428*, 851–854. [CrossRef]

15. Thomas, S.C. Photosynthetic capacity peaks at intermediate size in temperate deciduous trees. *Tree Physiol.* **2010**, *30*, 555–573. [CrossRef] [PubMed]

16. Bond, B.J. Age-related changes in photosynthesis of woody plants. *Trends Plant Sci.* **2000**, *5*, 349–353. [CrossRef]

17. Niinemets, U. Stomatal conductance alone does not explain the decline in foliar photosynthetic rates with increasing tree age and size in *Picea abies* and *Pinus sylvestris*. *Tree Physiol.* **2002**, *22*, 515–535. [CrossRef] [PubMed]

18. Thomas, S.C.; Ickes, K. Ontogenetic changes in leaf size in Malaysian rain forest trees. *Biotropica* **1995**, *27*, 427–434. [CrossRef]

19. Cao, M.; Zhang, J. Tree species diversity of tropical forest vegetation in Xishuangbanna, SW China. *Biodivers. Conserv.* **1997**, *6*, 995–1006. [CrossRef]

20. Cao, M.; Zhu, H.; Wang, H.; Lan, G.; Hu, Y.; Deng, S.Z.X.; Cui, J. *Xishuangbanna Tropical Seasonal Rainforest Dynamics Plot: Tree Distribution Maps, Diameter Tables and Species Documentation*; Yunnan Science and Technology Press: Kunming, China, 2008.

21. Van der Velden, N.; Ferry Slik, J.W.; Hu, Y.H.; Lan, G.; Lin, L.; Deng, X.; Poorter, L. Monodominance of *Parashorea chinensis* on fertile soils in a Chinese tropical rain forest. *J. Trop. Ecol.* **2014**, *30*, 311–322. [CrossRef]

22. Ying, S.; Shuai, J. Study on fruiting behavior, seedling establishment and population age classes of *Parashorea chinensis*. *Acta Bot. Yunnanica* **1990**, *12*, 415–420.

23. Tang, J.W.; Shi, J.P.; Zhang, G.M.; Bai, K.J. Density, structure and biomass of *Papashorea chinensis* population in different patches in Xishuangbanna, SW. *J. Plant Ecol.* **2018**, *32*, 40–54.

24. Yan, X.F.; Cao, M. Effects of light intensity on seed germination and seedling early growth of *Shorea wantianshuea*. *Chin. J. Appl. Ecol.* **2007**, *18*, 23–29.

25. He, C.X.; Li, J.Y.; Zhou, P.; Guo, M.; Zheng, Q.S. Changes of leaf morphological, anatomical structure and carbon isotope ratio with the height of the Wangtian tree (*Parashorea chinensis*) in Xishuangbanna, China. *J. Integr. Plant Biol.* **2008**, *50*, 168–173. [CrossRef] [PubMed]

26. Engelbrecht, B.M.J.; Herz, H.M. Evaluation of different methods to estimate understorey light conditions in tropical forests. *J. Trop. Ecol.* **2001**, *17*, 207–224. [CrossRef]

27. Atkin, O.K.; Evans, J.R.; Siebke, K. Relationship between the inhibition of leaf respiration by light and enhancement of leaf dark respiration following light treatment. *Funct. Plant Biol.* **1998**, *25*, 437–443. [CrossRef]

28. Florez-Sarasa, I.; Araújo, W.L.; Wallström, S.V.; Rasmusson, A.G.; Fernie, A.R.; Ribas-Carbo, M. Light-responsive metabolite and transcript levels are maintained following a dark-adaptation period in leaves of *Arabidopsis Thaliana*. *New Phytol.* **2012**, *195*, 136–148. [CrossRef]

29. Meng, L.Z.; Zhang, J.L.; Cao, K.F.; Xu, Z.F. Diurnal changes of photosynthetic characteristics and chlorophyll fluorescence in canopy leaves of four diptocarp species under ex-situ conservation. *Acta Phytoecol. Sin.* **2005**, *29*, 976–984. [CrossRef]

30. Woodruff, D.R.; Meinzer, F.C.; Lachenbruch, B.; Johnson, D.M. Coordination of leaf structure and gas exchange along a height gradient in a tall conifer. *Tree Physiol.* **2009**, *29*, 261–272. [CrossRef]

31. Zhang, J.L.; Meng, L.Z.; Cao, K.F. Sustained diurnal photosynthetic depression in uppermost-canopy leaves of four dipterocarp species in the rainy and dry seasons: Does photorespiration play a role in photoprotection? *Tree Physiol.* **2009**, *29*, 217–228. [CrossRef]

32. Santiago, L.S.; Mulkey, S.S. A test of gas exchange measurements on excised canopy branches of ten tropical tree species. *Photosynthetica* **2003**, *41*, 343–347. [CrossRef]

33. Bassman, J.H.; Zwier, J.C. Gas exchange characteristics of *Populus trichocarpa*, *Populus deltoides* and *Populus trichocarpa* x *P. deltoides* clones. *Tree Physiol.* **1991**, *8*, 145–159. [CrossRef]

34. Wright, I.J.; Reich, P.B.; Cornelissen, J.H.C.; Falster, D.S.; Garnier, E.; Hikosaka, K.; Lamont, B.B.; Lee, W.; Oleksyn, J.; Osada, N.; et al. Assessing the generality of global leaf trait relationships. *New Phytol.* **2005**, *166*, 485–496. [CrossRef] [PubMed]

35. Coble, A.P.; Cavaleri, M.A. Light acclimation optimizes leaf functional traits despite height-related constraints in a canopy shading experiment. *Oecologia* **2015**, *177*, 1131–1143. [CrossRef] [PubMed]

36. De'ath, G. Multivariate regression trees: A new technique for modeling species–environment relationships. *Ecology* **2002**, *83*, 1105–1117. [CrossRef]

37. Breiman, L.; Friedman, J.H.; Olshen, R.A.; Stone, C.I. *Classification and Regression Trees*; Wadsworth International Group: Belmont, CA, USA, 1984.

38. Ishii, H.T.; Jennings, G.M.; Sillett, S.C.; Koch, G.W. Hydrostatic constraints on morphological exploitation of light in tall *Sequoia sempervirens* trees. *Oecologia* **2008**, *156*, 751–763. [CrossRef]

39. R Core Team. *A Language and Environment for Statistical Computing*; R for Statistical Computing: Vienna, Austria, 2018.

40. Givnish, T.J. Adaptation to sun and shade: A whole-plant perspective. *Funct. Plant Biol.* **1988**, *15*, 63–92. [CrossRef]

41. Craine, J.M.; Reich, P.B. Leaf-level light compensation points in shade-tolerant woody seedlings. *New Phytol.* **2005**, *166*, 710–713. [CrossRef]

42. Niinemets, Ü. A review of light interception in plant stands from leaf to canopy in different plant functional types and in species with varying shade tolerance. *Ecol. Res.* **2010**, *25*, 693–714. [CrossRef]

43. Amthor, J.S. *Respiration and Crop Productivity*; Springer: New York, NY, USA, 1989.

44. Kenzo, T.; Yoneda, R.; Sano, M.; Araki, M.; Shimizu, A.; Tanaka-Oda, A.; Chann, S. Variations in leaf photosynthetic and morphological traits with tree height in various tree species in a Cambodian tropical dry evergreen forest. *JARQ* **2012**, *46*, 167–180. [CrossRef]

45. Oguchi, R.; Hikosaka, K.; Hirose, T. Leaf anatomy as a constraint for photosynthetic acclimation: Differential responses in leaf anatomy to increasing growth irradiance among three deciduous trees. *Plant Cell Environ.* **2005**, *28*, 916–927. [CrossRef]

46. Baltzer, J.L.; Thomas, S.C. Determinants of whole-plant light requirements in Bornean rain forest tree saplings. *J. Ecol.* **2007**, *95*, 1208–1221. [CrossRef]

47. Yan, X.F.; Cao, M. The endangered causes and protective strategies for *Shorea wantianshuea*, a tropical rain forest tree species in Xishuangbanna. *J. Fujian Sci. Technol.* **2008**, *35*, 187–191.

48. Field, C. Allocating leaf nitrogen for the maximization of carbon gain: Leaf age as a control on the allocation program. *Oecologia* **1983**, *56*, 341–347. [CrossRef] [PubMed]

49. Kogami, H.; Hanba, Y.T.; Kibe, T.; Terashima, I.; Masuzawa, T. CO_2 transfer conductance, leaf structure and carbon isotope composition of *Polygonum cuspidatum* leaves from low and high altitudes. *Plant Cell Environ.* **2001**, *24*, 529–538. [CrossRef]

50. Ishida, A.; Yazaki, K.; Hoe, A.L. Ontogenetic transition of leaf physiology and anatomy from seedlings to mature trees of a rain forest pioneer tree, *Macaranga gigantea*. *Tree Physiol.* **2005**, *25*, 513–522. [CrossRef]

51. Yan, X.F.; Cao, M. Seeding growth and survival of the endangered tree species *Shorea wantianshuea* after a mast-fruiting event. *J. Plant Ecol.* **2008**, *32*, 55–64.

52. Clark, D.A.; Clark, D.B. Life history diversity of canopy and emergent trees in a neotropical rain forest. *Ecol. Monogr.* **1992**, *62*, 315–344. [CrossRef]

53. Connell, J.H.; Green, P.T. Seedling dynamics over thirty-two years in a tropical rain forest tree. *Ecology* **2000**, *81*, 568–584. [CrossRef]

54. Still, M.J. Rates of mortality and growth in three groups of dipterocarp seedlings in Sabah, Malaysia. In *The Ecology of Tropical Forest Tree Seedlings (Man & The Biosphere Series)*; Swaine, M.D., Ed.; UNESCO: Paris, France, 1996; Volume 18, pp. 315–332.

55. Su, W.H. Preliminary study on the dynamics of *Pometia tomentosa* population in the tropical seasonal rain forest of Xishuangbanna. *Acta Bot. Yunnanica* **1997**, *9*, 92–96.

56. Silvertown, J. *Introduction to Plant Population Ecology*; Longman Scientific & Technical: New York, NY, USA, 1987.

57. Kitajima, K. Relative importance of photosynthetic traits and allocation patterns as correlates of seedling shade tolerance of 13 tropical trees. *Oecologia* **1994**, *98*, 419–428. [CrossRef]

58. Kitahashi, Y.; Ichie, T.; Maruyama, Y.; Kenzo, T.; Kitaoka, S.; Matsuki, S.; Chong, L.; Nakashizuka, T.; Koike, T. Photosynthetic water use efficiency in tree crowns of *Shorea beccariana* and *Dryobalanops aromatica* in a tropical rain forest in Sarawak, East Malaysia. *Photosynthetica* **2008**, *46*, 151–155. [CrossRef]

59. Kenzo, T.; Ichie, T.; Yoneda, R.; Kitahashi, Y.; Watanabe, Y.; Ninomiya, I.; Koike, T. Interspecific variation of photosynthesis and leaf characteristics in canopy trees of five species of Dipterocarpaceae in a tropical rain forest. *Tree Physiol.* **2004**, *24*, 1187–1192. [CrossRef]

60. Kenzo, T.; Ichie, T.; Watanabe, Y.; Yoneda, R.; Ninomiya, I.; Koike, T. Changes in photosynthesis and leaf characteristics with tree height in five dipterocarp species in a tropical rain forest. *Tree Physiol.* **2006**, *26*, 865–873. [CrossRef]

61. Martin, A.R.; Thomas, S.C. Size-dependent changes in leaf and wood chemical traits in two Caribbean rainforest trees. *Tree Physiol.* **2013**, *33*, 1338–1353. [CrossRef]

62. Thomas, S.; Appanah, S. On the statistical analysis of reproductive size thresholds in dipterocarp forests. *J. Trop. For. Sci.* **1995**, *27*, 412–418.

63. Patankar, R.; Thomas, S.C.; Smith, S.M. A gall-inducing arthropod drives declines in canopy tree photosynthesis. *Oecologia* **2011**, *167*, 701–709. [CrossRef]

64. Hossain, S.M.Y.; Caspersen, J.P. In-situ measurement of twig dieback and regrowth in mature *Acer saccharum* trees. *For. Ecol. Manag.* **2012**, *270*, 183–188. [CrossRef]

65. Sillett, S.C.; Van Pelt, R.; Koch, G.W.; Ambrose, A.R.; Carroll, A.L.; Antoine, M.E.; Mifsud, B.M. Increasing wood production through old age in tall trees. *For. Ecol. Manag.* **2010**, *259*, 976–994. [CrossRef]

66. Parkhurst, D.F.; Loucks, O.L. Optimal leaf size in relation to environment. *J. Ecol.* **1972**, *60*, 505–537. [CrossRef]

Publisher's Note: MDPI stays neutral with regard to jurisdictional claims in published maps and institutional affiliations.

Article

How Do Mediterranean Pine Trees Respond to Drought and Precipitation Events along an Elevation Gradient?

Sonja Szymczak [1,*], Martin Häusser [1], Emilie Garel [2,3], Sébastien Santoni [2,3], Frédéric Huneau [2,3], Isabel Knerr [4], Katja Trachte [5], Jörg Bendix [4] and Achim Bräuning [1]

[1] Institute of Geography, Friedrich-Alexander University Erlangen-Nürnberg, Wetterkreuz 15, 91058 Erlangen, Germany; martin.haeusser@fau.de (M.H.); achim.braeuning@fau.de (A.B.)
[2] Faculté des Sciences et Techniques, Laboratoire d'Hydrogéologie, Campus Grimaldi, Université de Corse Pascal Paoli, BP 52, 20250 Corte, France; garel_e@univ-corse.fr (E.G.); santoni_s@univ-corse.fr (S.S.); huneau_f@univ-corse.fr (F.H.)
[3] Centre national de la recherche scientifique, UMR 6134, SPE, 20250 Corte, France
[4] Laboratory of Climatology and Remote Sensing, Philipps-University of Marburg, Deutschhausstraße 12, 35037 Marburg, Germany; isabel.knerr@geo.uni-marburg.de (I.K.); bendix@mailer.uni-marburg.de (J.B.)
[5] Institute for Environmental Scienes, Brandenburg University of Technology (BTU) Cottbus-Senftenberg, Burger Chaussee 2, 03044 Cottbus, Germany; katja.trachte@b-tu.de
*   Correspondence: sonja.szymczak@fau.de

Received: 5 June 2020; Accepted: 10 July 2020; Published: 14 July 2020

Abstract: Drought is a major factor limiting tree growth and plant vitality. In the Mediterranean region, the length and intensity of drought stress strongly varies with altitude and site conditions. We used electronic dendrometers to analyze the response of two native pine species to drought and precipitation events. The five study sites were located along an elevation gradient on the Mediterranean island of Corsica (France). Positive stem increment in the raw dendrometer measurements was separated into radial stem growth and stem swelling/shrinkage in order to determine which part of the trees' response to climate signals can be attributed to growth. Precipitation events of at least 5 mm and dry periods of at least seven consecutive days without precipitation were determined over a period of two years. Seasonal dynamics of stem circumference changes were highly variable among the five study sites. At higher elevations, seasonal tree growth showed patterns characteristic for cold environments, while low-elevation sites showed bimodal growth patterns characteristic of drought prone areas. The response to precipitation events was uniform and occurred within the first six hours after the beginning of a precipitation event. The majority of stem circumference increases were caused by radial growth, not by stem swelling due to water uptake. Growth-induced stem circumference increase occurred at three of the five sites even during dry periods, which could be attributed to stored water reserves within the trees or the soils. Trees at sites with soils of low water-holding capacity were most vulnerable to dry periods.

Keywords: dendrometer; stem circumference changes; climate response; Mediterranean; *Pinus nigra*; *Pinus pinaster*

---

1. Introduction

Water availability is the most limiting abiotic factor for plant growth and productivity, especially in arid or semi-arid environments [1,2]. Tree dieback can often be related to water shortage and enhanced drought stress [3–5]. The identification of available water sources and the trees' response to short-term changes in environmental conditions is therefore of great importance to evaluate the vulnerability of forest ecosystems to current climate change.

A tree's water status is controlled by atmospheric conditions, the depth and distribution of the root system, and by soil water status. If the tree's transpiration exceeds the water uptake from the soil, the tree is able to relocate water from internal reservoirs to maintain the transpiration process and to adjust its water use [6] The hydraulic conductance of roots is highly variable and depends on soil water content, salinity of the soil, and on the demands for water from the transpiring shoot [7]. Soil water content is highly variable in space and time. Depending on the rooting and soil depth, trees can have access to different soil water layers and hence may respond differently to single precipitation events [8–10]. Beside the amount of water provided by precipitation, its seasonal distribution is of great relevance for regulating plant ecological processes. Several studies underline the importance of winter precipitation in summer-dry climates, which may stimulate an entirely different plant response than a summer rain event as shown for shrubs in desert environments [11,12]. This is particularly the case in climates where deeper soil water reservoirs are replenished by winter precipitation [9].

An indicator for the whole tree water status is the diurnal cycle of shrinking and swelling of the stem [6], which is caused by imbalances between transpiration and root water uptake [13], and by processes altering osmotic water potentials [14]. The daily cycle of water depletion and replenishment can be influenced by several factors, such as rainfall and soil water content, atmospheric vapor pressure deficit, and ambient air temperature [15–17]. Hence, stem circumference changes can be used to assess a tree's climate sensitivity, i.e., the response to extreme meteorological events [18,19], and its water status on short time scales [15].

Stem circumference changes (SCCs) are commonly measured with point or band dendrometers, a powerful tool providing data in unmatched quality and resolution [20]. Dendrometer measurements allow to investigate the influence of site conditions and meteorological factors on growth, as well as species-specific responses to changing climate conditions [19,21,22]. Dendrometers have been used to investigate the seasonal dynamics and growth phenology in a broad range of forest ecosystems in different climate zones, ranging from tropical [19] to subtropical [17,22] or alpine conditions [15,21]. For example, [17] determined water availability as the main factor influencing growth cessation in *Pinus pinaster* at the west coast of Portugal under Mediterranean climate conditions. However, the interpretation of dendrometer data can be complex, as SCCs depend on various factors. Beside the water-related swelling and shrinkage of phloem, xylem, and bark, increasing stem circumference can also be caused by irreversible radial growth due to newly formed sapwood and bark tissue cells, including cambial division and cell expansion [23,24]. Hence, it is often not possible to accurately differentiate "growth" from "stem water increase" [25]. Models to separate these two parameters require additional physiological data, which might not be available or which are difficult to measure over longer time periods [20]. Therefore, some studies use a combination of dendrometers and wood formation monitoring methods, as, e.g., [22] in the montane conifer *Cedrus libani* in Turkey.

Little is known about how fast trees respond to short-term fluctuations of water availability, although already [26] stressed the importance of single precipitation events for long-term plant functioning and survival. This is particularly important in climate regimes where the time of the highest rainfall amount and the vegetation period do not coincide, e.g., in the Mediterranean climate. Additionally, time lags in the signal transfer from water uptake to the tree-ring archive can occur [1,27–30], further complicating the interpretation of tree-ring proxy data related to water uptake, e.g., stable oxygen isotope ratios. Hence, the determination of the response time to discrete climatic events is of high importance. In this study, we used dendrometer measurements to analyze the response of pine trees to discrete climate events in Mediterranean forest ecosystems along elevation gradients on the island of Corsica. In contrast to other studies, we investigated the response to climate events not only during dry periods, but also to precipitation events. Due to the high spatial variability of ecological conditions on Corsica, we were able to study the tree response under both Mediterranean and subalpine climate conditions. We investigated the response of two pine species, which are widespread over the Mediterranean basin, i.e., *Pinus pinaster* Aiton growing at low to middle elevations, and *Pinus nigra* J.F. Arn subsp. *laricio* (Poiret) Maire var. Corsicana Hyl. growing at middle to high elevations, up to the upper tree limit. Despite its economic and ecological importance on Corsica, there are no studies on the intra-annual growth dynamics of *P. nigra* so far. The main aims of our study are (i) to determine the trees' response to dry periods and precipitation events by separating the growth- and stem water-related signals in SCC measurements, and (ii) to identify changes in tree response along an elevation gradient in order to distinguish tree growth from stem increment due to physiological responses under various local climate conditions. Hence, this study contributes to a better understanding of the productivity and resilience of both pine species under future climate conditions.

2. Materials and Methods

2.1. Study Sites

The mountainous island of Corsica (France) is located in the Western Mediterranean basin between 41–43°N and 8–10°E. It is characterized by a main mountain range extending in the north–south direction, with several peaks exceeding 2000 m above sea level (asl) (e.g., Mte. Cinto: 2706 m asl; Mte. Renoso: 2352 m asl). Due to steep elevation gradients, different microclimates are found, ranging from typical Mediterranean climates with dry hot summers and temperate wet winters close to sea level, to temperate and alpine conditions in the mountain areas, with a continuous winter snow cover above 1500 m asl [31].

Our five study sites are located along two elevation gradients from the Mediterranean coast to the high mountain zone along the western and eastern slopes of the Renoso Massif in central Corsica (Figure 1). For better readability, the study site IDs refer to the location of the sites, where "E" and "W" stands for "east" and "west", and "L", "M", and "H" for "low", "middle", and "high" elevation. The two studied pine species, *P. nigra* and *P. pinaster*, are widely distributed on Corsica, however, none of the species covers the full range of our transect (0–1600 m asl). A comparison of the cumulative growth rates in the transitional elevation belt in which both species coexist revealed no significant differences between the species [32], making it possible to compare tree data derived from *P. nigra* with those derived from *P. pinaster*. Tree-ring width chronologies from two additional sites support the similarity between the growth pattern of both species in elevations between 570 and 1240 m asl (Supplementary Figure S1). All five sites were very similar in terms of acidic soils, the dominance of granites and metamorphic rocks in the underground, and an average tree coverage below 30%. The low-elevation sites Ajaccio (WL) and Ghisonaccia (EL) are small forest patches of pure *P. pinaster* stands close to the coastlines. While the west coast is characterized by a complex topography with cliffs and bays, the east coast around EL is a large alluvial plain with sandy beaches and lagoons. On these sandy alluvial deposits, the soils are well developed Cambisols (WL) and podzols (EL). The mid-elevation sites Bocognano (WM) and Vivario (EM) are located within the mountain forest belt at 790 (WM) and 1000 m asl (EM), respectively, where *P. pinaster* and *P. nigra* co-occur. Soils are Cambisols of varying depths on granitic bedrock. The high-elevation site Capanelle (EH) is located at the upper timberline on the eastern slopes of Monte Renoso. The site is located close to the crest line of the mountain chain and is affected by air masses from the east as well as from the west, so that this high elevation site is representative for both slopes of the transect. The open forest at site EH consists of old-grown *P. nigra* and *Fagus sylvatica* trees. The soil is a shallow podzolic Cambisol on granitic bedrock with low water-holding capacity.

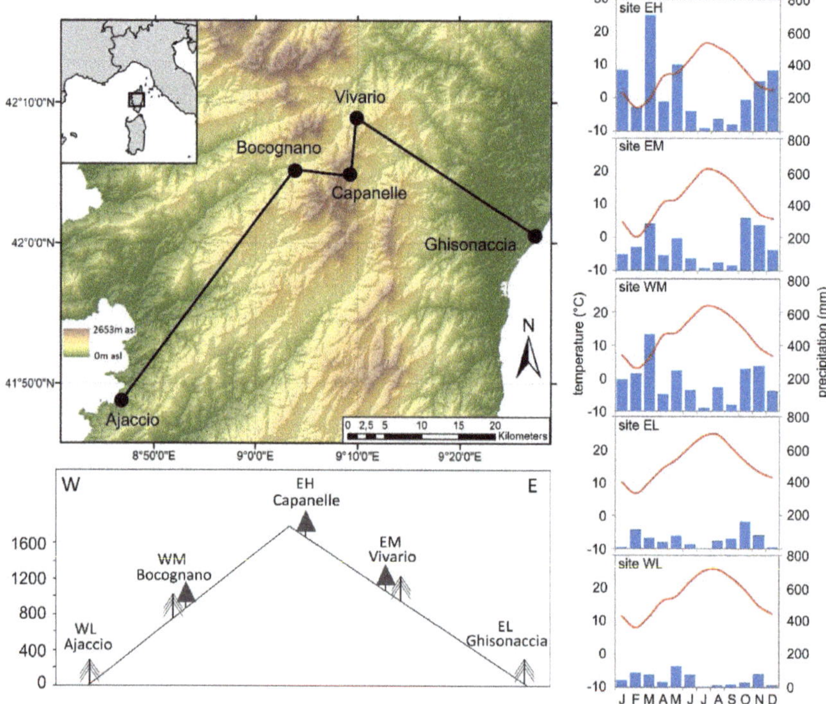

Figure 1. Map showing the location of the study sites and a schematic view of the study transect with site names and site codes explained in the main text. The two different tree signatures indicate the species *Pinus nigra* (filled symbol) and *Pinus pinaster* (dashed symbol). The digital elevation model is based on SRTM (Shuttle Radar Topography Mission) data [33]. Climate characteristics are expressed as the mean monthly temperature (red lines) and monthly precipitation sums (blue bars) in 2018. Data gaps from our own climate stations are filled with data from nearby MétéoFrance climate stations [34].

2.2. Meteorological Data

All sites were equipped with automatic weather stations (Campbell Scientific stations at EH, EM, and WM; Metek at EL and WL), which measured air temperature, precipitation, wind speed and direction, relative humidity, global radiation, soil moisture, and soil temperature in high resolution (5 min for EH, EM, and WM and 1 min for EL and WL). Soil moisture was recorded with a Campbell CS650 sensor, placed horizontal at 5–10 cm depth. Due to the rocky terrain, it was not possible to measure soil parameters at site EH. Mean temperature decreases with altitude, however, the western coastal site is warmer than the eastern site (Figure 1). Precipitation shows a more diverse pattern (Figure 1). The highest site (EH) is the wettest site, but at mid-elevation, the western site is wetter than the eastern site, although site EM is located at slightly higher elevation than site WM. In contrast, the eastern coast is more humid than the western coast.

To investigate the effect of single precipitation events on tree growth, we extracted all precipitation events between 01 May 2017 and 30 April 2019 from the climate data, thus covering a period of two complete years. We restricted the analysis to precipitation events with a precipitation sum of more than 5 mm, since lower precipitation amounts do not penetrate deeply into the soils [35]. To separate discrete precipitation events, we chose a threshold of one hour without any precipitation, i.e., all precipitation recorded with interruptions of less than one hour was regarded as one precipitation event.

To analyze their impact on tree behavior, we extracted dry periods of at least seven consecutive days without precipitation. The analysis of dry periods was restricted to the time with tree growth at all sites (March–November), i.e., dry periods in winter DJF were excluded because trees at higher elevations are dormant during this period and hence are not expected to respond to moisture changes.

2.3. Dendrometer Data

At each study site, six trees of similar tree age and stem size per study species were equipped with logging band dendrometers with a built-in thermometer (DRL26, EMS Brno) in April 2017, resulting in a total number of 42 study trees (Table 1). Stem circumference was registered in 30-min intervals. To avoid damages by grazing cattle, the dendrometers were installed at a stem height of 2 m. Since Mediterranean pine species have a very thick bark for fire protection, we removed parts of the outer bark without injuring the cambial zone to minimize the influence of bark swelling and shrinking on the dendrometer data.

Table 1. Overview of the tree data set. The tree species refers to *Pinus pinaster* (PIPI) and *Pinus nigra* (PINI), respectively. Site codes refer to the eastern (E) and western (W) side of Corsica, L, M, and H stand for low-, middle-, and high-elevation sites, respectively. The ages and stem circumferences are expressed as mean values ± 1 standard deviation (STD).

	Number of Trees	Tree Species	Mean Age (yr) ± 1 STD	Mean Stem Circumference (cm) ± 1 STD
WL	6	PIPI	32 ± 10	131 ± 24
WM	12	PIPI, PINI	42 ± 7	116 ± 24
EH	6	PINI	59 ± 10	109 ± 15
EM	12	PIPI, PINI	53 ± 10	120 ± 30
EL	6	PIPI	47 ± 20	139 ± 19

We analyzed the trees' responses to precipitation events and dry periods at the site level, i.e., we aggregated the dendrometer data from the individual trees to a mean curve for each site. This eliminates possible disturbance effects of individual trees and provides a representative picture of the common climate response at a specific site. Stem circumference change (SCC) measured by dendrometers contains information about the irreversible stem expansion related to cambial growth and cell expansion, as well as the reversible tree water deficit-induced shrinking and swelling of the stem. To distinguish between SCC caused by growth and stem shrinking/swelling, we used the growth definition of Zweifel [20], who defined growth as a current stem radius value exceeding a precedent maximum. By using this definition, each value of the dendrometer data was classified into one of the three categories of growth, swelling, or shrinkage (Figure 2). Afterwards, we calculated SCC curves free from the growth component by subtracting the values of all changes classified as growth from the raw data, i.e., all values classified as growth were set to the level of the value prior to the growth. Both SCC curves, the raw dendrometer data and the growth-free curves, were used for analyzing the response to precipitation events and dry periods, in order to determine which portion of the detected tree response to climate can be attributed to growth.

For each precipitation event, we extracted the mean SCC for each site from the starting point of the event over the following 48 h. For analyzing effects of dry periods, we extracted the SCC from the first day without precipitation and the following 21 days. We defined this threshold because longer dry periods occurred only at three sites and only once or twice in the investigated time period. To compare the tree responses between sites, the dendrometer data were standardized to the starting point of the precipitation event/dry period. We additionally investigated SCC after precipitation events for different seasons and different precipitation amounts. The precipitation events were divided into six classes regarding their amount. As events with low rainfall totals were more common than events with high rainfall totals, the thresholds to separate rainfall totals were not chosen with regular intervals, i.e., the class with the lowest rainfall totals included 5–9.9 mm/event precipitation, and the following four classes each had an increment of 10 mm/event (10–19.9, 20–29.9 mm/event, etc.). The class with the highest precipitation amount included totals of 50–99.9 mm/event.

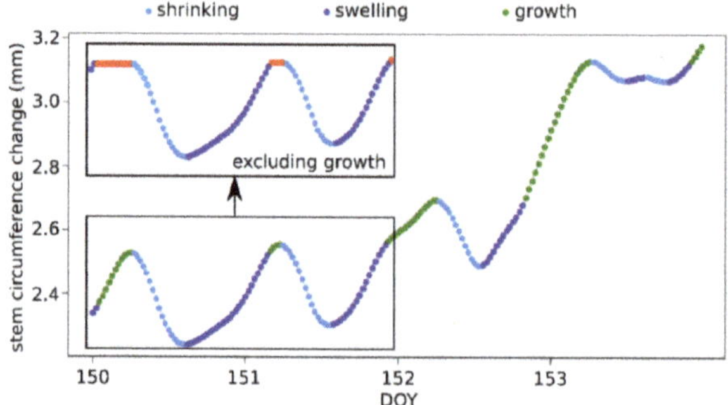

Figure 2. Schematic figure explaining the classification of the stem circumference change into shrinking, swelling, and growth, and the exclusion of the growth component, created with the R-program dendRoAnalyst [36]. Each dot represents a 30-min measurement recorded in 2017.

3. Results and Discussion

3.1. Site-Specific Stem Circumference Changes and Growing Periods

For the year 2018, the mean SCC curves from the coastal locations (EL and WL) can be clearly distinguished from the other sites by the much earlier start of the growing season, which did not begin before mid/end of April at mid- to high-elevation sites (Figure 3). A remarkable stem shrinkage is visible at site EH before the start of the growing season. The strong stem circumference increase in spring is followed by a decline in the growth rates in July. A second but less intense phase of increasing stem size occurred in October, followed by a cessation of stem diameter increment. The reduction in tree growth during summer did not occur simultaneously at all sites. The lowest cumulative increment sums occurred at sites EH and WM.

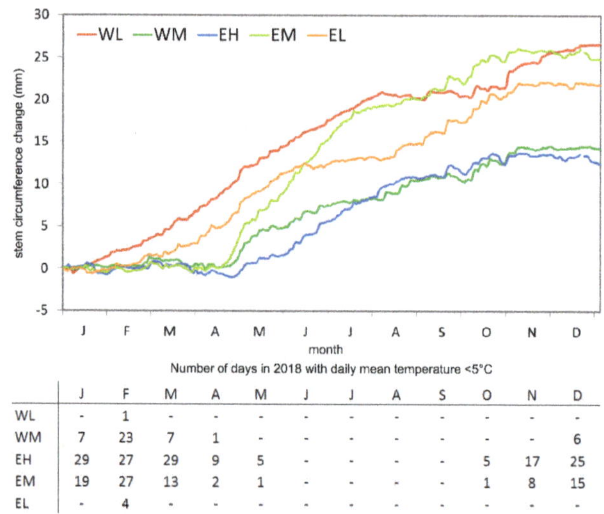

	J	F	M	A	M	J	J	A	S	O	N	D
WL	-	1	-	-	-	-	-	-	-	-	-	-
WM	7	23	7	1	-	-	-	-	-	-	-	6
EH	29	27	29	9	5	-	-	-	-	5	17	25
EM	19	27	13	2	1	-	-	-	-	1	8	15
EL	-	4	-	-	-	-	-	-	-	-	-	-

Figure 3. Mean cumulative stem circumference changes from the five study sites for the year 2018 reported as daily mean values from all trees (6 at WL, EH and EL, 12 at WM and EM) per site. Site codes refer to the east (E) and west (W) sides of Corsica, L, M, and H stand for low-, middle-, and high-elevation sites, respectively. The inserted table indicates the number of days in 2018 with daily mean temperatures <5 °C per site. Data gaps from our own climate stations are filled with data from nearby MétéoFrance climate stations [34].

The dendrometer data revealed that at all sites except WL, longer periods without growth occurred during several months in autumn/winter. The length of this dormancy period depended on the elevation of the site. At site WL, it was not possible to determine a dormancy season in autumn/winter from the dendrometer data because radial growth showed only short interruptions. Instead, prolonged periods without growth occurred during summer. The shortest dormancy occurred at site EL (from December to February), and the longest at site EH (from October/November to mid-May). Dormancy lasted from November/December to April at site EM, and from January to April at site WM.

As the climatic conditions vary strongly along the studied elevation transect, we expected different growth patterns and durations of the growing season. Bud break and growth onset after winter dormancy are highly responsive to temperature in temperate and boreal trees, as well as at drought-prone sites [37–40]. Rossi et al. [41] defined a mean daily air temperature of 4–5 °C as a critical value to reactivate xylem cell production of conifers in cold environments. This critical temperature is exceeded earlier at lower sites, leading to an earlier start of the growing season. Temperatures are lower at site EL than at site WL (Figure 1), so that the critical temperatures occur only during few days during a year, or in the case of a mild winter, even not at all, so that no clear dormancy season occurred at site WL in 2018. For WL, the mean temperature in JFM (ND) 2018 was 10.8 °C (13.1 °C), with 1% (0%) of days below the critical value of 5 °C. While the onset of radial growth is controlled by air temperature, several factors can cause the termination of radial growth, e.g., soil water availability [37,40], site conditions, stand age, and tree density [42]. A bimodal stem radial growth pattern with two growth peaks in the transitional seasons (spring-early summer and autumn), and a decreased growth rate in summer as a strategy for coping with harsh environmental conditions in dry periods is a typical pattern observed in climates with summer drought [17,21,43]. The summer growth suppression can either be caused by water deficit or by high temperatures above a certain threshold, which lead to a reduction in net photosynthesis [43]. A bimodal growth pattern is visible at the coastal locations on Corsica. As shown for other conifer species in Mediterranean regions [43–45], trees at our study sites also have the capability to resume cambial activity after periods of minimum growth or dormancy during summer droughts.

3.2. Precipitation Events and Dry Periods

The number of identified precipitation events over the two-year observation period ranged from 25 (site WL) to 138 (site WM), with a significantly higher number of rainfall events at higher elevations (Table 2). The same trends were visible in the mean precipitation sums and mean duration of the events. At the higher sites, single precipitation events could last for two days (sites EM and WM) or even longer (site EH), while they were usually shorter than one day at the coastal locations. The precipitation events were unevenly distributed over the seasons. The lowest number of events occurred at all sites in JJA, the highest in either MAM, SON, or DJF. It should be kept in mind that the total number of precipitation events is underestimated at some stations due to data gaps, especially at the coastal locations. However, the observed trend (stronger precipitation events and a higher number of events with increasing elevation) is confirmed by long-term climate stations from MétéoFrance [31].

The number of dry periods was less variable between the sites and ranged from 14 (site EL) to 20 (site WM) (Table 2). The maximum duration of rainless periods was 62 days at site WL, followed by 33 days at site EM. Even at the wettest sites, 19 and 21 consecutive dry days occurred. Coastal sites showed the longest mean duration of dry periods. Except for site WL, dry periods were most common in JJA, but they occurred in all seasons. The low number of dry periods in JJA at site WL was caused by a long dry period in July–August 2017, which lasted for 62 consecutive days.

Table 2. Characteristics of precipitation events and dry periods between May 2017 and April 2019. Mean precipitation sum was calculated independent from the mean duration of the events. Site codes refer to the eastern (E) and western (W) side of Corsica, L, M, and H stand for low-, middle-, and high-elevation sites, respectively. The seasons are expressed as abbreviations for the corresponding months (DJF: December, January, February; MAM: March, April, May; JJA: June, July, August; SON: September, October, November). Data gaps at individual climate stations: WL: January–February 2018, April–June 2018; EL: April–June 2018, mid November 2018–mid April 2019; EM: 15 days in March 2018.

Precipitation Events						
		WL	WM	EH	EM	EL
mean precipitation sum per event (mm)		11.5	20.8	33.0	18.6	14.3
mean duration of precipitation events (h)		6	9	14	10	6
mean intensity per event (mm/h)		2.6	2.7	2.9	2.3	3.3
number of events per season	DJF	6	42	41	31	7
	MAM	8	43	48	29	5
	JJA	1	19	13	12	1
	SON	10	34	24	26	24
number of events per amount class	5–9.9 mm	14	47	38	38	14
	10–19.9 mm	7	45	26	37	14
	20–29.9 mm	4	17	24	9	7
	30–39.9 mm	-	13	10	5	1
	40–49.9 mm	-	5	3	1	1
	50–99.9 mm	-	10	16	7	-
	> 100 mm	-	-	9	1	-
total number of precipitation events		25	138	126	98	37
Dry periods with at least 7 days without precipitation						
		WL	WM	EH	EM	EL
maximum duration (days)		62	21	19	33	27
mean duration (days)		18	10	11	13	14
number of dry periods per season	DJF	3	2	3	3	1
	MAM	3	5	3	4	2
	JJA	3	7	6	7	8
	SON	7	6	4	5	3
total number of dry periods		16	20	16	19	14

Precipitation on Corsica is very unevenly distributed throughout the year, with maximum precipitation rates in autumn/winter [31]. Single precipitation events can be followed by prolonged periods without any precipitation of up to more than one month at all sites, except WM and EH. As these periods occur mainly during the time of tree growth, it becomes obvious that the trees need to develop adaptation strategies to cope with drought stress related to this precipitation regime.

3.3. Tree Response to Precipitation Events

More than 50% of the stem circumference increase after precipitation events occurred within the first 12 h after the event (Figure 4). The strongest increase was found at the coastal locations and diminished with increasing elevation. The SCC increase was much steeper in the raw dendrometer data. The difference between the reaction including and excluding growth indicates that the response to precipitation events is a mixed effect of stem swelling and growth, with growth being the largest part at all sites (between around 60 and 80%), except EH (around 40%). The largest change within the 48 h after the event occurred at site EL, followed by WL and EM. The smallest stem increase after precipitation events occurred at the high-elevation site EH. Stem swelling followed immediately after the precipitation event, while stem increment due to growth extended over a longer time. This finding is in concordance with the soil water content (SWC), which increased mainly directly after the event.

At all sites except WL, the standard deviation was remarkably reduced when growth was extracted from the raw dendrometer measurements.

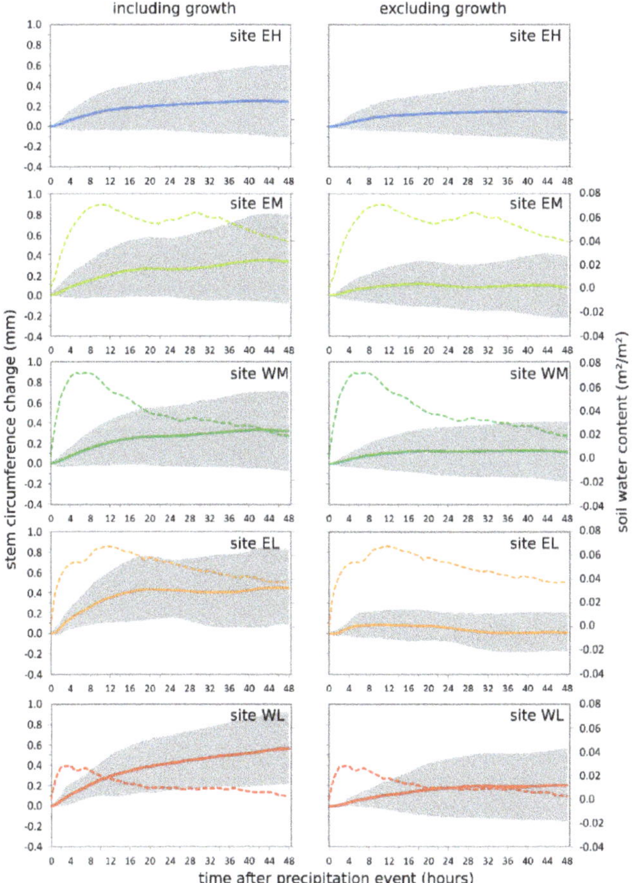

Figure 4. Mean stem circumference changes (solid lines; grey shaded areas indicate the range expressed as mean ± 1 standard deviation) and change of soil water content (dashed lines) over 48 h after precipitation events per site including (**left column**) and excluding (**right column**) tree growth. Precipitation events are extracted from the time period May 2017 to April 2019. Site codes refer to the east (E) and west (W) sides of Corsica, L, M, and H stand for low-, middle-, and high-elevation sites, respectively. Soil water measurements are not available for site EH.

Soil water content measurements were only available for four of the five sites (Figure 4). At EH, the stone content in the soil profile was too high for a correct measurement of the soil moisture. SWC at site WL is probably underestimated because the climate station was located on a nearly bare rocky surface, while the trees were located in a valley floor with a higher soil depth. All sites responded very quickly to precipitation events since measurements are obtained in a soil depth of 10–20 cm. The highest soil water content was reached three to twelve hours after the beginning of the precipitation event.

We observed a fast tree response to precipitation events at all five sites. The time lag between the beginning of the precipitation events and a visible response in the dendrometer data is very short, indicating a quick water uptake after an event. This is a faster response as found by [21], who suggested that replenishment of usable stem water reservoirs occurs within a few days after scattered rainfall events. They observed a stronger relationship between environmental variables and daily radial stem increments, when a time lag of one day was considered. The SCC increase in our study is a mixed effect of stem swelling and radial growth, with growth causing the larger part of the increase. The SCC increase due to growth is highest at the coastal locations, indicating that trees at coastal sites are more effective in using the water of precipitation events directly for growth. They are also more effective in using even small precipitation events.

Although trees at our sites grow under different climate conditions and show altitude-specific differences in the seasonal growth patterns, the response to precipitation events was very uniform among trees, sites, and did not vary with precipitation amounts. In the majority of cases (88%), SCC increased within 48 h after the onset of precipitation events. These similar responses are in concordance with the results of [46] and [22], who studied stem circumference changes of *Picea abies* and *Cedrus libani* along elevation gradients. Further, in the Italian Alps, positive responses of stem increment to precipitation were found [15]. In approximately 12% of events, we observed stem shrinkage following precipitation. This shrinkage occurred only at the higher sites and between December and April, i.e., during winter dormancy (Figure 3). A marked rehydration of the stem before the beginning of the growing season was observed in trees of cold environments [47,48] because water is withdrawn from the living cells to avoid frost-induced cavitation [25]. At sites EH and WM, temperatures commonly drop below zero between December and April.

On a seasonal basis, the SCC responses to precipitation events showed site-specific patterns (Figure 5). In summer (JJA), a pronounced bimodal to multimodal pattern was apparent, which was not or only weakly developed in other seasons. This diurnal cycle occurring with a temporal shift at site WL can be explained by the fact that only one precipitation event was recorded in JJA (Table 2). Since the values were standardized to the starting point of the precipitation event, a different starting time between events was responsible for the observed shift. SCC response including tree growth was highest in JJA and SON, and lowest in DJF. The steepest SCC increase directly following precipitation events was apparent in JJA, while the lowest SCC occurred in DJF at all sites. Only at site WL, the change in JJA was on a comparably low level. Except for JJA, EH showed the lowest stem diameter increase in all seasons. In all seasons, the response excluding growth was smaller than the reaction including growth, especially in JJA. Furthermore, the response was more homogenous among the sites, except in SON. Surprisingly, the strongest response occurred during SON at site WL. In all other seasons, the maximum increase was rarely above 0.2 mm.

The fluctuations of SCC over 48 h reflect the typical pattern of the daily stem cycle observed in dendrometer data by several authors [16,17,49]. The stem contracts during the day due to transpiration and photosynthesis and expands during the night as a result of replenishment of the stem internal water reserves. The amplitude of this daily cycle is most pronounced in summer as observed in alpine [15,16] as well as in Mediterranean environments [17] because the increased temperatures in combination with increased day length reduce the duration of the recovery phase, while increasing the water loss by transpiration [50].

The mean SCC curves after precipitation events classified by precipitation amount are shown in Figure 6. In the three classes with the lowest precipitation amounts per event (5–29.9 mm), site EH showed the smallest increase. In the two classes with the lowest precipitation amounts per event, both coastal locations showed a similar increase in the beginning, but their behavior diverged within the first day after the event. In contrast, in the following event class, site EL showed a much stronger SCC increase than site WL over the whole 48 h period. Site EM benefited more from precipitation events with high amounts (event class 50–99.9 mm) than sites EH and WM. The response after excluding growth was lower for all event classes. This was especially true for the two event classes of low amounts, indicating that even low precipitation events provide enough water for provoking a growth reaction.

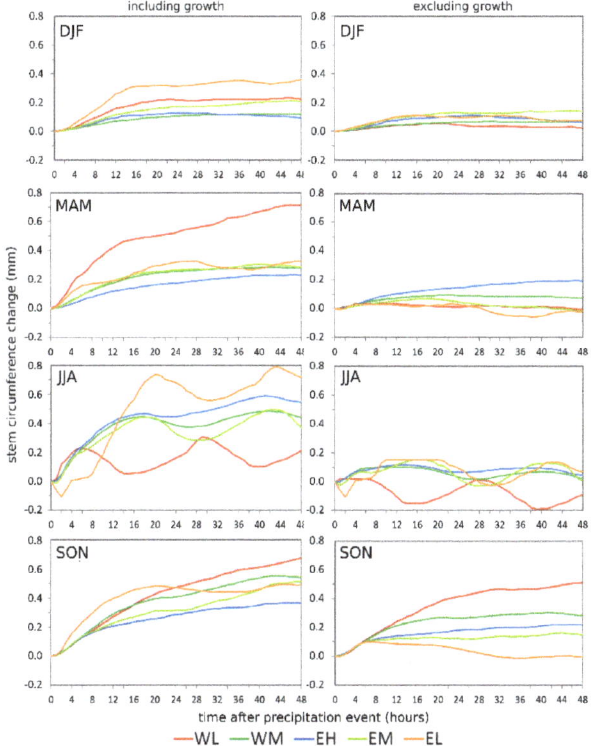

Figure 5. Mean stem circumference changes after precipitation events during different seasons, including (**left column**) and excluding (**right column**) tree growth. Precipitation events are extracted from the time period May 2017 to April 2019. Site codes refer to the east (E) and west (W) sides of Corsica, L, M, and H stand for low-, middle-, and high-elevation sites, respectively.

It becomes apparent that the strongest SCC increase is not correlated with the highest precipitation amount. Large amounts of precipitation are more common during winter than during summer (Table 2) due to passages of frontal systems, i.e., the time where no or less growth occurs at the high-elevation sites. The response to events with a low amount of precipitation in summer is therefore superimposed by growth. Another reason is that events with high precipitation amounts often show high precipitation intensity, so we assume that a large part of the precipitation is lost by surface runoff and does not infiltrate into the soils. The correlation factors between precipitation and soil water content support this assumption, as they are rather weak, ranging between 0.32 at site EL and 0.47 at site WL. Additionally, these events often occur during phases with several rainfall events on consecutive days. Soils can already be water-saturated prior to the event and the stems are already replenished with water, so that additional rainfall does not trigger additional stem swelling. This stem internal water storage can provide a significant proportion of the total diurnal and even seasonal water used by a plant [50].

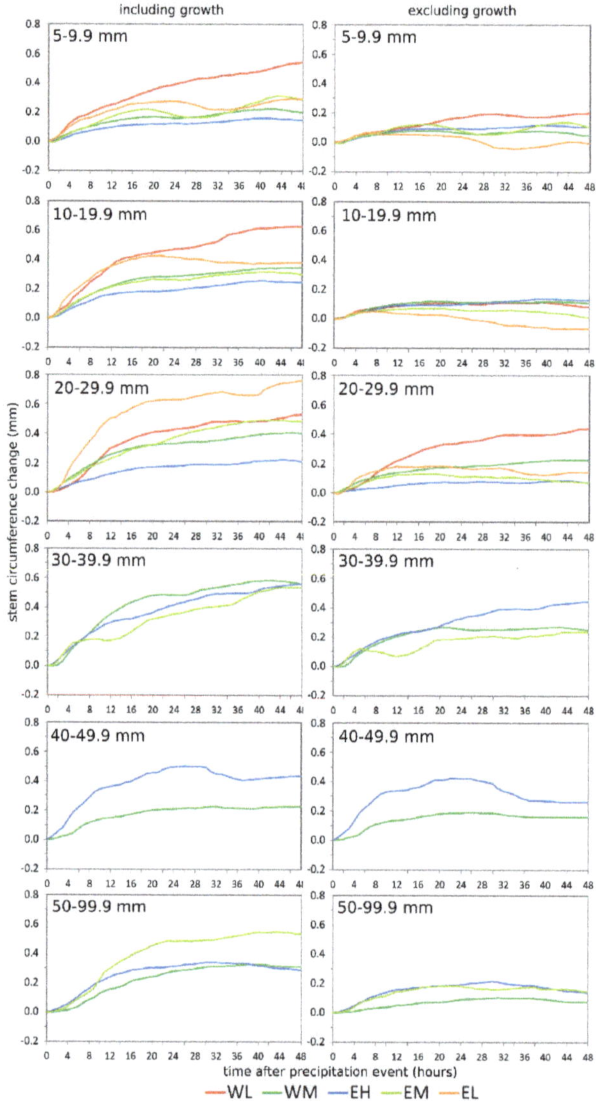

Figure 6. Mean stem circumference changes after precipitation events classified by precipitation amount, including (**left column**) and excluding (**right column**) tree growth. Stem circumference changes (SCC) curves are only shown for classes with at least two events per site. Precipitation events are extracted from the time period May 2017 to April 2019. Site codes refer to the east (E) and west (W) sides of Corsica, L, M, and H stand for low-, middle-, and high-elevation sites, respectively.

3.4. Tree Response to Dry Periods

The mean daily circumference changes during dry periods between 7 and 22 days are summarized in Figure 7. Dry periods longer than 11 days occurred more frequently at the coastal locations. In contrast to the rather homogenous response to precipitation events, the response to dry periods was site-specific, when considering raw dendrometer data including growth. Interestingly, sites EL and WL did not show any stem shrinkage, but instead an increase in stem circumference. The same was found at site EM, but less pronounced and only during the first seven dry days. In contrast, sites EH and WM showed a shrinkage, which was most severe at site WM. The steep increase in the later days was caused by the lower number of events, however, it indicated that growth occurred at least during single dry periods. After excluding tree growth from the data, the response to dry periods became more uniform among the sites. All sites showed a decreasing trend during the first days of the dry periods. The standard deviation was remarkably reduced when growth was extracted from the curves, especially at the coastal sites.

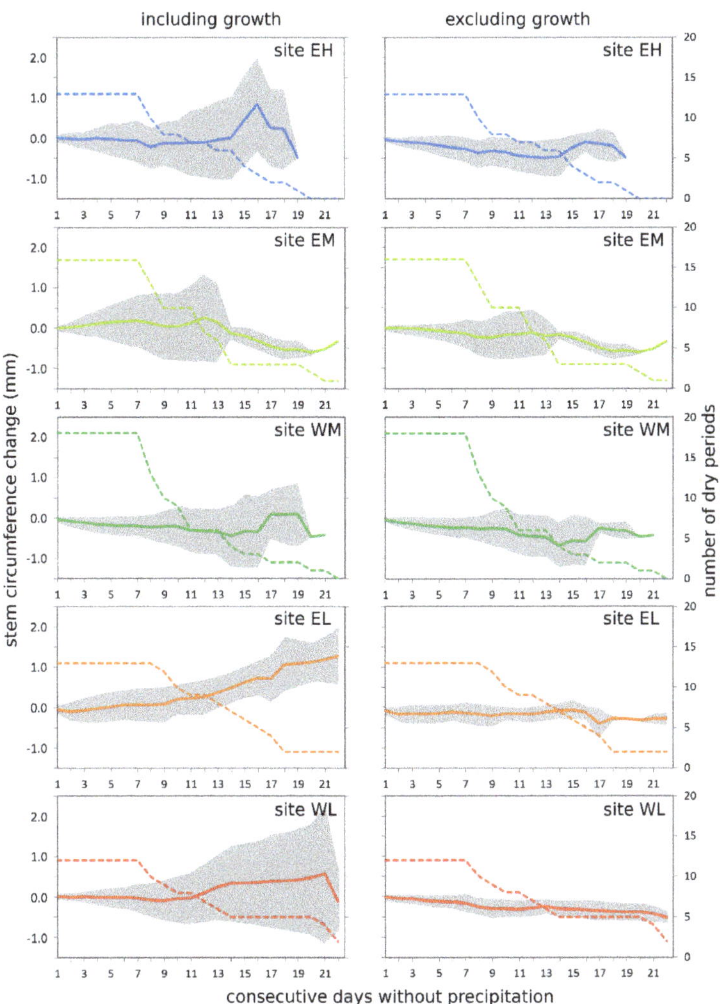

Figure 7. Mean stem circumference changes (solid lines; grey shaded areas indicate the range expressed as mean ± 1 standard deviation) during dry periods of at least seven consecutive days without precipitation including (**left column**) and excluding (**right column**) tree growth. Dashed lines indicate the number of dry periods. Dry periods are extracted from the time period May 2017 to April 2019. Site codes refer to the east (E) and west (W) sides of Corsica, L, M, and H stand for low-, middle-, and high-elevation sites, respectively.

The continuous stem circumference increases over consecutive dry days at sites WL, EL, and EM could be attributed to cambial growth because the trend disappeared after eliminating growth from the dendrometer data. An enlargement of freshly formed tracheids even during an extended period of drought was also observed by [21], suggesting that conifers can draw upon water reserves stored in sapwood and bark [51,52], more so at dry than at moist sites [13].

Another important factor influencing SCC during dry periods is soil water availability. The water-holding capacity of the soils differs among our study sites. The coastal sites are located in flat areas, where water can be stored after precipitation events. Additionally, soils at site WL can be affected by a small episodic creek, so water from precipitation events in the hilly surrounding becomes available for the trees. At the mid-elevation sites (EM and WM), soil conditions and hence water availability are different. At site EM, the soil is deeper and has a higher water-holding capacity, so water from single precipitation events is longer available. At site WM, the shape of the SWC curve (Figure 4) indicates that most of the precipitation is lost via surface runoff because SWC shows an earlier peak and a faster decrease after the beginning of a precipitation event compared with sites EL and EM. Therefore, although site WM is generally more humid than site EM, only a smaller amount of water from each precipitation event becomes available for the trees. The inferences about water availability derived from dendrometer data are corroborated by oxygen-stable isotope data derived from needle water, stem water, and precipitation from the same sites [53]. Sites EH and WM have limited or no access to deeper soil water, while the other three sites have access to deeper soil water and/or groundwater. *P. pinaster* is not a Mediterranean floral element in the strict sense, as it needs a certain air humidity and summer precipitation amounts of at least 100 mm [54]. The summer precipitation amount at our sites is lower than 100 mm, hence the vitality of *P. pinaster* at our coastal study sites indicates that these trees have access to deeper soil water or ground water pools, which are refilled in autumn/winter.

4. Conclusions

Dendrometers are suitable for studying the response of individual trees to weather (extreme) events. Analyzing the response to single precipitation events instead of comparing standard parameters like tree-ring width with seasonal or annual total precipitation is a promising approach to better understand the linkage of tree growth and tree physiological processes to environmental conditions on short timescales with a high temporal resolution. These studies can improve forecasts of tree growth under changing climate conditions, thus providing scenarios of future forest productivity, ecosystem functioning, and tree mortality risk, which are valuable sources of information to make forest management decisions on climate-adapted tree species composition.

In our study, we observed different adaptation strategies of pine trees to deal with the Mediterranean precipitation regime, including a bimodal growth pattern, the use of deeper water sources, and the capability to take up water quickly in the case of moisture availability. The different response to dry phases and precipitation events indicates that the water transfer in pine trees on Corsica is very complex. At sites with deeper soils, trees can use deeper soil water sources under unfavorable rainfall conditions. Thus, trees at sites with soils of low water-holding capacity are most vulnerable to dry periods.

At the ecosystem level, our results can help to improve models of the future distribution of *P. nigra* and *P. pinaster* along humidity gradients. *P. nigra* is the more water-demanding species, so declining water availability may reduce its potential distribution range. This can have serious impacts on silvicultural activities on Corsica as there is a high demand for *P. nigra* wood.

Supplementary Materials: The following figure is available online at http://www.mdpi.com/1999-4907/11/7/758/s1, Figure S1: Comparison of tree-ring width chronologies of *Pinus nigra* and *Pinus pinaster* at four study sites located between 570 and 1240 m asl.

Author Contributions: Conceptualization, S.S. (Sonja Szymczak) and A.B.; methodology, S.S. (Sonja Szymczak), M.H. and A.B.; formal analysis, S.S. (Sonja Szymczak) and M.H.; investigation, S.S. (Sonja Szymczak); writing—original draft preparation, S.S. (Sonja Szymczak); writing—review and editing, S.S. (Sonja Szymczak),

M.H., E.G., S.S. (Sébastien Santoni), F.H., A.B., I.K., K.T., J.B.; visualization, S.S. (Sonja Szymczak) and M.H.; All authors have read and agreed to the published version of the manuscript.

Funding: This research was funded by the German Research Foundation (DFG) (grant numbers BE 1780/45-1, BR 1895/27-1, SZ 356/1-1 and TR 1201/2-1) in the scope of the interdisciplinary research program CorsicArchive (Altitudinal Gradients and Forest Response: Climate, Hydrology and Isotope Variability of a Mediterranean Ecosystem) DFG-PAK 927/1.

Acknowledgments: We are thankful to the whole CorsicArchive team for sampling and field support.

Conflicts of Interest: The authors declare no conflict of interest.

References

1. Meinzer, F.C.; Brooks, J.R.; Gartner, B.L.; Warren, J.M.; Wodruff, D.R.; Bible, K.; Shaw, D.C. Dynamics of water transport and storage in conifers studied with deuterium and heat tracing techniques. *Plant Cell Environ.* **2006**, *29*, 105–114. [CrossRef] [PubMed]

2. Battipaglia, G.; DeMicco, V.; Brand, W.A.; Saurer, M.; Aronne, G.; Linke, P.; Cherubini, P. Drought impact on water use efficiency and intra-annual density fluctuations in *Erica arborea* on Elba (Italy). *Plant Cell Environ.* **2013**, *37*, 382–391. [CrossRef]

3. Anderegg, W.R.L.; Flint, A.; Huang, C.-Y.; Flint, L.; Berry, J.A.; Davis, F.W.; Sperry, J.S.; Field, C.B. Tree mortality predicted from drought-induced vascular damage. *Nat. Geosci.* **2015**. [CrossRef]

4. Hember, R.A.; Kurz, W.A.; Coops, N.C. Relationships between the individual-tree mortality and water-balance variables indicate positive trends in water stress-induced tree mortality across North America. *Glob. Chang. Biol.* **2017**, *23*, 1691–1710. [CrossRef] [PubMed]

5. Choat, B.; Brodribb, T.J.; Brodersen, C.R.; Duursma, R.A.; López, R.; Medlyn, B.E. Triggers of tree mortality under drought. *Nature* **2018**, *558*, 531–539. [CrossRef]

6. Schäfer, C.; Rötzer, T.; Thurm, E.A.; Biber, P.; Kallenbach, C.; Pretzsch, H. Growth and tree water deficit of mixed Norway spruce and European beech at different heights in a tree and under heavy drought. *Forests* **2019**, *10*, 577. [CrossRef]

7. Steudle, E. Water uptake by roots: Effects of water deficit. *J. Exp. Bot.* **2000**, *51*, 1531–1542. [CrossRef]

8. Jackson, R.; Moore, L.; Hoffmann, W.; Pockman, W.; Linder, C. Ecosystem rooting depth determined with caves and DNA. *Proc. Natl. Acad. Sci. USA* **1999**, *96*, 11387–11392. [CrossRef]

9. Ogle, K.; Reynolds, J.F. Plant responses to precipitation in desert ecosystems: Integrating functional types, pulses, thresholds, and delays. *Oecologia* **2004**, *141*, 282–294. [CrossRef] [PubMed]

10. Eggemeyer, K.D.; Awada, T.; Harvey, F.E.; Wedin, D.A.; Zhou, X.; Zanner, C.W. Seasonal changes in depth of water uptake for encroaching trees *Juniperus virginiana* and *Pinus ponderosa* and two dominant C_4 grasses in a semiarid grassland. *Tree Physiol.* **2009**, *29*, 157–169. [CrossRef]

11. Reynolds, J.F.; Virginia, R.A.; Kemp, P.R.; de Soyza, A.G.; Tremmel, D.C. Impact of drought on desert shrubs: Effects of seasonality and degree of resource island development. *Ecol. Monogr.* **1999**, *69*, 69–106. [CrossRef]

12. Gebauer, R.L.W.; Ehleringer, J.R. Water and nitrogen uptake patterns following moisture pulses in a cold desert community. *Ecology* **2000**, *81*, 1415–1424. [CrossRef]

13. Zweifel, R.; Zimmermann, L.; Newbery, D.M. Modeling tree water deficit from microclimate: An approach to quantifying drought stress. *Tree Physiol.* **2005**, *25*, 147–156. [CrossRef] [PubMed]

14. Mencuccini, M.; Hölttä, T.; Sevanto, S.; Nikinmaa, E. Concurrent measurements of change in the bark and xylem diameters of trees reveal a phloem-generated turgor signal. *New Phytol.* **2013**, *198*, 1143–1154. [CrossRef]

15. Deslauriers, A.; Rossi, S.; Anfodillo, T. Dendrometer and intra-annual tree growth: What kind of information can be inferred? *Dendrochronologia* **2007**, *25*, 113–124. [CrossRef]

16. King, G.; Fonti, P.; Nievergelt, D.; Büntgen, U.; Frank, D. Climatic drivers of hourly to yearly tree radius variations along a 6 °C natural warming gradient. *Agric. Meteorol.* **2013**, *168*, 36–46. [CrossRef]

17. Vieira, J.; Rossi, S.; Campelo, F.; Freitas, H.; Nabais, C. Sesaonal and daily cycles of stem radial variation of *Pinus pinaster* in a drought-prone environment. *Agric. Meteorol.* **2013**, *180*, 173–181. [CrossRef]

18. Siegmund, J.F.; Sanders, T.G.M.; Heinrich, I.; van der Maaten, E.; Simard, S.; Helle, G.; Donner, R.V. Meteorological drivers of extremes in daily stem radius variations of beech, oak, and pine in northeastern Germany: An event coincidence analysis. *Front. Plant Sci.* **2016**, *7*, 733. [CrossRef]

19. Raffelsbauer, V.; Spannl, S.; Pena, K.; Pucha-Cofrep, D.; Steppe, K.; Bräuning, A. Tree circumference changes and species-specific growth recovery after extreme dry events in a montane rainforest in southern Ecuador. *Front. Plant Sci.* **2019**, *10*. [CrossRef]

20. Zweifel, R. Radial stem variations—A source of tree physiological information not fully exploited yet. *Plant Cell Environ.* **2016**, *39*, 231–232. [CrossRef]

21. Oberhuber, W.; Gruber, A.; Kofler, W.; Swirdak, I. Radial stem growth in response to microclimate and soil moisture in a drought-prone mixed coniferous forest at an inner Alpine site. *Eur. J. Res.* **2014**, *133*, 467–479. [CrossRef] [PubMed]

22. Güney, A.; Gülsoy, S.; Sentürk, Ö.; Niessner, A.; Küppers, M. Environmental control of daily stem radius increment in the montane conifer *Cedrus libani*. *J. Res.* **2019**. [CrossRef]

23. Chan, T.; Hölttä, T.; Berninger, F.; Mäkinen, H.; Nöjd, P.; Mencuccini, M.; Nikinmaa, E. Separating water-potential induced swelling and shrinking from measured radial stem variations reveals a cambial growth and osmotic concentration signal. *Plant Cell Environ.* **2016**, *39*, 233–244. [CrossRef]

24. Mencuccini, M.; Salmon, Y.; Mitchell, P.; Hölttä, T.; Choat, B.; Meir, P.; O'Grady, A.; Tissue, D.; Zweifel, R.; Sevanto, S.; et al. An empirical method that separates irreversible stem radial growth from bark water content changes in trees: Theory and case studies. *Plant Cell Environ.* **2017**, *40*, 290–303. [CrossRef] [PubMed]

25. Zweifel, R.; Häsler, R. Dynamics of water storage in mature subalpine *Picea abies*: Temporal and spatial patterns of change in stem radius. *Tree Physiol.* **2001**, *21*, 561–569. [CrossRef]

26. Noy-Meir, I. Desert ecosystems: Environment and producers. *Annu. Rev. Ecol. Syst.* **1973**, *4*, 25–51. [CrossRef]

27. Barnard, R.L.; Salmon, Y.; Kodama, N.; Sörgel, K.; Holst, J.; Rennenberg, H.; Gessler, A.; Buchmann, N. Evaporative enrichment and time lags between $\delta^{18}O$ of leaf water and organic pools in a pine stand. *Plant Cell Environ.* **2007**, *30*, 539–550. [CrossRef] [PubMed]

28. Brandes, E.; Wenninger, J.; Koeninger, P.; Schindler, D.; Rennenberg, H.; Leibundgut, C.; Mayer, H.; Gessler, A. Assessing environmental and physiological controls over water relations in a Scots pine (*Pinus sylvestris* L.) stand through analyses of stable isotope composition of water and organic matter. *Plant Cell Environ.* **2007**, *30*, 113–127. [CrossRef]

29. Vicente-Serrano, S.M.; Gouveia, C.; Camarero, J.J.; Beguería, S.; Trigo, R.; López-Moreno, J.L.; Azorín-Molina, C.; Pasho, E.; Lorenzo-Lacruz, J.; Revuelto, J.; et al. Response of vegetation to drought time-scales across global land biomes. *Proc. Natl. Acad. Sci. USA* **2013**, *110*, 52–57. [CrossRef]

30. Wu, D.H.; Zhao, X.; Liang, S.L.; Zhou, T.; Huang, K.C.; Tang, B.J.; Zhao, W.Q. Time-lag effects of global vegetation responses to climate change. *Glob. Chang. Biol.* **2015**, *21*, 3520–3531. [CrossRef]

31. Bruno, C.; Dupre, G.; Giorgetti, G.; Borne, D. *Chì Tempu Face? Météorologie, Climat et Microclimats de la Corse*; CNDP-CRDP de Corse/MétéoFrance: Ajaccio, France, 2001.

32. Häusser, M.; Szymczak, S.; Garel, E.; Santoni, S.; Huneau, F.; Bräuning, A. Growth variability of two native pine species on Corsica as a function of altitude. *Dendrochronologia* **2019**, *54*, 49–55. [CrossRef]

33. Shuttle Radar Topography Mission of the United States Geological Survey. Available online: https://lta.cr.usgs.gov/SRTM/ (accessed on 15 January 2018).

34. MétéoFrance. Climate Data from Stations Ajaccio, Bocognano, Evisa, Sampolo and Solenzara. Available online: https://www.meteofrance.fr (accessed on 20 May 2020).

35. Cheng, X.L.; An, S.Q.; Li, B.; Chen, J.Q.; Lin, G.H.; Liu, Y.H.; Luo, Y.; Liu, S. Summer rain pulse size and rainwater uptake by three dominant desert plants in a desertified grassland ecosystem in northwestern China. *Plant Ecol.* **2006**, *184*, 1–12. [CrossRef]

36. Aryal, S.; Häusser, M.; Grießinger, J.; Fan, Z.-X.; Bräuning, A. DendRoAnalyst: A Complete Tool for Processing and Analyzing Dendrometer Data. R Package Version 0.1.0. 2020. Available online: https://CRAN.R-project.org/package=dendRoAnalyst (accessed on 7 May 2020).

37. Thabeet, A.; Vennetier, M.; Gadbin-Henry, C.; Denelle, N.; Roux, M.; Caraglio, Y.; Vila, B. Response of *Pinus sylvestris* L. to recent climatic events in the French Mediterranean region. *Trees* **2009**, *23*, 843–853. [CrossRef]

38. Hänninen, H.; Tannino, K. Tree seasonality in a warming climate. *Trends Plant. Sci.* **2011**, *16*, 412–416. [CrossRef]

39. Rossi, S.; Morin, H.; Deslauriers, A.; Plourde, P.Y. Predicting xylem phenology in black spruce under climate warming. *Glob. Chang. Biol.* **2011**, *17*, 614–625. [CrossRef]

40. Swidrak, I.; Gruber, A.; Kofler, W.; Oberhuber, W. Effects of environmental conditions on onset of xylem growth in *Pinus sylvestris* under drought. *Tree Physiol.* **2011**, *31*, 483–493. [CrossRef]

41. Rossi, S.; Deslauriers, A.; Gricar, J.; Seo, J.-W.; Rathgeber, C.B.K.; Anfodillo, T.; Morin, H.; Levanic, T.; Oven, P.; Jalkanen, R. Critical temperatures for xylogenesis in conifers of cold climates. *Glob. Ecol. Biogeogr.* **2008**, *17*, 696–707. [CrossRef]
42. Güney, A.; Kerr, D.; Sökücü, A.; Zimmermann, R.; Küppers, M. Cambial activity and xylogenesis in stems of *Cedrus libani* A. Rich at different altitudes. *Bot. Stud.* **2015**. [CrossRef]
43. Liu, X.; Nie, Y.; Wen, F. Seasonal dynamics of stem radial increment of *Pinus taiwanensis* Hayata and its response to environmental factors in the Lushan mountains, southeastern China. *Forests* **2018**, *9*, 387. [CrossRef]
44. De Luis, M.; Gričar, J.; Čufar, K.; Raventós, J. Sesaonal dynamics of wood formation in *Pinus halepensis* from dry and semi-arid ecosystems in Spain. *IAWA J.* **2007**, *28*, 389–404. [CrossRef]
45. Camarero, J.J.; Olano, J.M.; Parras, A. Plastic bimodal xylogenesis in conifers from continental Mediterranean climates. *New Phytol.* **2010**, *185*, 471–480. [CrossRef] [PubMed]
46. Cocozza, C.; Palombo, C.; Tognetti, R.; La Porta, N.; Anichini, M.; Giovannelli, A.; Emiliani, G. Monitoring intra-annual dynamics of wood formation with microcores and dendrometers in *Picea abies* at two different altitudes. *Tree Physiol.* **2016**, *36*, 832–846. [CrossRef] [PubMed]
47. Deslauriers, A.; Morin, H.; Begin, Y. Cellular phenology of annual ring formation of *Abies balsamea* in the Quebec boreal forest (Canada). *Can. J. For. Res.* **2003**, *33*, 190–200. [CrossRef]
48. Turcotte, A.; Morin, H.; Krause, C.; Deslauriers, A.; Thibeault-Martel, M. The timing of spring rehydration and its relation with the onset of wood formation in black spruce. *Agric. Meteorol.* **2009**, *149*, 1403–1409. [CrossRef]
49. Klippel, L.; Hartl-Meier, C.; Lindén, J.; Kochbeck, M.; Emde, K.; Esper, J. Hourly resolved climate response of *Picea abies* beyond its natural distribution range. *Balt. For.* **2017**, *23*, 556–563.
50. Čermak, J.; Kučerá, J.; Bauerle, W.L.; Phillips, N.; Hinckley, T.M. Tree water storage and its diurnal dynamics related to sap flow and changes in stem volume in old-growth Douglas-fir trees. *Tree Physiol.* **2007**, *27*, 181–198. [CrossRef]
51. Holbrook, N.M. Stem water storage. In *Plant Stems: Physiology and Functional Morphology*; Gartner, B.L., Ed.; Academic Press: San Diego, CA, USA, 1995; pp. 151–175.
52. Grip, H.; Hällgren, J.E. Water cycling in coniferous forest ecosystems. In *Ecosystems of the World. Volume 6. Coniferous Forests*; Andersson, F., Ed.; Elsevier: Amsterdam, The Netherlands, 2005; pp. 385–426.
53. Szymczak, S.; Barth, J.; Bendix, J.; Huneau, F.; Garel, E.; Häusser, M.; Juhlke, T.; Knerr, I.; Santoni, S.; Trachte, K.; et al. Tracking the oxygen isotope composition from source to sink in pine trees along an elevation gradient in a Mediterranean ecosystem. *Chem. Geol.* **2020**. under review. [CrossRef]
54. Schütt, P.; Weisgerber, H.; Schuck, H.J.; Lang, U.; Stimm, B.; Roloff, A. *Lexikon der Nadelbäume*; Nikol: Hamburg, Germany, 2004; p. 639.

 *forests*

Article

An ERF Transcription Factor Gene from *Malus baccata* (L.) Borkh, *MbERF11*, Affects Cold and Salt Stress Tolerance in *Arabidopsis*

Deguo Han, Jiaxin Han, Guohui Yang, Shuang Wang, Tianlong Xu and Wenhui Li *

Key Laboratory of Biology and Genetic Improvement of Horticultural Crops (Northeast Region), Ministry of Agriculture and Rural Affairs, College of Horticulture & Landscape Architecture, Northeast Agricultural University, Harbin 150030, China; deguohan_neau@126.com (D.H.); A02140301@163.com (J.H.); yangguohui_neau@126.com (G.Y.); ws18045312966@163.com (S.W.); 18846829610@163.com (T.X.)
* Correspondence: lwh_neau@126.com; Tel.: +86-4515-5190-781

Received: 2 April 2020; Accepted: 29 April 2020; Published: 2 May 2020

Abstract: Apple, as one of the most important economic forest tree species, is widely grown in the world. Abiotic stress, such as low temperature and high salt, affect apple growth and development. Ethylene response factors (ERFs) are widely involved in the responses of plants to biotic and abiotic stresses. In this study, a new ethylene response factor gene was isolated from *Malus baccata* (L.) Borkh and designated as *MbERF11*. The *MbERF11* gene encoded a protein of 160 amino acid residues with a theoretical isoelectric point of 9.27 and a predicated molecular mass of 17.97 kDa. Subcellular localization showed that MbERF11 was localized to the nucleus. The expression of *MbERF11* was enriched in root and stem, and was highly affected by cold, salt, and ethylene treatments in *M. baccata* seedlings. When *MbERF11* was introduced into *Arabidopsis thaliana*, it greatly increased the cold and salt tolerance in transgenic plant. Increased expression of *MbERF11* in transgenic *A. thaliana* also resulted in higher activities of superoxide dismutase (SOD), peroxidase (POD), and catalase (CAT), higher contents of proline and chlorophyll, while malondialdehyde (MDA) content was lower, especially in response to cold and salt stress. Therefore, these results suggest that *MbERF11* probably plays an important role in the response to cold and salt stress in *Arabidopsis* by enhancing the scavenging capability for reactive oxygen species (ROS).

Keywords: *Malus baccata*; *MbERF11*; cold stress; salt stress; transgenic plant

---

1. Introduction

During growth and development, plants are frequently exposed to various abiotic stresses, such as drought, cold, salt, heat, and nutrient deprivation. Under different environmental stresses, plants have developed different adaptable mechanisms to ensure their normal growth and development [1,2]. The response mechanism of plants to abiotic stresses was regulated by multiple signaling pathways [3]. In these processes, transcription factors (TFs) play a key role in the interaction of these signaling pathways [4–7]. TFs, also known as trans-acting factors, are a class of DNA-binding proteins that specifically bind to cis-acting elements. Interactions between transcription factors and cis-acting elements or other proteins can be transcriptional activation or inhibition. Studies have clarified that TF genes are abundantly present in plants, and can regulate plant growth and metabolism [8,9].

The APETALA2/ethylene responsive factor (AP2/ERF) TFs are widely distributed in different types of plants [10]. About 145 AP2/ERF TF genes have been isolated from *A. thaliana* [11], and 167 AP2/ERF TFs were also found in Oryza sativa [12]. The AP2/ERF TF family may have evolved from HNHAP2 endonucleases in bacteria and viruses [13]. The AP2/ERF genes were widely isolated and studied from different plants, such as *APETALA2, AINTEGUMENTA* and *AtCBF1* from *Arabidopsis* [14–16],

ThCRF1 from *Tamarix hispida* [17], and *CsERF025* from cucumber [18]. The AP2/ERF TFs were found to participate in almost every process of plant growth and development, especially in response to biotic or abiotic stresses in plants [19].

As a class of TFs, AP2/ERF TFs usually contain an AP2/ERF domain consisting of about 60 amino acid residues [20]. This domain was firstly discovered from APETALA2 homologues of *A. thaliana* [21], a similar sequence was also found in tobacco [22]. According to the different numbers of AP2/ERF domains contained in AP2/ERF, the AP2/ERF TFs were divided into two subfamilies. Among them, one subfamily is called ERF containing one AP2/ERF domain. The other subfamily is called the AP2 subfamily containing two AP2/ERF domains. The ERF subfamily can be further divided into DREB (dehydration responsive element binding protein), ERF and other three categories [23]. In the AP2/ERF domain of DREB subfamily, the 14th and 19th amino acids are valine (V) and glutamic acid (E), respectively, while they are alanine and aspartate in the AP2/ERF domain of ERF subfamily [24].

ERF TFs are widely involved in biotic and abiotic stress responses, which play important roles in drought, high salt and low temperature tolerance, as well as plant development, hormone response and other regulatory networks [24,25]. Previous researches found ERF TFs were also involved in organ development, cell division, differentiation, flower development and fruit maturation in plants [26,27]. Overexpression of *Sl-ERF2* gene in tomato could activate *Sl-Man2*, a mannanase-encoding gene, and result in the premature germination of tomato immature seeds [28]. The *Sub1A/C* gene in *O. sativa* participates in plant growth and metabolism [29]. The *MdERF1/2* genes in *Malus domestica* are associated with fruit ripening [30]. However, these studies mostly focused on model plants or crops, and the roles of the ERF TFs genes in *Malus* plant stress responses were less well known.

M. baccata is widely used as an apple rootstock in northern China, and also as a source of forestry wood or greening tree species. Due to its high resistance to low temperature and drought, M. baccata is also used as a breeding material for cold and drought resistance [31]. From the transcriptome analysis of *M. baccata* seedlings under cold and/or drought stresses (results not presented here), we found the *MbERF11* level was significantly up-regulated under both stresses. More importantly, through NCBI blast (https://blast.ncbi.nlm.nih.gov/Blast.cgi) of *MbERF11* gene, we found that the closest *Arabidopsis* *ERF* gene is *AtERF7*, which is a famous ERF TF gene involved in drought stress through ABA signal transduction [32]. To better understand the role of ERF TFs genes involved in low temperature and salt stress, and to provide potentially genetic resources for the improvement of the drought resistance of *Malus* plant, a new ERF TFs gene was isolated from *M. baccata* and designated as *MbERF11*. Moreover, it was found that the tolerance of transgenic *A. thaliana* to cold and salt stress was increased because of the overexpression of *MbERF11*.

2. Materials and Methods

2.1. Plant Material and Growth Conditions

The *M. baccata* test-tube seedlings were rapidly propagated in MS growth medium (containing 0.6 mg/L IBA + 0.6 mg/L 6-BA) for 30 days. Then they were transplanted to MS + 1.2 mg/L IBA for 45 days for rooting. Finally, the seedlings were transferred to Hoagland solution for 40 days for growth. The solution was changed three times per week. The temperature of the culture chamber was maintained at 20 °C and the relative humidity was maintained at about 85%. When the test-tube seedlings grew to 8–9 leaves (completely expanded), a part of them was placed in Hoagland solution with a NaCl concentration of 200 μM and pH 5.8 for salt stress treatment. The other part of them was placed in Hoagland solution at 4 °C for cold stress treatment. Test-tube seedlings incubated in Hoagland solution at 20 °C were used as control. The unexpanded young leaf, the completely unfolded mature leaf, the phloem at the second and third node stem segments, and the newly emerged root were taken as samples. The samples of all control and treated plants were sealed after treatments of respectively 0, 2, 4, 8, 12, 24, and 48 h, immediately frozen in liquid nitrogen, and then stored at −80 °C for RNA extraction.

2.2. The qRT-PCR Expression Analysis of MbERF11

Total RNA was respectively extracted from young leaf, mature leaf, new root, and stem using the EasyPure Plant RNA Kit (TransGen, Beijing). Synthesis of cDNA first strands with TransGen's Trans Script® First-Strand cDNA Synthesis Super Mix (TransGen, Beijing). The whole sequence of *MbERF11* was obtained by PCR, with the first strand cDNA of *M. baccata* as a template. A pair of primers (*MbERF11*-F and *MbERF11*-R, Table 1) were designed based on the homologous regions of *MdERF011* (MDP0000258562) to amplify the full-length cDNA sequence. The obtained DNA fragments were gel purified and cloned into the pEasy-T1 vector (TransGen) and sequenced (BGI, Beijing).

Table 1. Primers used in this study.

Name of Primer	Sequence of Primer (5′→3′)	Purpose
MbERF11-F	ATGGAAGGAGATTACTGCTGCT	Cloning
MbERF11-R	TTAACTTTCATCGGAGTTTTCTGGG	Cloning
ACTIN-F	ACACGGGGAGGTAGTGACAA	qPCR
ACTIN-R	CCTCCAATGGATCCTCGTTA	qPCR
GAP-F	GTCGTACTACTGGTATCGTT	qPCR
GAP-R	TCATAGTCAAGAGCAATGTA	qPCR
MF	TGGAGAAGCGTAAGCATCCC	qPCR
MR	CGTCGTCTTGGAATACAAGCT	qPCR
Site-F	GAGCTCATGGAAGGAGATTACTGCT	Add *Sac*I site
Site-R	CCTAGGACTTTCATCGGAGTTTTCTG	Add *Bam*HI site

The qRT-PCR expression analysis of *MbERF11* was performed according to method of Han et al. [33]. The *Malus ACTIN* gene (AB638619.1) amplified from *M. baccata* tissues was as control, which was stably expressed under various conditions [34]. We designed the primers (*ACTIN*-F and *ACTIN*-R, Table 1) from the sequences and published in the GenBank databases. The primers (MF and MR, Table 1) of *MbERF11* were designed from partial sequences cloned in this study for qRT-PCR detection. The thermal cycling program was one initial cycle of 94 °C for 30 s, followed by 40 cycles of 94 °C for 15 s, and 55 °C for 30 s. The relative transcription level data was analyzed by the Pfaffl method [35].

2.3. Subcellular Localization Analysis of MbERF11 Protein

The *MbERF11* ORF was cloned into the *Sac*I and *Bam*HI sites of the pSAT6-GFP-N1 vector. This vector contains a modified red-shifted green fluorescent protein (GFP) at *Sac*I–*Bam*HI sites. The *MbERF11*-GFP construct was transformed into onion (*Allium cepa*) epidermal cells by particle bombardment [36]. The DAPI staining was used as a nucleus marker for nucleus detection. The transient expression of the MbERF11–GFP fusion protein was observed by confocal microscopy.

2.4. A. Thaliana Transformation

To construct an expression vector for the transformation of *A. thaliana*, restriction enzyme cut sites of *Sac*I and *Bam*HI at 5′ and 3′ ends of the *MbERF11* cDNA was respectively added by PCR with the primers (*Site*-F, *Site*-R, Table 1). To construct the PCAM3301-*MbERF11* vector, PCAM3301 and the products of PCR were digested by *Sac*I and *Bam*HI, and linked together by T4 DNA ligase. The *MbERF11* gene driven by the CaMV 35S promoter and the vecror (only PCAM3301) were introduced into *A. thaliana* by Agrobacterium-mediated LBA4404 transformation [37]. Columbia ecotype *A. thaliana* plants were transformed using the vacuum infiltration method. Transformants (transgenic lines and vector line) were selected on MS medium containing 6 mg/mL glufosinate. The transgenic lines (roots used as materials) were confirmed by qRT-PCR analysis with wild type (WT) and vacant-vector line (VL) as control. T3 generation plants were used for further analysis.

2.5. Determination Survival Rates Under Cold and/or Salt Stress Treatment in Transgenic Arabidopsis

Wild-type *A. thaliana* (WT), vacant-vector line (VL, the line only transformed with vacant vector) and three *MbERF11* transgenic lines (S2, S6, S7) were respectively sown in culture medium, and transferred to nutrient soil for two weeks after 10 days. The WT, VL and T3 transgenic *A. thaliana* were cultured under control condition, low temperature treatment (−4 °C) for 12 h, and high salinity stress (200 mM NaCl) for 7 d, respectively, after which their survival rates were recorded with 15 nutrition pots.

2.6. Detection of the Contents of Chlorophyll, MDA and Proline and the Activity of SOD, POD and CAT

All the materials of different lines above were collected for measurements. The chlorophyll content was determined with method of Li et al. [38]. The proline content was measured according to the spectrophotometric method [39]. The MDA content and the activities of SOD, POD, and CAT were measured according to the protocol described by Shin et al. [40].

2.7. Statistical Analysis

DPS 7.05 data processing system software was used for one-way analysis of variance (ANOVA). All experiments were repeated for three times and the standard errors (±SE) were measured, respectively. Statistical differences were referred to as significant * $p \leq 0.05$, ** $p \leq 0.01$.

3. Results

3.1. Isolation of MbERF11 Gene from M. Baccata

The ProtParam analysis (http://www.expasy.org/tools/protparam.html) showed that the *MbERF11* gene encodes 160 amino acids (Figure 1). The theoretical molecular mass of MbERF11 is 17.97 kDa, with theoretical isoelectric point 9.27 and the average hydrophilicity coefficient −0.995. The underlined part of Figure 1 is the conserved sequence of the AP2/ERF family, which contains two conserved elements, namely YRG and RAYD. The 14th and 19th amino acids of the conserved sequence are valine and glutamic respectively, indicating that it belongs to the DREB subfamily in the AP2/ERF family.

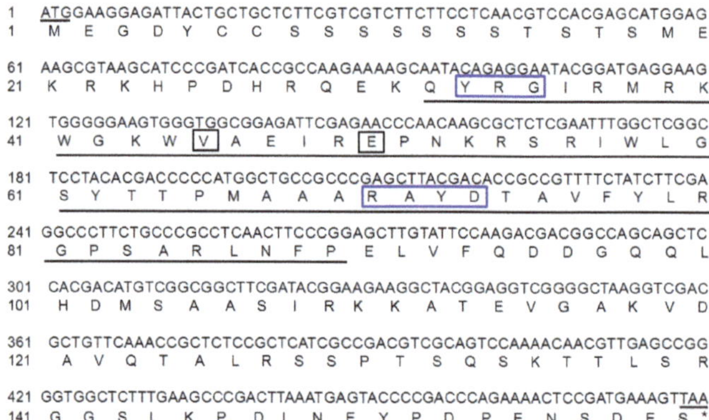

Figure 1. Nucleotide and deduced amino acid sequences of *MbERF11* gene. Underlines indicate conserved amino acid sequences. Black boxes indicate specific amino acids of the AP2/ERF domain. Blue boxes indicate conserved elements.

3.2. Phylogenetic Relationship of MbERF11 with Other ERF Proteins

To explore the evolutionary relationship among plant ethylene response factors, DNAman was used to compare MbERF11 with other 13 ERF proteins of different species. The phylogenetic tree showed that MbERF11 contains an AP2/ERF conserved domain consisting of 58 amino acid residues. This conserved sequence is the characteristic sequence of the ERF TF in plants (Figure 2A). As shown in Figure 2A, the TF of ERF family has higher homology in the conserved domain. There are also many changes in the non-conserved domain, which is consistent with the characteristics of the TF.

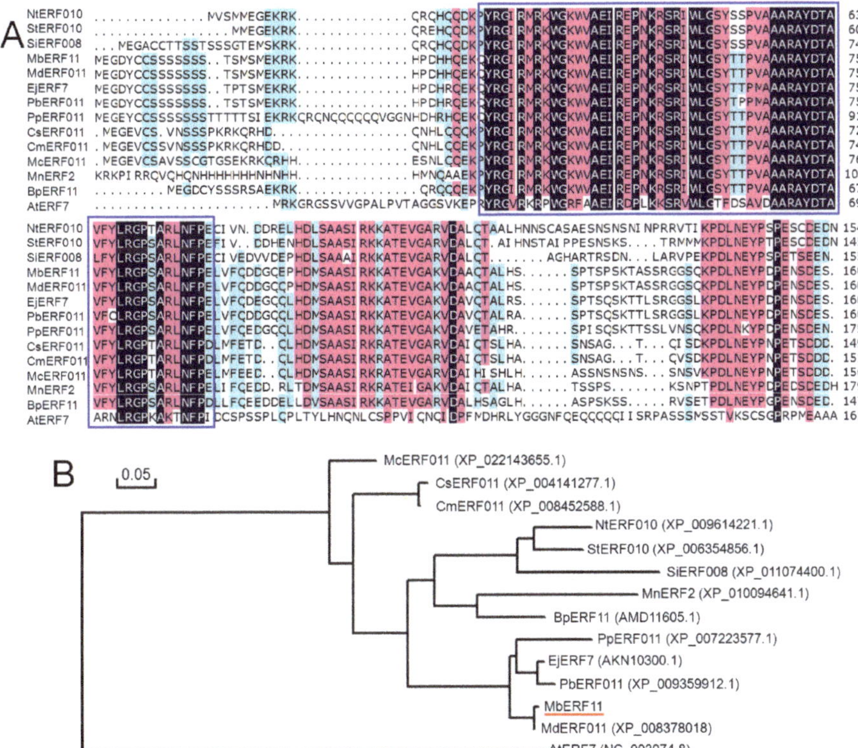

Figure 2. Comparison and phylogenetic relationship of MbERF11 with ethylene-responsive transcription factor proteins from other species. (**A**) Comparison completes alignment of MbERF11 with other plant ethylene-responsive transcription factor proteins. Conserved domains are shown in blue boxes. Positions containing identical residues are shaded in navy blue, while conservative residues are shown in green. (**B**) Phylogenetic tree analysis of MbERF11 and other plant ethylene-responsive transcription factor proteins. The tree was constructed by the neighbour-joining method with MEGA 7 (http://www.megasoftware.net). The gene accession numbers are listed in Figure 2B.

The homologous phylogenetic tree showed that MbERF11, MdERF011 (XP_008378018, *Malus domestica*), EjERF7 (AKN10300.1, *Eriobotrya japonica*), PbERF011 (XP_009359912.1, *Pyrus bretschneideri*) and PpERF011 (XP_007223577.1, *Prunus persica*) clustered together. NtERF010 (XP_009614221.1, *Nicotiana tomentosiformis*), StERF010 (XP_006354856.1, *Solanum tuberosum*), SiERF008 (XP_011074400.1, *Sesamum indicum*), MnERF2 (XP_010094641.1, *Morus notabilis*), and BpERF11 (AMD11605.1, from *Betula platyphylla*) were grouped into the second cluster, followed by CsERF011 (XP_004141277.1, *Cucumis sativus*), CmERF011 (XP_008452588.1, *Cucumis melo*), and McERF011 (XP_022143655.1, *Momordica charantia*). However, AtERF7 (NC_003074.8) was grouped into another cluster on its own (Figure 2B).

3.3. MbERF11 was Localized to the Nucleus

As shown in Figure 3, the MbERF11–GFP fusion protein is targeted to nucleus (Figure 3E) with 4′, 6-diamidino-2-phenylindole (DAPI) staining (Figure 3F), whereas the control GFP alone is distributed throughout the cytoplasm (Figure 3B). These results determined that MbERF11 is a nucleus localized protein.

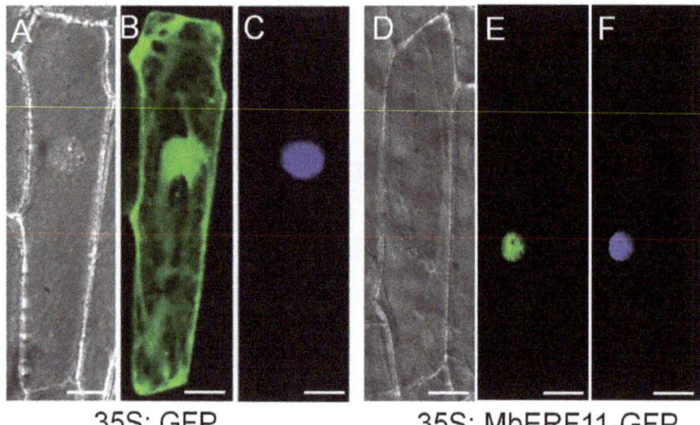

Figure 3. Subcellular localization of MbERF11 protein. The transient vector harboring 35S-GFP and 35S-*MbERF11*-GFP cassettes were transformed into onion epidermal cells by particle bombardment. Transient expressions of green fluorescent protein 35S-GFP (**B**) and 35S-*MbERF11*-GFP (**E**) translational product were visualized in onion epidermal cells by fluorescence microscopy. Onion epidermal cells of 35S-GFP (**C**) and 35S-*MbERF11*-GFP (**F**) stained with DAPI for 24 h. (**A**,**D**) were taken in the bright light. Scale bar corresponds to 5 μm.

3.4. Expression Analysis of MbERF11 in M. Baccata

As shown in Figure 4A, in control condition, the expression level of *MbERF11* in *M. baccata* seedlings was higher in root and stem, while very low in leaf. When dealt with salt (200 mM NaCl), cold (4 °C), and ethephon treatments (500 μL/L, the ratio of ethylene:water is 1:2000), the expression level of *MbERF11* in young leaf of *M. baccata* increased quickly, reached maximum at 12 h, 24 h, and 4 h, respectively, and then decreased (Figure 4B). The expression level of *MbERF11* in root had a similar trend, which reached the maximum at 8 h, 12 h, and 2 h, respectively, then decreased slightly (Figure 4C).

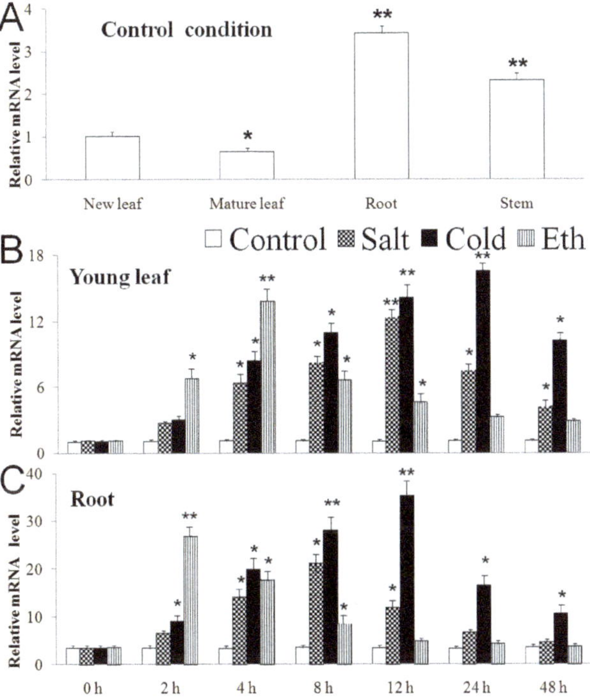

Figure 4. Time-course expression patterns of *MbERF11* in *Malus baccata* using qRT-PCR methord. (**A**) Expression patterns of *MbERF11* in young leaf (partly expanded), mature leaf (fully expanded), root and stem in normal condition (room temperature and normal nutrient solution). The expression level of young leaf was as control. (**B,C**) Expression patterns of *MbERF11* in control, salt (200 μM), cold (4 °C) and ethephon (500 μL/L) in young leaf (**B**), and root (**C**) at the following time points: 0 h, 2 h, 4 h, 8 h, 12 h, 24 h and 48 h. The reference genes *MdACTIN* and *MdGAPDH* were used as controls in this study. The error bars represent standard deviation. Asterisks above the error bars indicate a significant difference between the treatment and control (0 h) using Student's *t* test (* $p \leq 0.05$; ** $p \leq 0.01$).

3.5. Overexpression of MbERF11 in A. Thaliana Contributed to Low Temperature Stress Tolerance

To study the role of *MbERF11* in cold and salt stress responses, *MbERF11* gene was transformed into *A. thaliana* under the control of the CaMV 35S promoter. Among 12 transformed lines, seven of them (S1, S2, S4, S6, S7, S9, and S10) were confirmed by qRT-PCR analysis with wild type (WT) and vacant-vector line (VL) as control (Figure 5A).

As shown in Figure 5B, no significant difference in appearance was found among all the *A. thaliana* lines (WT, VL, S2, S6, and S7) under control condition (Cold 0h). However, when dealt with low temperature (−4 °C) stress for 12 h (Cold 12 h), the transgenic plants (S2, S6 and S7) look much healthier than WT and VL. There were no differences in appearance between WT and VL under control condition and cold stress.

Under control condition, there was no significant difference in the survival rates among all *A. thaliana* lines (WT, VL, S2, S6 and S7). However, when dealt with cold stress, the survival rates of WT and VL *A. thaliana* were only 16.7% and 18.3%, while the transgenic plants of S2, S6, and S7 were 78.7%, 75.1%, and 81.3%, respectively. The survival rates of transgenic plants were significantly higher than those of WT and VL lines under low temperature treatments (Figure 5C).

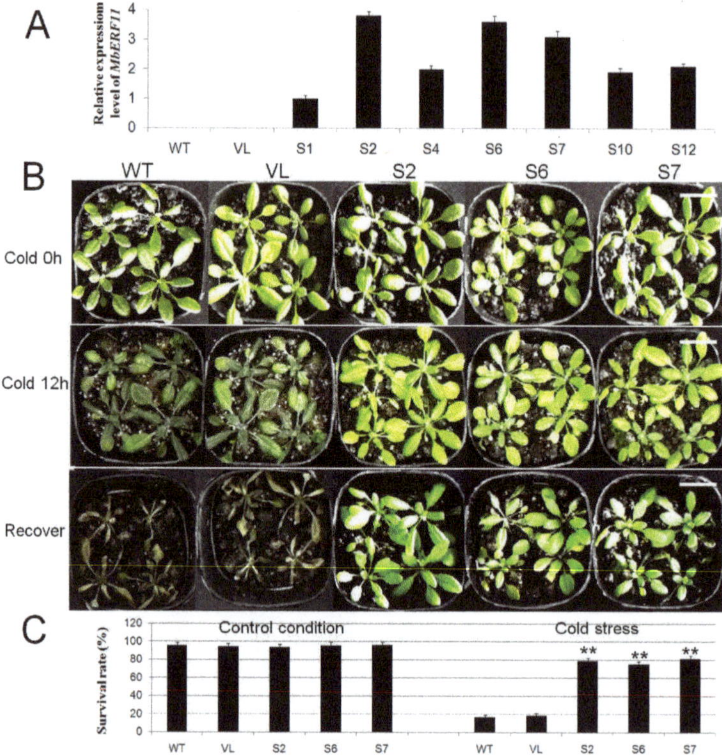

Figure 5. Overexpression of *MbERF11* in *Arabidopsis* improved cold tolerance. (**A**) Transgenic *Arabidopsis* qRT-PCR validation. (**B**) Phenotypic map of *Arabidopsis* under control condition (Cold 0 h), cold stress (Cold 12 h) and recover. All the test lines, including the T_3 transgenic *A. thaliana* (S2, S6, and S7), WT and VL were dealt with low temperature stress (−4 °C) for 12 h, and then with control temperature (20 °C) 1 h for recover. Scale bar corresponds to 3 cm. (**C**) The survival rates of transgenic and WT *Arabidopsis* under control condition and low temperature stress. Extremely significant differences between transgenic *Arabidopsis* (S2, S6, and S7), vacant-vector line (VL) and WT line were shown by the *t* test, ** $p \leq 0.01$.

As shown in Figure 6, for the contents of chlorophyll, MDA and proline, as well as the activities of SOD, POD, and CAT, these was no significant difference among all the *A. thaliana* lines, i.e., S2, S6, S7, WT and VL plants under control conditions (Cold 0 h). However, when dealt with cold treatment (−4 °C) for 12 h (Cold 12 h), the activities of SOD, POD, and CAT, the chlorophyll and proline contents were higher than those in WT and VL. However, the contents of MDA in transgenic *A. thaliana* (S2, S6, and S7) were significantly lower than those in WT and VL (Figure 6).

3.6. Overexpression of MbERF11 in A. Thaliana Contributed to Improved Salt Stress Tolerance

The WT, VL and transgenic lines (S2, S6, and S7) of *A. thaliana* were treated with 200 mM NaCl solution daily for seven days, and then the phenotype of each line was observed. As shown in Figure 7A, the transgenic lines (S2, S6, and S7), WT and VL plants all grew well under control condition (Salt 0 d). However, when dealt with salt stress for 7 days (Salt 7 d), the transgenic *A. thaliana* (S2, S6 and S7) had better appearance than WT and VL plants.

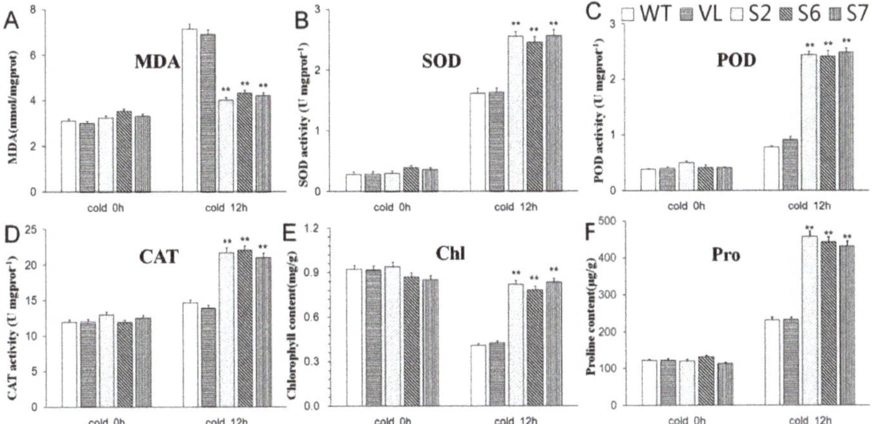

Figure 6. The levels of malondialdehyde content ((**A**) MDA); superoxide dismutase activity ((**B**), SOD); peroxidase activity ((**C**), POD); catalase activity ((**D**), CAT); chlorophyll content ((**E**), Chl) and proline content ((**F**), Pro) Proline in WT, VL (vacant-vector line) and *MbERF11*-OE (S2, S6, and S7) *Arabidopsis* under control condition (Cold 0 h) and cold stresses (Cold 12 h). Asterisks above the error bars indicate an extremely significant difference between the transgenic and WT plants with Student's *t* test (** $p \leq 0.01$). The level of each index in WT was as control.

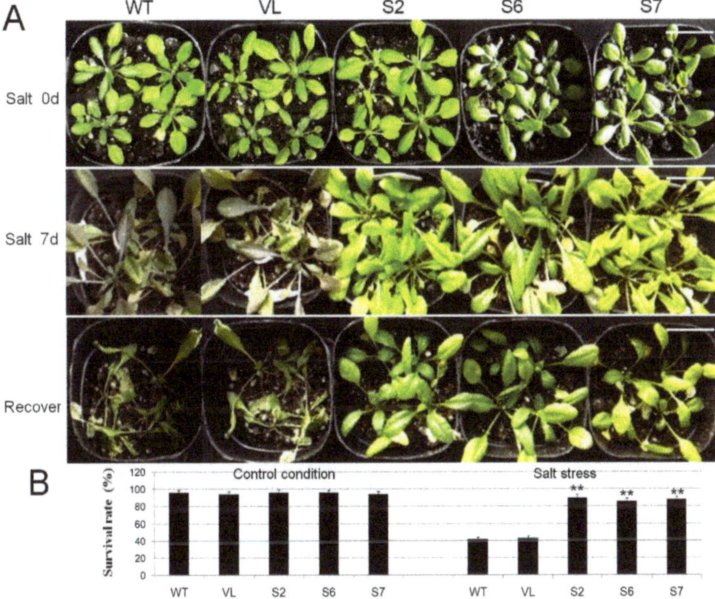

Figure 7. Overexpression of *MbERF11* in *Arabidopsis* improved salt tolerance in transgenic plants. (**A**) Phenotypic map of *Arabidopsis* under control condition (Salt 0 d), salt stress (Salt 7 d) and recover. All the test lines, including the T₃ transgenic *A. thaliana* (S2, S6, and S7), WT and VL were dealt with salt stress (200 mM NaCl) for 7 d, and then with control water management 3 d for recover. Scale bar corresponds to 3 cm. (**B**) The survival rate of *Arabidopsis* under control condition and salt stress. Extremely significant differences between transgenic *Arabidopsis* (S2, S6, and S7), VL and WT line were shown by the *t* test, ** $p \leq 0.01$.

Under control condition, there was no significant difference in the survival rates of all *A. thaliana* lines (WT, VL, S2, S6, and S7). However, when dealt with salt stress for 7 days, the survival rates of WT and VL plants were only 41.9% and 43.2%, while the transgenic lines of S2, S6, and S7 were 89.7%, 85.8% and 87.6%, respectively. The survival rates of the transgenic plants under salt stress were significantly higher than those of WT and VL plants (Figure 7B).

As shown in Figure 8, when treated with salt stress (Salt 7 d), overexpression of *MbERF11* in transgenic *A. thaliana* resulted in lower MDA contents, higher levels of chlorophyll and proline contents, as well as higher activities of SOD, POD, and CAT than those of WT and VL plants. However, for the indices above, these was no significant difference among the entire test lines (WT, VL, S2, S6, and S7) under control condition (Salt 0 d).

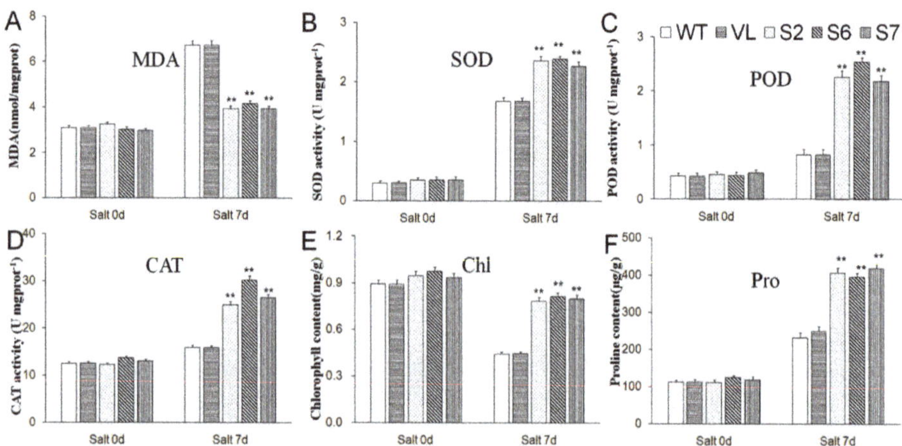

Figure 8. The levels of malondialdehyde content ((**A**), MDA); superoxide dismutase activity ((**B**), SOD); peroxidase activity ((**C**), POD); catalase activity ((**D**), CAT); chlorophyll content ((**E**), Chl) and proline content ((**F**), Pro) Proline in WT, VL (vacant-vector line) and *MbERF11*-OE (S2, S6, and S7) under control condition (Salt 0 d) and salt stresses (Salt 7 d). Asterisks above the error bars indicate an extremely significant difference between the transgenic and WT *Arabidopsis* with Student's *t* test (** $p \leq 0.01$). The level of each index in WT was as control.

4. Discussion

From the transcriptome analysis of *M. baccata* seedlings under cold and/or drought stresses, we found the *MbERF11* level was significantly up-regulated under both stresses. More importantly, through NCBI blast (https://blast.ncbi.nlm.nih.gov/Blast.cgi) of *MbERF11* gene, we found that the closest *Arabidopsis* ERF gene is *AtERF7*, which is a famous ERF TF gene involved in drought stress through ABA signal transduction [32]. Sequence homologous analysis showed that MbERF11 is a member of the ERF family (Figure 2A). All the ERF family includes one conserved ERF domain in the middle region [41]. These results showed that the ERF family genes are highly conserved during the evolution. ERF genes were widely distributed in apple, *A. thaliana*, pear, jujube, cucumber, tobacco, and rice, and were known to be involved in a variety of processes, including stress [10,12,18,21,32]. Subcellular localization has revealed that the MbERF11 is a nucleus localized protein (Figure 3), which is consistent with other ERF proteins [5,17,28,30,32]. Phylogenetic tree analysis shows that the relationship between MbERF11 and MdERF011 is the closest among 14 species. Among Arabidopsis ERF TFs, AtERF7 has the highest homology to MbERF11 (Figure 2B).

The expression level of *MbERF11* was more enriched in stem and root than in young leaf and mature leaf (Figure 4A). This expression pattern indicated that *MbERF11* may play an important role in organs that related to transport of stress signal. When dealt with cold, salt, and ethephon treatments, the expression level of *MbERF11* in *M. baccata* was markedly affected. It is possible that *MbERF11* plays a key role in regulating stress response in *M. baccata*. Ethylene is considered as a signal of stress in plants [17,19,25,41], and ethylene treatment affected the expression of *MbERF11*. The changed expression level of *MbERF11* induced by cold and salt probably depends on the synthesis of endogenous ethylene.

When dealt with salt, cold, and ethephon treatments, the *MbERF11* expression level in root of *M. baccata* reached the highest level at 8 h, 12 h, and 2 h, respectively (Figure 4C), while in leaf, which got the maximum at 12 h, 24 h, and 4 h, respectively (Figure 4B). The results showed that the response rate to stresses, such as low temperature, salt stress and ethephon in root is faster than in leaf, indicating that the expression of *MbERF11* in root is more sensitive than that in leaf. This expression profile indicated that *MbERF11* may play a key role in the plants' stress signal transportation. Ethylene is a gaseous plant hormone that regulates many aspects of the plant life cycle, including seed germination, leaf senescence, fruit ripening, abscission, as well as biotic and abiotic stress responses [42–44]. Consequently, a higher concentration of ethylene in plants may be a signal to regulate plant stress response. This conclusion was consistent with the cold resistance mechanism of *VaERF057* [45]. Ethylene had been proposed to protect mitochondrial activity in *A. thaliana* under cold stress [46]. *GmERF7* had been confirmed to regulate the expression of stress-related genes through regulating the content of ethylene, thereby improving the salt tolerance [47]. Therefore, the increased expression level of *MbERF11* induced by cold and salt stresses may depend on the biosynthesis of ethylene.

The expression of *AtERF7* can be induced by ethylene, ABA and JA [32]. Abiotic stresses could induce the expression of *AtERF71, AtERF73, RAP2.1, RAP2.2,* and *RAP2.3* [48–50]. These results were consistent with our research (*MbERF11* expression level can be induced by ethylene, salt, and cold treatments). The expression level of *GmERF3* increased when dealt with abiotic stresses such as drought and high salinity. External factors such as ethylene and other hormone treatments could also change the expression level of *GmERF3*. However, low temperature stress had little effect on *GmERF3* expression [51]. The expression level of *MbERF11* gene in young leaf and new root was also significantly affected by cold, salt, and ethephon treatments (Figure 4B,C). Low temperature and salt stress could also increase the ethylene levels and trigger cold and salt stress responses in plants [30,46]. Based on the previous studies and theories, we reckon that ethephon treatments induce stress responses, such as the increased expression of *MbERF11* in the above parts.

Overexpression of *MbERF11* enhanced the tolerance to both cold and salt stresses in transgenic *A. thaliana*. The levels of chlorophyll, proline, MDA and antioxidant enzymes can be used to indicate the damage extent from stress [52,53]. The higher MDA content indicates higher degree of membranous peroxidation of the plant cells and the more serious damage to the cell membrane [2]. The proline content in WT *A. thaliana* increased after exposure to low temperature stress. Low temperature stress caused the destruction of chloroplast and the yellowing of plants. Hence, chlorophyll content is one of the important indicators of whether the plant is subjected to adverse stress [54]. The antioxidant enzyme system in plant plays an important role in resisting external environmental stress. They can inhibit the accumulation of free radicals, thereby reducing the occurrence of oxidative damage and lethal effects. The above effects allowed a variety of biochemical metabolic activities in cells to proceed normally. Overexpression of *MbERF11* enhanced the tolerance to cold and salt stresses in transgenic *A. thaliana* (Figures 5B and 7A), also led to increased activities of SOD, POD, and CAT, contents of proline and chlorophyll, decreased MDA content, especially when dealt with stresses (Figures 6 and 8). It is possible that *MbERF11* could increase cold and salt tolerance through changing these physiological indicators in transgenic *A. thaliana* under stress.

These results indicate that the *MbERF11* may be an upstream regulatory gene for stress resistance, and the overexpression of *MbERF11* gene may enhance the cold and salt tolerance of *A. thaliana*. More works need to be done to further verify the function of *MbERF11* through heterologous expression in *Arabidopsis* mutants (*AtERF7* gene deletion). Clarifying the role of the different domains of *MbERF11* in stress response will be helpful in breeding stress-resistant *Malus* by gene transfer. Further experiments are required to identify other functions of *MbERF11* gene.

5. Conclusions

In the present study, a new ERF gene was isolated from M. baccata and named as MbERF11. Subcellular localization showed that MbERF11 protein was located to the nucleus. When MbERF11 was introduced into A. thaliana, it increased the levels of proline and chlorophyll, and improved the activities of SOD, POD, and CAT, but decreased MDA content, especially under cold and salt treatments. Taken together, our results suggest that MbERF11 plays an important role in response to cold and salt stress by enhancing the capability of scavenging ROS.

Author Contributions: D.H. and W.L. designed the experiments; D.H., J.H., G.Y., and S.W. performed the experiments; T.X. and J.H. analyzed the data; D.H. and W.L. wrote the manuscript. All authors have read and agreed to the published version of the manuscript.

Funding: This research was funded by by University Nursing Program for Young Scholars with Creative Talents in Heilongjiang Province (UNPYSCT-2017004), National Natural Science Foundation of China (31301757), Postdoctoral Scientific Research Development Fund of Heilongjiang Province, China (LBH-Q16020, LBH-Z15019), the Natural Science Fund Joint Guidance Project of Heilongjiang Province (LH2019C031) and the Open Project of Key Laboratory of Biology and Genetic Improvement of Horticultural Crops (Northeast Region), Ministry of Agriculture and Rural Affairs (neauhc201803).

Conflicts of Interest: The authors declare no conflicts of interest.

References

1. Tuskan, G.A.; Difazio, S.; Jansson, S.; Bohlmann, J.; Grigoriev, I.; Hellsten, U.; Hellsten, U.; Putnam, N.; Ralph, S.; Rombauts, S.; et al. The genome of black cottonwood, *Populus trichocarpa* (Torr. & Gray). *Science* **2006**, *313*, 1596–1604.

2. Han, D.; Du, M.; Zhou, Z.; Wang, S.; Li, T.; Han, J.; Xu, T.; Yang, G. Overexpression of a *Malus baccata* NAC transcription factor gene *MbNAC25* increases cold and salinity tolerance in *Arabidopsis*. *Int. J. Mol. Sci.* **2020**, *21*, 1198. [CrossRef]

3. Ryu, H.; Cho, Y.G. Plant hormones in salt stress tolerance. *J. Plant Biol.* **2015**, *58*, 147–155. [CrossRef]

4. Yamaguchishinozaki, K.; Shinozaki, K. Transcriptional regulatory networks in cellular responses and tolerance to dehydration and cold stresses. *Ann. Rev. Plant Biol.* **2006**, *57*, 781–803. [CrossRef]

5. Huang, Z.; Zhang, Z.; Zhang, X.; Zhang, H.; Huang, D.; Huang, R. Tomato *TERF1* modulates ethylene response and enhances osmotic stress tolerance by activating expression of downstream genes. *FEBS Lett.* **2004**, *573*, 110–116. [CrossRef] [PubMed]

6. Lucas, S.; Durmaz, E.; Akpınar, B.A.; Budak, H. The drought response displayed by a dre-binding protein from triticum dicoccoides. *Plant Physiol. Biochem.* **2011**, *49*, 346–351. [CrossRef]

7. Mcconn, M.; Creelman, R.A.; Bell, E.; Mullet, J.E.; Browse, J. Jasmonate is essential for insect defense in *Arabidopsis*. *Proc. Natl. Acad. Sci. USA* **1997**, *94*, 5473–5477. [CrossRef] [PubMed]

8. Liu, L.; White, M.J.; Macrae, T.H. Transcription factors and their genes in higher plants functional domains, evolution and regulation. *Eur. J. Biochem.* **1999**, *262*, 247–257. [CrossRef]

9. Singh, K.; Foley, R.C.; Oñatesánchez, L. Transcription factors in plant defense and stress responses. *Curr. Opin. Plant Biol.* **2002**, *5*, 430–436. [CrossRef]

10. Nakano, T.; Suzuki, K.; Fujimura, T.; Shinshi, H. Genome-wide analysis of the *ERF* gene family in *Arabidopsis* and rice. *Plant Physiol.* **2006**, *140*, 411–432. [CrossRef]

11. Feng, J.X.; Liu, D.; Pan, Y.; Gong, W.; Ma, L.G.; Luo, J.C.; Deng, X.W.; Zhu, Y.X. An annotation update via cdna sequence analysis and comprehensive profiling of developmental, hormonal or environmental responsiveness of the *Arabidopsis*, Ap2/EREBP transcription factor gene family. *Plant Mol. Biol.* **2005**, *59*, 853–868. [CrossRef] [PubMed]

12. Sun, B.; Zhan, X.D.; Cao, L.Y.; Cheng, S.H. Research Progress of Rice AP2/ERF Transcription Factors. *J. Agric. Biotech.* **2017**, *11*, 1860–1869.

13. Shigyo, M.; Hasebe, M.; Ito, M. Molecular evolution of the AP2 subfamily. *Gene* **2006**, *366*, 256–265. [CrossRef] [PubMed]

14. Jofuku, K.D.; den Boer, B.G.; Van, M.M.; Okamuro, J.K. Control of *Arabidopsis* flower and seed development by the homeotic gene *APETALA2*. *Plant Cell* **1994**, *6*, 1211–1225. [PubMed]

15. Elliott, R.C.; Betzner, A.S.; Huttner, E.; Oakes, M.P.; Tucker, W.Q.; Gerentes, D.; Perez, P.; Smyth, D.R. *AINTEGUMENTA*, an *APETALA2-Like* gene of *Arabidopsis* with pleiotropic roles in ovule development and floral organ growth. *Plant Cell* **1996**, *8*, 155–168. [PubMed]

16. Stockinger, E.J.; Gilmour, S.J.; Thomashow, M.F. *Arabidopsis thaliana cbf1* encodes an *AP2* domain-containing transcriptional activator that binds to the c-repeat/dre, a cis-acting dna regulatory element that stimulates transcription in response to low temperature and water deficit. *Proc. Natl. Acad. Sci. USA* **1997**, *94*, 1035–1040. [CrossRef]

17. Qin, L.; Wang, L.; Guo, Y.; Li, Y.; Ümüt, H.; Wang, Y. An ERF transcription factor from *Tamarix hispida*, ThCRF1, can adjust osmotic potential and reactive oxygen species scavenging capability to improve salt tolerance. *Plant Sci.* **2017**, *265*, 154–166. [CrossRef]

18. Wang, C.; Xin, M.; Zhou, X.; Liu, C.; Li, S.; Liu, D.; Xu, Y.; Qin, Z.W. The novel ethylene-responsive factor *CsERF025* affects the development of fruit bending in cucumber. *Plant Mol. Biol.* **2017**, *95*, 519–531. [CrossRef]

19. Erpen, L.; Devi, H.S.; Grosser, J.W.; Dutt, M. Potential use of the DREB/ERF, MYB, NAC and WRKY transcription factors to improve abiotic and biotic stress in transgenic plants. *Plant Cell Tissue Organ Cult.* **2017**, *132*, 1–25. [CrossRef]

20. Hao, D.; Ohmetakagi, M.; Sarai, A. Unique mode of gcc box recognition by the dna-binding domain of ethylene-responsive element-binding factor (ERF domain) in plant. *J. Biol. Chem.* **1998**, *273*, 26857–26861. [CrossRef]

21. Liu, W.; Li, Q.; Wang, Y.; Wu, T.; Yang, Y.; Zhang, X.; Han, Z.; Xu, X. Ethylene response factor *AtERF72* negatively regulates *Arabidopsis thaliana* response to iron deficiency. *Biochem. Biophys. Res. Commun.* **2017**, *491*, 862–868. [CrossRef] [PubMed]

22. Son, G.H.; Wan, J.; Kim, H.J.; Nguyen, X.C.; Chung, W.S.; Hong, J.C.; Stacey, G. Ethylene-responsive element-binding factor 5, *ERF5*, is involved in chitin-induced innate immunity response. *Mol. Plant* **2012**, *25*, 48–60. [CrossRef] [PubMed]

23. Lee, S.Y.; Hwang, E.Y.; Seok, H.Y.; Tarte, V.N.; Jeong, M.S.; Jang, S.B. *Arabidopsis* AtERF71/HRE2 functions as transcriptional activator via cis-acting gcc box or dre/crt element and is involved in root development through regulation of root cell expansion. *Plant Cell Rep.* **2015**, *34*, 223–231. [CrossRef] [PubMed]

24. Sakuma, Y.; Liu, Q.; Dubouzet, J.G.; Abe, H.; Shinozaki, K.; Yamaguchishinozaki, K. DNA-binding specificity of the ERF/AP2 domain of *Arabidopsis* drebs, transcription factors involved in dehydration- and cold-inducible gene expression. *Biochem. Biophys. Res. Commun.* **2002**, *290*, 998–1009. [CrossRef] [PubMed]

25. Zhang, Z.; Huang, R. Enhanced tolerance to freezing in tobacco and tomato overexpressing transcription factor *TERF2/LeERF2* is modulated by ethylene biosynthesis. *Plant Mol. Biol.* **2010**, *73*, 241–249. [CrossRef]

26. El-Sharkawy, I.; Sherif, S.; Mila, I. Molecular cliaracterization of seven genes encoding ethylene-responsive transcriptional factors during plum fruit development and ripening. *J. Exp. Bot.* **2009**, *60*, 907–922. [CrossRef]

27. Nakano, T.; Fujisawa, M.; Shima, Y.; Ito, Y. The AP2/ERF transcription factor *SlERF52* functions in flower pedicel abscission in tomato. *J. Exp. Bot.* **2014**, *65*, 1–4. [CrossRef]

28. Pirrello, J.; Jaimesmiranda, F.; Sanchezballesta, M.T.; Tournier, B.; Khalilahmad, Q.; Regad, F.; Latché, A.; Pech, J.C.; Bouzayen, M. Sl-ERF2, a tomato ethylene response factor involved in ethylene response and seed germination. *Plant Cell Physiol.* **2006**, *47*, 1195–1205. [CrossRef]

29. Fukao, T.; Xu, K.; Ronald, P.C.; Baileyserres, J. A variable cluster of ethylene response factor-like genes regulates metabolic and developmental acclimation responses to submergence in rice. *Plant Cell* **2006**, *18*, 2021–2034. [CrossRef]

30. Wang, A.; Tan, D.; Takahashi, A.; Li, T.Z.; Harada, T. MdERFs, two ethylene-response factors involved in apple fruit ripening. *J. Exp. Bot.* **2007**, *58*, 3743–3748. [CrossRef]

31. Xiao, H.; Yin, L.; Xu, X.; Li, T.; Han, Z. The iron-regulated transporter, mbnramp1, isolated from *Malus baccata* is involved in fe, mn and cd trafficking. *Ann. Bot.* **2008**, *102*, 881–889. [CrossRef] [PubMed]

32. Song, C.P.; Agarwal, M.; Ohta, M.; Guo, Y.; Halfter, U.; Wang, P.; Zhu, J.K. Role of an *Arabidopsis* AP2/EREBP-type transcriptional repressor in abscisic acid and drought stress responses. *Plant Cell* **2005**, *17*, 2384–2396. [CrossRef] [PubMed]

33. Han, D.; Yufang Wang, Y.; Zhang, Z.; Pu, Q.; Ding, H.; Han, J.; Fan, T.; Bai, X.; Yang, G. Isolation and functional analysis of *MxCS3*: A gene encoding a citrate synthase in *Malus xiaojinensis*, with functions in tolerance to iron stress and abnormal flower in transgenic *Arabidopsis thaliana*. *Plant Growth Regul.* **2017**, *82*, 479–489. [CrossRef]

34. An, J.P.; Qu, F.J.; Yao, J.F.; Wang, X.N.; You, C.X.; Wang, X.F.; Hao, Y.J. The bZIP transcription factor MdHY5 regulates anthocyanin accumulation and nitrate assimilation in apple. *Hortic. Res.* **2017**, *4*, 17023. [CrossRef]

35. Pfaffl, M.W. A new mathematical model for relative quantification in real time RT-PCR. *Nucleic Acids Res.* **2001**, *29*, e45. [CrossRef]

36. Yang, G.H.; Li, J.; Liu, W.; Yu, Z.Y.; Shi, Y.; Lv, B.Y.; Wang, B.; Han, D.G. Molecular cloning and characterization of *MxNAS2*, a gene encoding nicotianamine synthase in *Malus xiaojinensis*, with functions in tolerance to iron stress and misshapen flower in transgenic tobacco. *Sci. Hortic.* **2015**, *183*, 77–86. [CrossRef]

37. An, G.; Watson, B.D.; Chiang, C.C. Transformation of tobacco, tomato, potato, and *Arabidopsis thaliana* using a binary ti vector system. *Plant Physiol.* **1986**, *81*, 301–305. [CrossRef] [PubMed]

38. Liu, S.Y.; Liu, Y.F.; Xu, J.; Liu, T.; Dong, S.Z. Effectiveness of lysozyme coatings and 1-MCP treatments on storage and preservation of kiwifruit. *Food Chem.* **2019**, *288*, 201–217.

39. Jiménez-Bremont, J.F.; Becerra-Flora, A.; Hernandéz-Lucero, E.; Rodriguéz-Kessler, M.; Acosta-Gallegos, J.A.; Ramiréz-Pimentel, J.G. Proline accumulation in two bean cultivars under salt stress and the effect of polyamines and ornithine. *Biol. Plant* **2006**, *50*, 763–766. [CrossRef]

40. Shin, L.J.; Lo, J.C.; Yeh, K.C. Copper chaperone antioxidant protein1 is essential for copper homeostasis. *Plant Physiol.* **2012**, *59*, 1099–1110. [CrossRef]

41. Ohta, M.; Matsui, K.; Hiratsu, K.; Shinshi, H.; Ohme-Takagi, M. Repression domains of class II ERF transcriptional repressors share an essential motif for active repression. *Plant Cell* **2001**, *13*, 1959–1968. [CrossRef] [PubMed]

42. Achard, P.; Baghour, M.; Chapple, A.; Hedden, P.; Van Der Straeten, D.; Genschik, P.; Harberd, N.P. The plant stress hormone ethylene controls floral transition via DELLA-dependent regulation of floral meristem-identity genes. *Proc. Natl. Acad. Sci. USA* **2007**, *104*, 6484–6489. [CrossRef] [PubMed]

43. Chen, H.; Xue, L.; Chintamanani, S.; Germain, H.; Lin, H.; Cui, H. Ethylene insensitive3 and ethylene insensitive3-like1 repress salicylic acid induction deficient2 expression to negatively regulate plant innate immunity in *Arabidopsis*. *Plant Cell* **2009**, *21*, 2527–2540. [CrossRef] [PubMed]

44. Yin, C.C.; Ma, B.; Collinge, D.P.; Pogson, B.J.; He, S.J.; Xiong, Q. Ethylene responses in rice roots and coleoptiles are differentially regulated by a carotenoid isomerase-mediated abscisic acid pathway. *Plant Cell* **2015**, *27*, 1061–1081. [CrossRef]

45. Kendrick, M.D.; Chang, C. Ethylene signaling: New levels of complexity and regulation. *Curr. Opin. Plant Biol.* **2008**, *11*, 479–485. [CrossRef]

46. Sun, X.; Zhao, T.; Gan, S.; Ren, X.; Fang, L.; Karungo, S.K. Ethylene positively regulates cold tolerance in grapevine by modulating the expression of ethylene response factor 057. *Sci. Rep.* **2016**, *6*, 24066. [CrossRef]

47. Catalá, R.; Lópezcobollo, R.; Mar, M.C.; Angosto, T.; Alonso, J.M.; Ecker, J.R. The *Arabidopsis* 14-3-3 protein rare cold inducible 1a links low-temperature response and ethylene biosynthesis to regulate freezing tolerance and cold acclimation. *Plant Cell* **2014**, *26*, 3326–3342. [CrossRef]

48. Zhai, Y.; Wang, Y.; Li, Y.; Lei, T.; Yan, F.; Su, L. Isolation and molecular characterization of *GmERF7*, a soybean ethylene-response factor that increases salt stress tolerance in tobacco. *Gene* **2013**, *513*, 174–183. [CrossRef]

49. Park, H.Y.; Seok, H.Y.; Woo, D.H.; Lee, S.Y.; Tarte, V.N.; Lee, E.H. *AtERF71/HRE2* transcription factor mediates osmotic stress response as well as hypoxia response in *Arabidopsis*. *Biochem. Biophys. Res. Commun.* **2011**, *414*, 135–141. [CrossRef]

50. Papdi, C.; Perez-Salamo, I.; Joseph, M.P.; Giuntoli, B.; Bögre, L.; Koncz, C.; Szabados, L. The low oxygen, oxidative and osmotic stress responses synergistically act through the ethylene response factor vii genes rap2.12, rap2.2 and rap2.3. *Plant J.* **2015**, *82*, 772–784. [CrossRef]

51. Yang, C.Y.; Hsu, F.C.; Li, J.P.; Wang, N.N.; Shih, M.C. The AP2/ERF transcription factor *AtERF73/HRE1* modulates ethylene responses during hypoxia in *Arabidopsis*. *Plant Physiol.* **2011**, *156*, 202–212. [CrossRef] [PubMed]

52. Zhang, G.Y.; Chen, M.; Li, L.C.; Xu, Z.S.; Chen, X.P.; Guo, J.M. Overexpression of the soybean *GmERF3* gene, an AP2/ERF type transcription factor for increased tolerances to salt, drought, and diseases in transgenic tobacco. *J. Exp. Bot.* **2009**, *60*, 3781–3796. [CrossRef] [PubMed]

53. Han, D.; Ding, H.; Chai, L.; Liu, W.; Zhang, Z.; Hou, Y.; Yang, G. Isolation and characterization of *MbWRKY1*, a WRKY transcription factor gene from *Malus baccata* (L.) Borkh involved in drought tolerance. *Can. J. Plant Sci.* **2018**, *98*, 1023–1034. [CrossRef]
54. Han, D.; Zhang, Z.; Ding, H.; Chai, L.; Liu, W.; Li, H.; Yang, G. Isolation and characterization of *MbWRKY2* gene involved in enhanced drought tolerance in transgenic tobacco. *J. Plant Interact.* **2018**, *13*, 163–172. [CrossRef]

Article

The Role of Climate Niche, Geofloristic History, Habitat Preference, and Allometry on Wood Density within a California Plant Community

Rebecca A. Nelson [1,*], Emily J. Francis [2], Joseph A. Berry [3], William K. Cornwell [4] and Leander D. L. Anderegg [3,5]

[1] Department of Biology, School of Humanities and Sciences, Stanford University, Stanford, CA 94305, USA
[2] Department of Integrative Biology, College of Natural Sciences, University of Texas at Austin, Austin, TX 78712, USA; emily.francis@austin.utexas.edu
[3] Department of Global Ecology, Carnegie Institution for Science, Stanford, CA 94305, USA; jberry@carnegiescience.edu (J.A.B.); leanderegg@gmail.com (L.D.L.A.)
[4] School of Biological, Earth and Environmental Sciences, University of New South Wales, Sydney, NSW 2052, Australia; w.cornwell@unsw.edu.au
[5] Department of Integrative Biology, College of Letters & Science, University of California, Berkeley, CA 94720, USA
*   Correspondence: rnelson3@stanford.edu

Received: 1 December 2019; Accepted: 7 January 2020; Published: 14 January 2020

Abstract: *Research Highlights:* To better understand within-community variation in wood density, our study demonstrated that a more nuanced approach is required beyond the climate–wood density correlations used in global analyses. *Background and Objectives:* Global meta-analyses have shown higher wood density is associated with higher temperatures and lower rainfall, while site-specific studies have explained variation in wood density with structural constraints and allometry. On a regional scale, uncertainty exists as to what extent climate and structural demands explain patterns in wood density. We explored the role of species climate niche, geofloristic history, habitat specialization, and allometry on wood density variation within a California forest/chaparral community. *Materials and Methods:* We collected data on species wood density, climate niche, geofloristic history, and riparian habitat specialization for 20 species of trees and shrubs in a California forest. *Results:* We found a negative relationship between wood density and basal diameter to height ratio for riparian species and no relationship for non-riparian species. In contrast to previous studies, we found that climate signals had weak relationships with wood density, except for a positive relationship between wood density and the dryness of a species' wet range edge (species with drier wet range margins have higher wood density). Wood density, however, did not correlate with the aridity of species' dry range margins. Geofloristic history had no direct effect on wood density or climate niche for modern California plant communities. *Conclusions:* Within a California plant community, allometry influences wood density for riparian specialists, but non-riparian plants are 'overbuilt' such that wood density is not related to canopy structure. Meanwhile, the relationship of wood density to species' aridity niches challenges our classic assumptions about the adaptive significance of high wood density as a drought tolerance trait.

Keywords: wood density; allometry; functional traits; climate niches

---

1. Introduction

Wood density (the mass per unit volume of wood) is considered an essential functional trait associated with mechanical support, carbon and nutrient storage, drought tolerance, water transport, and pathogen defense [1–3]. Variation in wood functional traits often reflect ecological tradeoffs that are influenced by allometric, biogeographic, and phylogenetic factors [1]. Understanding wood density patterns can help elucidate spatial trends in aboveground biomass [4]. Moreover, wood density is a useful predictor of growth and mortality rates in the tropics [5] and is related to growth and performance in Mediterranean climates [6]. Finally, models of carbon stocks often rely on wood density, making estimates of wood density operationally as well as ecologically important [7,8].

Wood density often relates to mechanical stability [9–11] with crown architecture, diameter, and height correlating with wood density at small scales [8,12]. Tree height is a key trait for understanding patterns in wood density and allometry in Mediterranean communities [6]. Allometric equations that relate wood density to height and diameter have explained site-specific patterns across tropical ecosystems [8,13]. The relationship of wood density to diameter–height allometry at an individual site level has varied across studies. Whereas some studies have found that tree height at a given diameter is unrelated to wood density [14,15], Iida et al. (2012) found that, at a constant tree height, higher wood density is associated with smaller tree diameters [16], and wood density is positively related to diameter-corrected tree height in rainforests [17]. Simultaneously, in a California mixed evergreen forest, wood density correlated strongly with vessel traits and microsite soil moisture [18,19], indicating that moisture availability can be an environmental filter for wood density at very small scales. The variation in the wood density–allometry relationship across different climate regimes and site conditions may be partially explained by the availability of water at particular microsites, with some individuals and species at or close to drainages being able to access consistent water throughout the year, while other species experience strong seasonal fluctuations in water availability. In other words, riparian habitat specialization could influence the degree to which structural constraints and climate niche relate to wood density.

Global and regional meta-analyses suggest that higher wood density correlates with lower precipitation and higher mean annual temperature [6,11,20]. Across a latitudinal gradient in Central America, patterns in wood density more strongly correlated to precipitation and aridity than temperature [21]. Wood density is positively associated with aridity in New Caledonia, and a global meta-analysis [22,23] suggests that high wood density is adaptive in dry environments. However, precipitation and temperature explain variation in tree height–diameter relationships across North America as well, possibly influencing wood density indirectly through structural constraints [24]. The findings that rainfall amount can modify wood density–allometry relationships at a regional level in water-stressed climates suggests at least some direct effect of climate on wood density [25], and denser wood is correlated with lower drought mortality [26]. Variations in shade tolerance and altitude add further nuance to relationships between wood density and rainfall patterns in tropical and Mediterranean climates [12,27].

While many studies have examined the relationship between climate and wood density across a broad geographic scale, and a number of tropical studies have highlighted the importance of structural or allometric constraints on wood density at a site level, few studies have assessed both structural and climatic constraints on wood density in temperate communities. Moreover, phylogenetic signal, biogeography, and geofloristic history (plant assemblages recorded in the fossil record) can strongly structure the composition and functional traits of temperate plant communities [11,22,27,28]. As a result, community composition that is the contingent outcome of past climates could drive functional trait patterns that are not mechanistically linked to the contemporary climate [28]. In water-stressed ecosystems, uncertainty exists as to whether mechanical demands, habitat specialization, geofloristic history, or macroclimatic niche best explain variation in wood density.

The California floristic province, with its high plant diversity and endemism, is vulnerable to increasing drought frequency and intensity exacerbated by anthropogenic climate change. Microsite variation in soil moisture influences the wood density and tree height of the local community across microsites in northern California [18,19]. This suggests environmental filtering for higher wood density species in drier sites. The current species composition of California mixed evergreen forests resulted from the confluence of several geofloras (plant community assemblages identified in the fossil record), especially the temperate Arcto-Tertiary geoflora and the subtropical Madro-Tertiary geoflora [29,30]. The Madro-Tertiary and Arcto-Tertiary geofloras, both originating in the Tertiary Period, are distinguished by their biogeographic origin and differing community compositions. The Madro-Tertiary geoflora, consisting mostly of subtropical, sclerophyll species, is associated with the development of a Mediterranean climate in California and is the ancestor of many extant chaparral specialists [29,30]. The Arcto-Tertiary geoflora, compromised of mixed deciduous forest species, advanced into California from more northerly climates [29,30].

We investigated whether structure, climatic niche, geofloristic history, and habitat specialization related to wood density variation among members of a regional woody plant community, sampling 20 species from a northern California plant community across multiple sites. We tested four hypotheses: (1) We hypothesize that the structural demand of supporting a plant canopy will constrain wood density values such that species with tall, thin stems must have higher wood density to support their canopies. We predict that 'taper' (the ratio of plant basal diameter to plant height) will be negatively related to wood density. (2) We predict that, within a community, more arid adapted species will have higher wood density. Thus, we predict that wood density and metrics of water availability will be negatively related for the dry and wet edges of species' aridity niche, with stronger relationships at the dry edge. (3) We hypothesize that species with a Madro-Tertiary geofloristic history would have stronger wood density–climate relationships than species with an Arcto-Tertiary geofloristic history because the Madro-Tertiary geoflora arose in conjunction with California's dry Mediterranean climate. (4) Finally, we hypothesize that riparian habitat specialization may greatly influence the relationship between wood density and structure as well as wood density and climate niche.

2. Materials and Methods

2.1. Study Site

We sampled naturally occurring vegetation at three sites near Stanford, CA: Jasper Ridge Biological Preserve (37.40311, −122.24428), a 481-ha preserve in the eastern foothills of the Santa Cruz Mountains; the Stanford Quarry (37.39999, −122.16154); and the Stanford main campus (37.42875, −122.17919) (Figure 1). The climate is Mediterranean with a mean annual temperature of 14 °C, mean annual precipitation of 570 mm, and most rainfall occurring from November to April. The vegetation communities present at all three sites included chaparral and broadleaf, mixed evergreen forest. All three sites have loam to gravely clay loam soils derived from similar mixed sedimentary and metamorphic residuum and alluvium [31].

Figure 1. Map of study sites. Green markers represent the geographic location of each site.

2.2. Species Selection

We selected woody species native to northern California. In the spring and summer of 2017, we sampled five individuals from 13 native species growing on the Stanford campus or the Stanford Quarry. We sampled reasonably abundant species (five or more individuals in a ~200 m search area) and focused sampling on the largest individuals of each species present based on a visual survey. We included comparable data of wood density and plant allometry from Cornwell and Ackerly (2009) collected at the Jasper Ridge Biological Preserve, making for a total of 20 species of California trees and shrubs [19]. A range of habitats including riparian forests and chaparral/oak woodlands were sampled at all three sites to capture the dominant woody species across multiple microsites. We only sampled species that grew naturally at our sites. All three sites have similar climate and native plant communities. See Table S1 for a full list of species sampled and which datasets these species occurred in. For data on the heights, diameters, wood density values, and other traits measured of the study species, see Tables S2 and S3.

2.3. Trait Measurement

We obtained three branches per individual for at least five individuals sampled per species to use for measuring branch wood density via a volumetric method. We randomly selected mid- to upper-canopy, sun-exposed branches from each individual using hand pruners or a 4.5 m pole pruner. Branches of ~1 cm diameter and ~3 cm length were peeled of bark, their wet volume measured via mass displacement, and their dry weight measured following 72+ h drying at 65 °C. We measured individual height with an inclinometer. We obtained basal diameter for larger trees as diameter at breast height. We then calculated 'taper' as the basal diameter to height ratio, back calculating the basal diameter from diameter at breast height measurements assuming a conic stem where necessary. We averaged branch wood density per individual, and then averaged wood density and taper of individuals per species. We used branch wood density in order make our results comparable to wood density values obtained by Cornwell and Ackerly (2009) [19], who used data from three-year old branches from five individuals per species (see Preston et al. (2006) [18] for details of wood density measurements).

2.4. Habitat Preferences

We used information in the National Wildlife Federation and Audubon field guides as well as Calscape.org to determine whether species in our study were riparian specialists [32–34]. If the literature described a species as preferring rivers, streambanks, or riparian areas, we recorded these species as riparian specialists. Our dataset contained 11 riparian species and 9 non-riparian species with a mix of riparian and non-riparian species sampled at each of the three sites (Figure S1). Table S2 provides information on whether species were riparian and the sample size per species.

2.5. Paleohistory

We conducted a literature search in order to determine whether each species had ancestors in the Madro-Tertiary or Arcto-Tertiary geoflora. We assigned a geoflora category based off the most taxonomically specific level possible; for most species, we obtained geofloristic information at the genus level. For chaparral species, we placed them in the Madro-Tertiary group based on literature stating that the current chaparral community evolved from the Madro-Tertiary geoflora.

2.6. Climate Data

We determined climatic parameters describing each species climate niche using the WorldClim database and geo-referenced specimens from North America in the Global Biodiversity Information Facility (GBIF) database [35]. We used the full climate range for each species. We obtained 1960–1990 long-term averages of the following annual climatic parameters at 2.5 arc-minutes from the WorldClim

1.4 database [36]: Minimum temperature (Tmin), maximum temperature (Tmax), mean annual temperature (MAT), mean annual precipitation (MAP), and precipitation seasonality (coefficient of variation of precipitation across months, Pseason).We also obtained aridity data from the Global Aridity and PET Database [37], which is derived from WorldClim input data, including potential evapotranspiration (PET) and climate moisture deficit (CMD). CMD was calculated as the amount of water by which PET exceeds precipitation. Climate data were extracted for all GBIF specimen locations for each species. The extent of each species' climate niche was calculated based on the 5th and 95th percentile for each of the above climate variables.

2.7. Data Analysis

We averaged all wood density and taper values per species (taper did not show variation with plant height within-species) and constructed a series of linear models for the data, using R version 3.3.3. First, we constructed linear models relating wood density to taper, habitat preference, the interaction between the two, and the interaction without the habitat preference main effect. We selected the most parsimonious model based on the second-order Aikike information criterion (AICc, or AIC for small sample sizes) using the 'stats' package and the 'MuMIn' [38]. We also tested for univariate relationships between wood density and climate niche parameters related to water availability (CMD, MAP, PET) for the driest edge and wettest edge of each species (based on the 5th and 95th percentile of their species range), and then selected the best single climate variable by built linear models for niche parameters that were initially found to be significant using AICc. Finally, we combined the best structural model and the best climate model to build a single full model and all potential nested models, and then used AICc to select the most parsimonious of these models. The equation for our final model was as follows: WD ~5th% CMD + Riparian Specialization × Taper, where WD is wood density and CMD is climate moisture deficit. Finally, we verified that this inference was not purely driven by phylogenetic non-independence by refitting the final model using a phylogenetic generalized least squares (PGLS) model. We used the Smith and Brown 2018 [39] phylogeny (which contained all species except *Baccharis pilularis*, which was replaced with a congener to estimate branch lengths) and a Brownian motion covariance structure [40] using the *ape* [41], *geiger* [42,43], and *nlme* [44] R packages. See Figure S2 for the results of the phylogenetic analysis.

3. Results

A higher taper (basal diameter to height ratio) refers to woody plants that are wider in basal diameter at a given height (Figure 2). For riparian habitat specialist species, branch wood density decreased with increasing taper (Figure 2). For non-riparian species, no correlation existed between branch wood density and taper (Figure 2). The taper model with the lowest AICc had a main effect of taper and a taper x riparian interaction (but no riparian main effect), though this model was only a marginal improvement over a model with only a riparian specialization main effect (ΔAICc = 1.1, likelihood ratio test $p = 0.06017$) (Table 1).

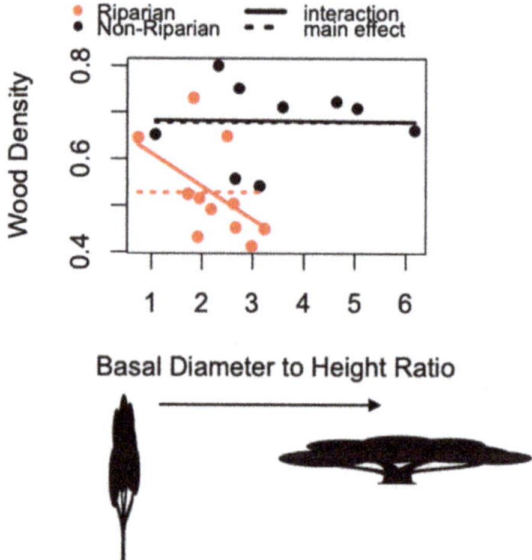

Figure 2. Branch wood density (g/cm^3) vs basal diameter to height ratio for California trees and shrubs. Trees with a low basal diameter to height ratio are tall and skinny, while trees with a high basal diameter to height ratio are short and wide, as represented by tree symbols. The arrow represents increasing basal diameter to height ratio. Red points/lines indicate riparian specialists and black points/lines non-riparian specialists. Dotted lines show the model fit for the habitat only linear model (Table 1), while solid lines show the model fit for the taper × habitat model.

Table 1. Model selection for structure and habitat specialization. WD refers to branch wood density. Taper is basal diameter to height ratio. Habitat refers to whether or not species are riparian specialists. AIC, Aikike information criterion.

Model	Degrees of Freedom	AICc
Null (WD~1)	2	−24.22
WD~taper	3	−22.29
WD~habitat	3	−31.84
WD~taper + habitat	4	−29.22
WD~taper + habitat + taper/habitat	5	−29.49
WD~taper + taper/habitat	4	−32.95

Few correlations existed between wood density and species driest range edge, on the contrary to the existing relationships between wood density and site water availability in global meta-analyses (Figure 3). However, positive relationships between wood density and the dryness of a species' wettest range boundary were found for all metrics of moisture availability (5th percentile CMD, PET, and MAT; 95th percentile MAP), with 5th percentile CMD being the best climate predictor of wood density (Table 2). Riparian specialists had generally lower wood density than non-specialists for the same wet edge aridity, but showed a similar positive relationship between wood density and wet edge aridity (Figure 3). Adding a main effect of habitat specialization improved the best climate model (ΔAICc = 6.86, Table 2), but adding an interaction between climate and habitat specialization did not improve the model. These results were qualitatively similar if climate niches were constructed from only California occurrence records in the California Consortium of Herbaria [45].

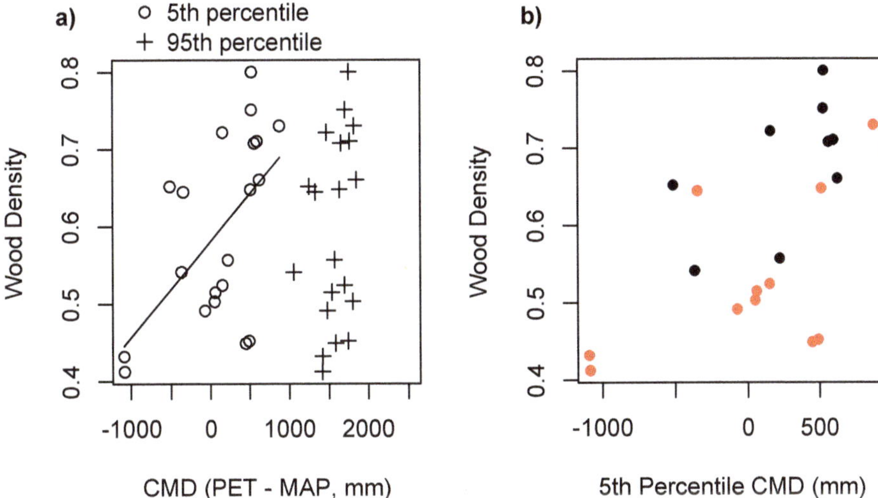

Figure 3. Branch wood density vs. (**a**) species 5% and 95% climate moisture deficit (CMD) and (**b**) species 5% CMD for riparian vs. non-riparian California trees and shrubs. Black dots represent non-riparian specialist species, while red dots represent species that are riparian habitat specialists. The black line indicates a significant relationship between wood density and CMD at the 5th percentile, but no significant effect at the 95th percentile of CMD. PET, potential evapotranspiration; MAP, mean annual precipitation.

Table 2. Model selection for wet edge climate effects. WD refers to branch wood density. Habitat refers to whether or not species are riparian specialists. CMD 0.05 is the 5th percentile climate moisture deficit. MAP 0.05 is the 5th percentile for mean annual precipitation. MAT 0.05 is the 5th percentile for mean annual temperature. PET 0.05 is the 5th percentile for potential evapotranspiration.

Model	Degrees of Freedom	AICc
Null	2	−21.20
WD~CMD 0.05	3	−29.21
WD~MAP 0.05	3	−22.04
WD~PET 0.05	3	−28.84
WD~MAT 0.05	3	−25.75
WD~CMD 0.05 + habitat	4	−36.07
WD~CMD 0.05 + habitat/CMD 0.05	4	−27.80
WD~CMD 0.05 + habitat + habitat/CMD 0.05	5	−32.58

Although geoflora suggested small differences between the climate niches of Madro-Tertiary and Arcto-Tertiary species, especially for the minimum temperature range (Figures 4 and 5), geoflora did not directly affect wood density (ΔAICc = 0.62918, likelihood ratio test p = 0.2794) or strongly differentiate the climate niches of modern California plant communities (Figures 4 and 5).

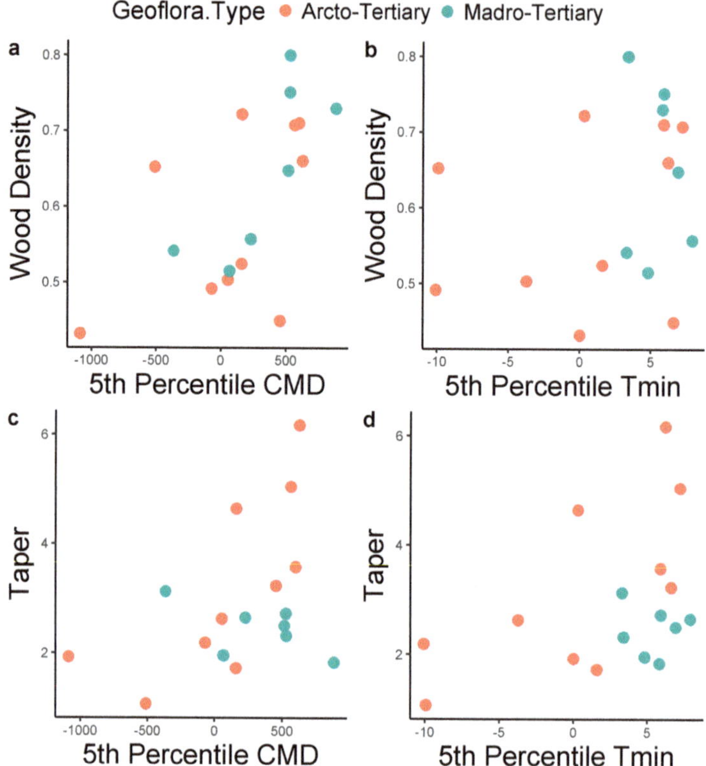

Figure 4. Branch wood density g/cm^3 vs. (**a**) 5th percentile climate moisture deficit (mm) and (**b**) 5th percentile Tmin (°C) for California trees and shrubs. Taper vs. (**c**) mean climate moisture deficit and (**d**) 5th percentile Tmin for California trees and shrubs.

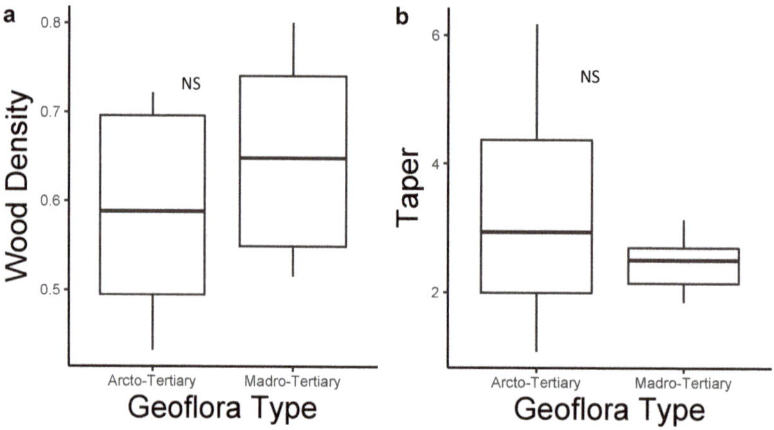

Figure 5. (**a**) Boxplot of wood density values and (**b**) taper (basal diameter to height ratio) values per geoflora type. NS denotes non-significance.

Combining all evidence, aridity of the wet range edge and, for riparian specialists, taper explained almost three quarters of the variation in wood density across this California plant community.

The most parsimonious full model overall included a main effect of 5th percentile CMD and a riparian specialization x taper interaction (Table 3, Figure 6) and had an R^2 of 0.74. On the basis of a decomposition of variance analysis (assessing the R^2 of nested climate and structural models), structure and habitat specialization explained more of the variance than climate, with species 5th percentile CMD explaining 32% of the total variance and the taper × riparian specialization interaction explaining 52% of the variance, with 10% of the variance shared between them. All model effects remained significant and qualitatively similar when controlling for phylogenetic non-independence using a PGLS model [39], though visual inspection of the congeneric species in the dataset suggested that the lack of relationship between taper and wood density for non-riparian species may be driven by the high wood density and high taper values of the four species of oaks in the dataset (Figure S2).

Table 3. Model selection for most parsimonious model overall. WD refers to branch wood density. Taper is basal diameter to height ratio. Habitat refers to whether or not species are riparian specialists. CMD 0.05 is the 5th percentile climate moisture deficit.

Model	Degrees of Freedom	AICc
Null (WD~1)	2	−21.20
WD~CMD 0.05	3	−29.21
WD~CMD 0.05 + taper + habitat + taper/habitat	6	−37.76
WD~CMD 0.05 + taper + taper/habitat	5	−41.74
WD~taper + habitat + taper/habitat	5	−29.49
WD~CMD 0.05 + habitat	4	−36.07
WD~habitat	3	−31.84
WD~taper	3	−22.29

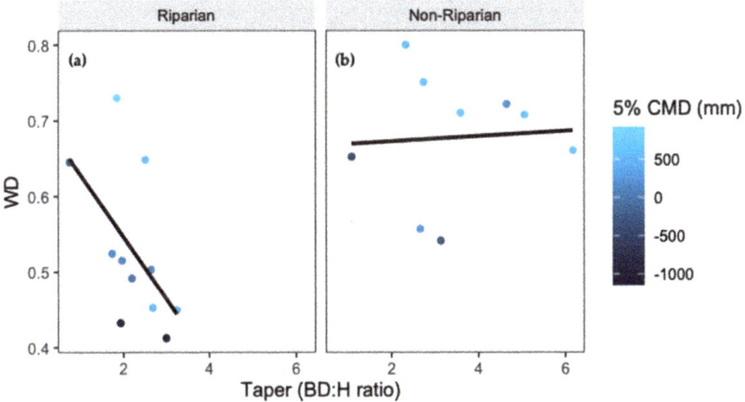

Figure 6. Branch wood density vs basal diameter to height ratio (taper) for (**a**) riparian and (**b**) non-riparian species, colored by the 5th percentile climate moisture deficit shown for each species (drier in lighter blue). Trend lines are only for the wood density (WD)~taper relationships.

4. Discussion

In a Mediterranean woody plant community, we found that canopy allometry was only related to wood density in riparian specialists, that geofloristic provenance had little influence on wood density, and that species climate niche had unexpected relationships with wood density. These results contrast with both global analyses of wood density and site-specific investigations of wood density variation within individual communities. We discuss the implications of our results for the context dependence of the climatic and structural controls on wood density, and the potential for disconnects between small-scale environmental filtering at microsites and large-scale environmental filtering of a regional species pool.

We found that allometry only relates to wood density in riparian contexts. In the California mixed evergreen forest, the relationship between wood density and allometry, however, does not appear as consistent as the wood density–allometry structural equations proposed by Chave et al. (2005) for tropical forests [13]. This lack of consistency may be explained by the differing functional strategies of riparian and non-riparian species. We found that riparian specialists had a negative relationship between wood density and taper consistent with Chave et al. (2005) [13], indicating that species with high wood density are taller at a given diameter than species with low wood density. This result is also consistent with Iida et al. (2012), who found that species with higher wood density have smaller diameters at a given height in Malaysia [16]. However, these findings contrast with those of Aiba & Nakashizuka (2009) as well as King, Davies, Tan, & Noor (2006), who found no relationship between wood density and height for a given diameter in tropical forests [14,15]. Predictions from engineering theory suggest that high wood density does not necessarily increase the mechanical stability of taller trees at a given diameter, which further contrasts with our results [46]. This negative relationship between wood density and taper is consistent with previous work at one of our field sites that showed strong relationships between wood density, allometry, and microsite soil moisture [18,19]. This allometric relationship with wood density for riparian plants may be similar to allometry–wood density relationships in the wet tropics, because both ecosystems are not water-limited.

Differences in allometry–wood density relationships between riparian and non-riparian species may reflect evolutionary processes. Chave et al. (2009) suggested that wood density is related to water transport efficiency [1]. Stronger selection on wood density–allometry relationships may, therefore, occur in wet climates where efficient transportation of water would be advantageous. Non-riparian species, however, lack this wood density–allometry relationship, suggesting that, in water-limited ecosystems, these species are 'overbuilt' for other reasons [47,48]. Riparian forests tend to be more light-limited than immediately adjacent non-riparian forests owing to their closed, tall canopies, so selection may occur in riparian habitats to favor taller, competitive trees, giving wood density more of a structural role in riparian than non-riparian ecosystems, but not in systems with less light competition. The high taper, non-riparian species were all oaks, including species with deep roots that can penetrate groundwater [49,50]. A large dataset covering more sites would be needed to determine if the oaks have a phylogenetic signal.

The California floristic province formed from the confluence of a diverse range of geofloras [29], but this history did not explain the variation in wood density. The Madro-Tertiary geoflora has been associated with the modern chaparral species in California and associated with the development of a Mediterranean climate in California [30]. Species with Madro-Tertiary ancestry did not significantly differ in climate niche than species with temperate Arcto-Tertiary history, suggesting that paleoclimatic influences do not clearly shape modern climate niches in this community. There were neither strong influences of paleoclimatic history on wood density directly nor on the effects of climate or taper on wood density. A more comprehensive study of the traits and climate niches of Madro-Tertiary versus Arcto-Tertiary geofloras could reveal more subtle legacies of their biogeographic origins, but we found little preliminary evidence of geofloristic legacies in this community.

Surprisingly, the wet edge of climate moisture deficit best predicted wood density when compared with other climate variables. This contrasts with meta-analyses conducted at regional and global scales where mean annual temperature and mean annual precipitation of the sampling site explained patterns in wood density across many species [11,20,21] and higher wood density was associated with decreased drought mortality [26]. The positive relationship between wood density and climate moisture deficit is consistent with previous literature, but it is surprising that this trend does not appear for the dry edge. For a Mediterranean plant community, Camarero (2019) found that the relationship between April precipitation and tree growth was the strongest in the species with the highest wood density and that the wet season particularly influenced radial growth rates [6]. In contrast, for a Mediterranean climate in South Africa, xylem density positively correlated with seasonal water stress, suggesting that the dry edge of climate niche was more of an influence [10]. For tropical and subtropical forests, wood density positively related to aridity across sites [22,25,27]. Meanwhile, in a tropical forest, deciduous versus evergreen leaf habit mediated the degree of variation in wood density in response to rainfall [51]. A water-stressed Mediterranean climate may select for lower wood density in wet environments rather than higher wood density in dry environments. Structural and competitive constraints may have more of an influence on wood density in wetter ecosystems, where it is more advantageous to grow tall and fast. Selection for extremely efficient water transport may occur in wetter climates where increased conductivity has an advantageous impact on plant fitness.

In order to better understand landscape-level variation in wood density, a more nuanced approach is required beyond the species-mean climate–wood density correlations used in global analyses. Moreover, examining relationships between climate and wood density alone may overlook the influence of structural constraints and habitat specialization on wood density. Structure influenced wood density only for riparian species, suggesting that the relationship between wood density and allometry is highly context-specific, and that differences in habitat specialization across a landscape should be accounted for. A disconnect often exists between the climate and wood density correlations observed at regional levels and the microsite influences on wood density at localized scales [6,11,18–20]. Our findings caution against the overgeneralization of how structural constraints influence wood density across ecosystems. Further research is needed to explore the complexity of factors, such as climate, allometry, and habitat specialization, that shape wood density variation at a community level. Understanding how these factors drive functional tradeoffs in wood traits can provide more nuanced insights into the biological mechanisms by which these tradeoffs occur. Elucidating the extent to which climate and local habitat explain wood functional traits is relevant in the context of increasing water stress resulting from anthropogenic climate change [26,52,53]. Furthermore, fire greatly shapes the structure and composition of California ecosystems and relates to wood density in tropical forests, yet the extent to which wood density relates to species' responses to fire remains unknown [54]. Finally, an exploration of within-species patterns of wood density and allometry between individuals might elucidate some of the fundamental structural constraints driving the patterns we found in riparian specialists in this comparative study.

5. Conclusions

In summary, we found that a negative relationship occurred between the basal diameter to height ratio and wood density only for riparian habitat specialists. Geofloristic history had little influence on relationships between wood density and climate in our woody plant community. The wet edge of climate moisture deficit further explained variation in wood density. These results suggest a more nuanced approach is required in order disentangle the drivers of wood density variation at a community level.

Supplementary Materials: The following are available online at http://www.mdpi.com/1999-4907/11/1/105/s1, Tables S1–S3 and Figures S1 and S2.

Author Contributions: Conceptualization, R.A.N., E.J.F., J.A.B., W.K.C., and L.D.L.A.; methodology and sampling design and data collection, R.A.N. and L.D.L.A.; analysis, R.A.N. and L.D.L.A.; data curation, R.A.N., W.K.C., and L.D.L.A.; writing—Original draft preparation, R.A.N., E.J.F., and L.D.L.A.; writing—Review and editing, R.A.N., E.J.F., J.A.B., W.K.C., and L.D.L.A.; visualization, R.A.N. and L.D.L.A.; supervision, L.D.L.A.; project administration, R.A.N. and L.D.L.A.; funding acquisition, R.A.N. and L.D.L.A. All authors have read and agreed to the published version of the manuscript.

Funding: This research was funded by a Stanford Vice Provost for Undergraduate Education small grant. L.D.L.A. received funding from the National Science Foundation (DBI-1711243) and a Climate and Global Change postdoctoral fellowship from the National Oceanic and Atmospheric Administration. Any opinions, findings, and conclusions or recommendations expressed in this material are those of the author(s) and do not necessarily reflect the views of the National Science Foundation.

Acknowledgments: Thank you to Rodolfo Dirzo for providing initial advice on this project.

Conflicts of Interest: The authors declare no conflict of interest. The funders had no role in the design of the study; in the collection, analyses, or interpretation of data; in the writing of the manuscript; or in the decision to publish the results.

References

1. Chave, J.; Coomes, D.; Jansen, S.; Lewis, S.L.; Swenson, N.G.; Zanne, A.E. Towards a worldwide wood economics spectrum. *Ecol. Lett.* **2009**, *12*, 351–366. [CrossRef]
2. Ogle, K.; Pathikonda, S.; Sartor, K.; Lichstein, J.W.; Osnas, J.L.D.; Pacala, S.W. A model-based meta-analysis for estimating species-specific wood density and identifying potential sources of variation. *J. Ecol.* **2014**, *102*, 194–208. [CrossRef]
3. Steffenrem, A.; Solheim, H.; Skrøppa, T. Genetic parameters for wood quality traits and resistance to the pathogens Heterobasidion parviporum and Endoconidiophora polonica in a Norway spruce breeding population. *Eur. J. For. Res.* **2016**, *135*, 815–825. [CrossRef]
4. Baker, T.R.; Phillips, O.L.; Malhi, Y.; Almeida, S.; Arroyo, L.; Di Fiore, A.; Erwin, T.; Killeen, T.J.; Laurance, S.G.; Laurance, W.F.; et al. Variation in wood density determines spatial patterns in Amazonian forest biomass. *Glob. Chang. Biol.* **2004**, *10*, 545–562. [CrossRef]
5. Wright, S.J.; Kitajima, K.; Kraft, N.J.B.; Reich, P.B.; Wright, I.J.; Bunker, D.E.; Condit, R.; Dalling, J.W.; Davies, S.J.; Díaz, S.; et al. Functional traits and the growth–mortality trade-off in tropical trees. *Ecology* **2010**, *91*, 3664–3674. [CrossRef]
6. Camarero, J.J. Linking functional traits and climate-growth relationships in Mediterranean species through wood density. *IAWA J.* **2019**, *40*, 215–240. [CrossRef]
7. Sarmiento, C.; Patiño, S.; Timothy Paine, C.E.; Beauchêne, J.; Thibaut, A.; Baraloto, C. Within-individual variation of trunk and branch xylem density in tropical trees. *Am. J. Bot.* **2011**, *98*, 140–149. [CrossRef]
8. Fayolle, A.; Doucet, J.L.; Gillet, J.F.; Bourland, N.; Lejeune, P. Tree allometry in Central Africa: Testing the validity of pantropical multi-species allometric equations for estimating biomass and carbon stocks. *For. Ecol. Manag.* **2013**, *305*, 29–37. [CrossRef]
9. Pratt, R.B.; Jacobsen, A.L.; Ewers, F.W.; Davis, S.D. Relationships among xylem transport, biomechanics and storage in stems and roots of nine Rhamnaceae species of the California chaparral. *New Phytol.* **2007**, *174*, 787–798. [CrossRef]
10. Jacobsen, A.L.; Agenbag, L.; Esler, K.J.; Pratt, R.B.; Ewers, F.W.; Davis, S.D. Xylem density, biomechanics and anatomical traits correlate with water stress in 17 evergreen shrub species of the Mediterranean-type climate region of South Africa. *J. Ecol.* **2007**, *95*, 171–183. [CrossRef]
11. Zhang, S.B.; Slik, J.W.F.; Zhang, J.L.; Cao, K.F. Spatial patterns of wood traits in China are controlled by phylogeny and the environment. *Glob. Ecol. Biogeogr.* **2011**, *20*, 241–250. [CrossRef]
12. Poorter, L.; Lianes, E.; Moreno-de las Heras, M.; Zavala, M.A. Architecture of Iberian canopy tree species in relation to wood density, shade tolerance and climate. *Plant Ecol.* **2012**, *213*, 707–722. [CrossRef]
13. Chave, J.; Andalo, C.; Brown, S.; Cairns, M.A.; Chambers, J.Q.; Eamus, D.; Fölster, H.; Fromard, F.; Higuchi, N.; Kira, T.; et al. Tree allometry and improved estimation of carbon stocks and balance in tropical forests. *Oecologia* **2005**, *145*, 87–99. [CrossRef]
14. King, D.A.; Davies, S.J.; Tan, S.; Noor, N.S.M. The role of wood density and stem support costs in the growth and mortality of tropical trees. *J. Ecol.* **2006**, *94*, 670–680. [CrossRef]
15. Aiba, M.; Nakashizuka, T. Architectural differences associated with adult stature and wood density in 30 temperate tree species. *Funct. Ecol.* **2009**, *23*, 265–273. [CrossRef]
16. Iida, Y.; Poorter, L.; Sterck, F.J.; Kassim, A.R.; Kubo, T.; Potts, M.D.; Kohyama, T.S. Wood density explains architectural differentiation across 145 co-occurring tropical tree species. *Funct. Ecol.* **2012**, *26*, 274–282. [CrossRef]
17. Read, J.; Evans, R.; Sanson, G.D.; Kerr, S.; Jaffré, T. Wood properties and trunk allometry of co-occurring rainforest canopy trees in a cyclone-prone environment. *Am. J. Bot.* **2011**, *98*, 1762–1772. [CrossRef]
18. Preston, K.A.; Cornwell, W.K.; DeNoyer, J.L. Wood density and vessel traits as distinct correlates of ecological strategy in 51 California coast range angiosperms. *New Phytol.* **2006**, *170*, 807–818. [CrossRef]

19. Cornwell, W.K.; Ackerly, D.D. Community assembly and shifts in plant trait distributions across an environmental gradient in coastal California. *Ecol. Monogr.* **2009**, *79*, 109–126. [CrossRef]

20. Swenson, N.G.; Enquist, B.J. Ecological and evolutionary determinants of a key plant functional trait: Wood density and its community-wide variation across latitude and elevation. *Am. J. Bot.* **2007**, *94*, 451–459. [CrossRef]

21. Martínez-Cabrera, H.I.; Jones, C.S.; Espino, S.; Jochen Schenk, H. Wood anatomy and wood density in shrubs: Responses to varying aridity along transcontinental transects. *Am. J. Bot.* **2009**, *96*, 1388–1398. [CrossRef] [PubMed]

22. Ibanez, T.; Chave, J.; Barrabé, L.; Elodie, B.; Boutreux, T.; Trueba, S.; Vandrot, H.; Birnbaum, P. Community variation in wood density along a bioclimatic gradient on a hyper-diverse tropical island. *J. Veg. Sci.* **2017**, *28*, 19–33. [CrossRef]

23. Nabais, C.; Hansen, J.K.; David-Schwartz, R.; Klisz, M.; López, R.; Rozenberg, P. The effect of climate on wood density: What provenance trials tell us? *For. Ecol. Manag.* **2018**, *408*, 148–156. [CrossRef]

24. Hulshof, C.M.; Swenson, N.G.; Weiser, M.D. Tree height-diameter allometry across the United States. *Ecol. Evol.* **2015**, *5*, 1193–1204. [CrossRef] [PubMed]

25. Onoda, Y.; Richards, A.E.; Westoby, M. The relationship between stem biomechanics and wood density is modified by rainfall in 32 Australian woody plant species. *New Phytol.* **2010**, *185*, 493–501. [CrossRef]

26. Greenwood, S.; Ruiz-Benito, P.; Martínez-Vilalta, J.; Lloret, F.; Kitzberger, T.; Allen, C.D.; Fensham, R.; Laughlin, D.C.; Kattge, J.; Bönisch, G.; et al. Tree mortality across biomes is promoted by drought intensity, lower wood density and higher specific leaf area. *Ecol. Lett.* **2017**, *20*, 539–553. [CrossRef]

27. Chave, J.; Muller-Landau, H.C.; Baker, T.R.; Easdale, T.A.; Hans Steege, T.E.R.; Webb, C.O. Regional and phylogenetic variation of wood density across 2456 neotropical tree species. *Ecol. Appl.* **2006**, *16*, 2356–2367. [CrossRef]

28. Ackerly, D.D. Adaptation, niche conservatism, and convergence: Comparative studies of leaf evolution in the California chaparral. *Am. Nat.* **2004**, *163*, 654–671. [CrossRef]

29. Axelrod, D.I. Evolution of the madro-tertiary geoflora. *Bot. Rev.* **1958**, *24*, 433–509. [CrossRef]

30. Axelrod, D.I. Axelrod. *Bull. Geol. Soc. Am.* **1957**, *68*, 19–46. [CrossRef]

31. Soil Survey Staff, Natural Resources Conservation Service, United States Department of Agriculture. Web Soil Survey. Available online: https://websoilsurvey.sc.egov.usda.gov/ (accessed on 16 December 2019).

32. Kershner, B.; Tufts, C.; Nelson, G. *National Wildlife Federation Field Guide to Trees of North America*; Sterling Publishing Company: New York, NY, USA, 2008.

33. Little, E. *The Audubon Society Field Guide to North American Trees: Western Region*; Knopf: New York, NY, USA, 1988.

34. Calscape. Available online: https://calscape.org/ (accessed on 21 December 2019).

35. GBIF Home Page. Available online: GBIF.org.

36. Hijmans, R.J.; Cameron, S.E.; Parra, J.L.; Jones, P.G.; Jarvis, A. Very high resolution interpolated climate surfaces for global land areas. *Int. J. Climatol.* **2005**, *25*, 1965–1978. [CrossRef]

37. Zomer, R.J.; Trabucco, A.; Bossio, D.A.; Verchot, L.V. Climate change mitigation: A spatial analysis of global land suitability for clean development mechanism afforestation and reforestation. *Agric. Ecosyst. Environ.* **2008**, *126*, 67–80. [CrossRef]

38. Barton, K. Package "MuMIn". *R Package Version* **2015**, *1*, 18.

39. Smith, S.A.; Brown, J.W. Constructing a broadly inclusive seed plant phylogeny. *Am. J. Bot.* **2018**, *105*, 302–314. [CrossRef] [PubMed]

40. Slater, G.J.; Harmon, L.J.; Wegmann, D.; Joyce, P.; Revell, L.J.; Alfaro, M.E. Fitting models of continuous trait evolution to incompletely sampled comparative data using approximate Bayesian computation. *Evolution* **2012**, *66*, 752–762. [CrossRef] [PubMed]

41. Paradis, E.; Schliep, K. Ape 5.0: An environment for modern phylogenetics and evolutionary analyses in R. *Bioinformatics* **2018**, *35*, 526–528. [CrossRef]

42. Pennell, M.W.; Eastman, J.M.; Slater, G.J.; Brown, J.W.; Uyeda, J.C.; FitzJohn, R.G.; Alfaro, M.E.; Harmon, L.J. Geiger v2.0: An expanded suite of methods for fitting macroevolutionary models to phylogenetic trees. *Bioinformatics* **2014**, *30*, 2216–2218. [CrossRef]

43. Harmon, L.J.; Weir, J.T.; Brock, C.D.; Glor, R.E.; Challenger, W. GEIGER: Investigating evolutionary radiations. *Bioinformatics* **2007**, *24*, 129–131. [CrossRef]

44. Pinheiro, J.; Bates, D.; DebRoy, S.; Sarkar, D.; Team, R.C. Nlme: Linear and Nonlinear Mixed Effects Models. *R Package Version* **2019**, *3*, 1–140.

45. Baldwin, B.G.; Thornhill, A.H.; Freyman, W.A.; Ackerly, D.D.; Kling, M.M.; Morueta-Holme, N.; Mishler, B.D. Species richness and endemism in the native flora of California. *Am. J. Bot.* **2017**, *104*, 487–501. [CrossRef]

46. Anten, N.P.R.; Schieving, F. The role of wood mass density and mechanical constraints in the economy of tree architecture. *Am. Nat.* **2010**, *175*, 250–260. [CrossRef] [PubMed]

47. Hacke, U.G.; Sperry, J.S.; Pockman, W.T.; Davis, S.D.; McCulloh, K.A. Trends in wood density and structure are linked to prevention of xylem implosion by negative pressure. *Oecologia* **2001**, *126*, 457–461. [CrossRef] [PubMed]

48. Lachenbruch, B.; McCulloh, K.A. Traits, properties, and performance: How woody plants combine hydraulic and mechanical functions in a cell, tissue, or whole plant. *New Phytol.* **2014**, *204*, 747–764. [CrossRef] [PubMed]

49. Mahall, B.E.; Tyler, C.M.; Cole, E.S.; Mata, C. A comparative study of oak (Quercus, Fagaceae) seedling physiology during summer drought in southern California. *Am. J. Bot.* **2009**, *96*, 751–761. [CrossRef]

50. Mclaughlin, B.C.; Zavaleta, E.S. Predicting species responses to climate change: Demography and climate microrefugia in California valley oak (Quercus lobata). *Glob. Chang. Biol.* **2012**, *18*, 2301–2312. [CrossRef]

51. Tarelkin, Y.; Hufkens, K.; Hahn, S.; Van den Bulcke, J.; Bastin, J.F.; Ilondea, B.A.; Debeir, O.; Van Acker, J.; Beeckman, H.; De Cannière, C. Wood anatomy variability under contrasted environmental conditions of common deciduous and evergreen species from central African forests. *Trees* **2019**, *33*, 893–909. [CrossRef]

52. Pretzsch, H.; Biber, P.; Schütze, G.; Kemmerer, J.; Uhl, E. Wood density reduced while wood volume growth accelerated in Central European forests since 1870. *For. Ecol. Manag.* **2018**, *429*, 589–616. [CrossRef]

53. Sapes, G.; Serra-Diaz, J.M.; Lloret, F. Species climatic niche explains drought-induced die-off in a Mediterranean woody community. *Ecosphere* **2017**, *8*, e01833. [CrossRef]

54. Brando, P.M.; Nepstad, D.C.; Balch, J.K.; Bolker, B.; Christman, M.C.; Coe, M.; Putz, F.E. Fire-induced tree mortality in a neotropical forest: The roles of bark traits, tree size, wood density and fire behavior. *Glob. Chang. Biol.* **2012**, *18*, 630–641. [CrossRef]

Article

Tree Growth and Water-Use Efficiency Do Not React in the Short Term to Artificially Increased Nitrogen Deposition

Francesco Giammarchi [1,*], Pietro Panzacchi [1,2], Maurizio Ventura [1] and Giustino Tonon [1]

[1] Faculty of Science and Technology, Free University of Bolzano-Bozen, Piazza Università 5, 39100 Bolzano, Italy; pietro.panzacchi@gmail.com (P.P.); maurizio.ventura@unibz.it (M.V.); giustino.tonon@unibz.it (G.T.)

[2] Centro di Ricerca per le Aree Interne e gli Appennini (ArIA), Università degli Studi del Molise, Via Francesco De Sanctis 1, 86100 Campobasso, Italy

* Correspondence: francesco.giammarchi@unibz.it

Received: 3 December 2019; Accepted: 27 December 2019; Published: 31 December 2019

Abstract: Increasing atmospheric CO_2 concentration and nitrogen deposition are, among the global change related drivers, those playing a major role on forests carbon sequestration potential, affecting both their productivity and water-use efficiency. Up to now, results are however contrasting, showing that the processes underlying them are far from being fully comprehended. In this study, we adopted an innovative approach to simulate the increase of N deposition in a sessile oak forest in North-Eastern Italy, by fertilizing both from above and below the canopy. We observed the dynamics of basal area increment, intrinsic water-use efficiency and of several leaf functional traits over 4 years, to evaluate how the added nitrogen and the two different fertilization system could affect them. We were not able, however, to detect any shift, besides a common yearly variability related to a prevailing background environmental forcing. To this end, we considered as relevant factors both the short time-span of the observation and the relatively low rate of applied nitrogen. Therefore, we stress the importance of long-term, manipulative experiments to improve the understanding of the C sequestration and mitigation ability of forests in response to increased N deposition.

Keywords: Sessile oak; deciduous forest; carbon sequestration

---

1. Introduction

Atmospheric carbon dioxide concentration (Ca) increase and nitrogen (N) depositions are considered among the climate change-related drivers that play a major role on forest productivity and carbon (C) sequestration potential [1,2]. However, evidences on the effect of rising CO_2 levels on forest productivity, as well as on the intrinsic water-use efficiency (iWUE) of trees are still contrasting. In fact, although several authors reported a positive relationship between forest growth and iWUE [3–5] especially in drought prone sites an increase in Ca was not followed by a corresponding increase in tree growth, though an increase in iWUE was observed [6,7]. In the last decades, nitrogen oxides (NO_x) emitted during fuel combustion and ammonia volatilization resulting from intensive agriculture have increased atmospheric N deposition, mostly as NO_3^- and NH_4^+, especially in the Northern Hemisphere [8]. Similarly to CO_2 increase, N deposition is thought to improve forest productivity and C storage [9], as both temperate and boreal forests are considered N-limited ecosystems [10,11]. However, also concerning the effect of atmospheric N deposition on forest ecosystem, contrasting results are reported in the literature. Some long-term experiments showed that chronic soil N addition could have negative effects on forest growth and soil organic matter mineralization due to the N-saturation process [12]. At the same time, other authors have shown an important stimulating effect of N deposition on forest biomass accumulation [13] and consequently on the overall ecosystem carbon (C) storage potential [9,14]. Erisman et al. [15], reviewing a number of studies based on experimental data and model simulations, reported C sequestration rates up to 160 kg of C per kg of N added, with most of the results ranging between 35 and 65 kg C/kg N. These conclusions are also supported by the results of long-term manipulation studies, where N has been added to the soil for periods of more than a decade [16]. In these studies, N deposition was found to exert an effect on the water use efficiency at the tree level by increasing either the leaf area [17] or the photosynthetic capacity [18]. Short-term studies showed different responses to increased N inputs, related both to an enhancement of assimilation rate (A) [19,20] and to a reduction of stomatal conductance (g_s) [21]. Overall, the discrepancies of the available research results suggest that more long-term experimental studies are needed for a better understanding of the response of forests to environmental changes and to elucidate the mechanisms behind the sensitivity of forest productivity to N addition. Furthermore, most of the manipulation experiments performed so far have simulated increase in N depositions providing N fertilization on the forest floor, bypassing the tree canopy [22]. However, it has been shown that tree canopy can absorb a large amount of N from atmospheric deposition [23,24]. The passage through the canopy can determine a change in the chemical form of the N input with deposition [24–26]. Results from previous fertilization experiments may therefore have been biased by the absence of the interaction with the canopy, which should be included in experimental designs if the real impact of N depositions on forest ecosystems has to be assessed. Nitrogen deposition is expected to increase in many regions and has been predicted to almost double globally by 2050 [27]. Therefore, to know the sensitivity of forest ecosystem to N deposition is crucial to guide forest managers' choices in times to come. In this study, we present the first results of a long-term experiment, where N was add to the forest both from above and below the canopy. Specifically, the aim was to assess the short-term effects of the differential N fertilization systems on forest growth, iWUE and on several leaf functional traits.

2. Results

The correlation matrix shows a significant relation between iWUE and leaf area index, (LAI), as well as between LAI and canopy N content and relative leaf N concentration (Table 1). The other parameters do not result to be significantly correlated. No significant correlations are evidenced when the dataset is split by single treatment.

Table 1. Correlation matrix for basal area increment (BAI), intrinsic water use efficiency (iWUE), relative leaf nitrogen content (Nleaves), leaf area index (LAI), leaf mass per area (LMA) and canopy nitrogen content (Ncanopy). For each variable in the first row are the Pearson correlation coefficients, whereas in the second row, in italic, are the significance values ($p < 0.05$). Significant correlations are highlighted in bold characters.

	BAI	iWUE	LAI	LMA	Ncanopy	Nleaves
BAI		0.0333 *0.8472*	0.2139 *0.2104*	−0.0269 *0.8674*	0.2658 *0.1171*	0.1880 *0.2722*
iWUE	0.0333 *0.8472*		−0.3310 ***0.0487***	−0.1426 *0.4068*	−0.2403 *0.1580*	0.3178 *0.0589*
LAI	0.2139 *0.2104*	−0.3310 ***0.0487***		−0.2953 *0.0804*	0.5895 ***0.0002***	−0.0401 *0.8165*
LMA	−0.0269 *0.8674*	−0.1426 *0.4068*	−0.2953 *0.0804*		0.3290 *0.2060*	0.0746 *0.6654*
Ncanopy	0.2658 *0.1171*	−0.2403 *0.1580*	0.5895 ***0.0002***	0.3290 *0.2060*		0.6114 ***0.0001***
Nleaves	0.1880 *0.2722*	0.3178 *0.0589*	−0.0401 *0.8165*	0.0746 *0.6654*	0.6114 ***0.0001***	

Regardless of the fertilization method, the basal area increments obtained from girth tape measurements present a clear inter-annual variability with higher values in 2016 and 2018, which are common to all the plots (Figure 1a). According to the general linear model (GLM) repeated measurements, the effect of the N addition was never significant, whereas the time, i.e., the years, resulted to be a highly significant factor (Table 2). Moreover, no significant interaction between year and treatment was detected for this variable. The same pattern emerges when taking into consideration the iWUE, the leaves N content, as well as the total canopy N content (Figure 1b–f), though with different trends among the single variable.

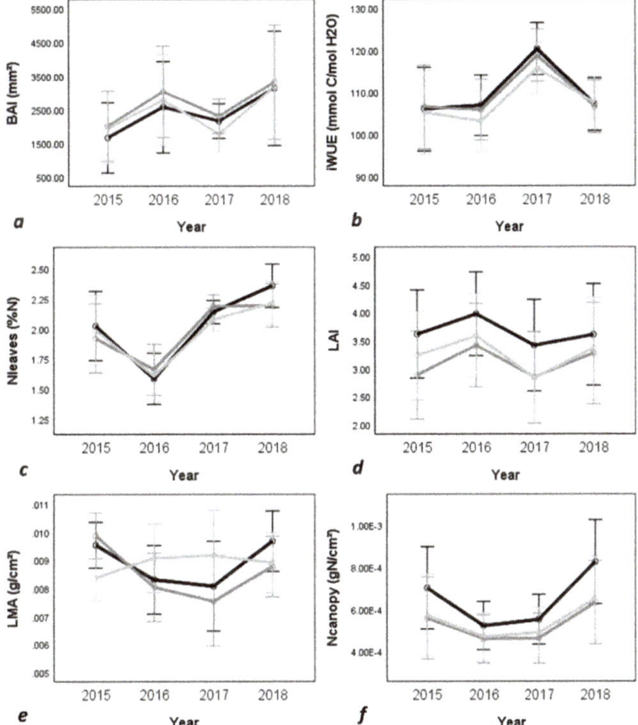

Figure 1. Trend of basal area increment (BAI, **a**), intrinsic water use efficiency (iWUE, **b**), relative leaf nitrogen content (Nleaves, **c**), leaf area index, (LAI, **d**), leaf mass per area (LMA, **e**) and canopy nitrogen content (Ncanopy, **f**), observed in the three different treatments during the 4 years observation period. Data points represent mean values of three plots. SE are shown in vertical bars. Black line = above canopy fertilization; dark grey line = below canopy fertilization; light grey = control).

Table 2. Results for the general linear model (GLM) repeated measures, considering the treatment as between-subject factor and years as within-subject factor. The interaction between the two is also shown. Significant statistics ($p < 0.05$) are highlighted in bold characters. Variables are: basal area increment (BAI), intrinsic water use efficiency (iWUE), relative leaf nitrogen content (Nleaves), leaf area index (LAI), leaf mass per area (LMA) and canopy nitrogen content (Ncanopy).

	Variable	F	p
Treatment	BAI	0.119	0.890
	iWUE	0.228	0.803
	Nleaves	0.180	0.839
	LAI	1.936	0.225
	LMA	2.060	0.208
	Ncanopy	2.060	0.208
Year	BAI	16.985	**0.000**
	iWUE	21.986	**0.000**
	Nleaves	62.191	**0.000**
	LAI	2.867	0.065
	LMA	2.925	0.062
	Ncanopy	11.638	**0.000**
Year × Treatment	BAI	0.358	0.896
	iWUE	0.310	0.923
	Nleaves	1.097	0.401
	LAI	0.156	0.985
	LMA	2.324	0.077
	Ncanopy	0.406	0.865

The iWUE displayed a particularly high value in 2017. The leaf N concentration is decreasing in 2016 and then constantly increasing in 2017 and 2018, whereas canopy N content decreased in 2016, remained constant in 2017 and increased in 2018. Values of LAI were slightly higher in 2016, while LMA was decreasing in 2016 and 2017 and increasing in 2018 in the treated plots, but remained more or less constant in the control plots throughout the whole period. However, for both LAI and LMA, the effect of year was not significant, as well as that of the treatment and the interaction between time and treatment (Table 2).

3. Discussion

The aim of this study was to evaluate the short-term effects of N addition by using two different N fertilization systems, above and below the canopy, on tree growth, intrinsic water-use efficiency at leaf level and on several leaves functional traits. To our knowledge, very few experimental studies explored the effects of N depositions on forest ecosystems through canopy fertilization [24,28,29]. Even less did it by comparing N fertilizer addition on the ground and above the canopy [22], or by investigating the iWUE changes after N fertilizations through stable isotope analyses [30]. Most probably, the reasons behind this deficiency are the costs, effort required and intrinsic difficulties of such an approach. However, this leads to a crucial lack of knowledge regarding the way that N depositions naturally occur. In fact, the amount of N intercepted by the canopy can represent a significant part of the overall N reaching the ecosystem. For instance, Gaige et al. [24] showed that, in a North-American coniferous forest, the canopy layer retains from 57% to 75% of the NO_3^- and from 73% to 83% of the NH_4^+. Thus, neglecting this portion of the overall N cycle can lead to relevant biases in the understanding of the forest ecosystem functioning.

Depending on tree species and original nutritional status, nitrogen deposition was found to affect plant transpiration and stomata conductance on leaf, tree and stand scales influencing in turn net carbon assimilation and climate change mitigation potential [31]. However, the existing studies reported often contrasting results and the mechanisms that are behind the response of forests to N depositions in terms of water balance and C gain are still matter of debate [15,32]. The balance between C gain and water loss by transpiration is well represents by the iWUE, being the ratio of assimilation rate to stomatal conductance. In this context, N deposition was found to increase iWUE in *Quercus velutina* and *Populus* × *euroamericana* [19,30] to have no effect in *Fagus sylvatica* L. and *Pinus sylvestris* L. [21,33], and to decrease iWUE of phosphorus-limited tropical forests [34]. N directly reaching the canopy is readily available for uptake by the leaves, particularly in the form of NH_4^+ [24] and its assimilation is expected to increase leaf N content, affecting photosynthesis [19] and ultimately the C sequestration potential. Therefore, the magnitude of the interception might have a greater impact on the C cycle and on the potential of forest ecosystems to act as carbon sink [9]. In fact, photosynthetic rate was found to be strictly correlated to leaf N content in many studies and generally associated with an increased rubisco activity [35]. Since it has been observed that the canopy reacts faster to N additions [20], in our study we were expecting to find already in the short-term a divergent response of the N_{AB} treated plots when compared to the N_{BL} plots. However, no clear effect of the treatment was found and only a temporal trend was detected, regardless of the treatment, which likely points to a still prevailing role of the local conditions in determining the variations of the investigated parameters. For instance, we observed a common and consistent trend of iWUE for all the treatments that could be ascribed to local climatic variability, while the influence of the N addition on the carboxylation [35], and hence on the *A* component of the iWUE, could still be masked by the former. The magnitude of the trees' reaction to N fertilization can be mostly ascribed to the stand species mixture and age, the background deposition as well as the amount and duration of N addition [36]. In this study, the absence of evident effects might be related to the time needed for an ecosystem such as an oak adult high-forest to shift its N cycle in response to added N, applied either above or below canopy. Among the studies that assessed the relationship between N, C and water cycles, Jennings et al. [30] investigated the behavior of both iWUE and tree growth on a 60 years old black oak forest following long-term manipulative fertilizations, and showed that there was a positive effect on both. Similar findings were also obtained by Cornejo-Oviedo et al. [37] on a 20 years old Douglas fir plantation. Here, N fertilization had a positive impact on iWUE, the increase of which promoted tree growth. The results of these studies, however, were gathered respectively after 23 years and after 7 years since the start of the experiment, whereas the time span of our study consists of only 4 years of observations after the beginning of the treatments. On the other hand, Guerrieri et al. [20] found a positive response of the iWUE in a Sitka spruce plantation already within 5 years. Nevertheless, this suggests that both the age of the stand and the ecology of the studied species might play a significant role. Our forest is, in fact, similar to the former case, being all the stands at around 80 years of age, whereas the latter two are young plantations of fast-growing species, in which the elevated N inputs applied (224 kg N/ha in the Douglas fir stand) likely triggered a much faster response. At this point, a second issue must be taken into consideration, i.e., the amount of added N. Though in this study this is not low if compared to the background N deposition levels in the Monticolo site, it could be a further possible explanation of the lack of a short-term response, especially when compared to similar manipulative experiments. Different studies reported that only high levels of N addition would result in higher leaf N content [38–40], with different species and forests being insensitive to low inputs of N deposition [41–43]. In line with the above-mentioned studies, both leaf N content and total canopy N content were unaffected by the external N addition in our experiment. This result helps to explain the absence of physiological response to nitrogen treatments in terms of both WUE and growth rate that we observed in our study. Besides the above-mentioned research of Cornejo-Oviedo et al. [37], other studies focusing on either below canopy fertilization or on both above and below addition normally applied greater quantities of N. For instance, Sheppard et al. [44] applied N at two different rates (48 and 96 Kg N ha^{-1} yr^{-1}); Jennings et al. [30] applied 50 and 150 kg N ha^{-1} yr^{-1} and Zhang et al. [22] 25 and 50 kg N ha^{-1} yr^{-1}. Only Gaige et al. [24] applied a quantity similar to the one added in this study (18–20 kg N ha^{-1} yr^{-1}), though for a longer period than in our experiment. This suggests that 4 years is still a short time to allow observing a clear reaction of the studied forest to the N addition, regardless of the fertilization system. To this end, our results show that the leaves N% was changing in a common way among the treatments almost all the time, with the N_{AB} showing a higher value only in 2018, though not significant. Moreover, not overly elevated N addition rates and a less fast responding forest ecosystem might have easily interacted and, combined, contributed to the observed lack of effects on both tree growth, iWUE and the leaves functional traits.

4. Material and Methods

4.1. Study Site and Experimental Design

The study area is located in Monticolo (46°25′35″ N; 11°17′55″ E), in the Autonomous Province of Bolzano, Italy, at about 550 m above sea level (a.s.l.). The forest stand in which the plots have been established is composed of Sessile oak (*Quercus petraea* L.), up to 97%. Minor species are Scots pine (*Pinus sylvestris* L.), sweet chestnut (*Castanea sativa* Mill.) and several others occurring sporadically, such as lime (*Tilia cordata* Mill.), European hop-hornbeam (*Ostrya carpinifolia* Scop.) and silver birch (*Betula pendula* Roth). The main biometric characteristics of the forest stand at the beginning of the experiment are reported in Table 3.

Table 3. Species mixture, total basal area and stand characteristics of the forest stand at the beginning of the experiment. Stand characteristics referred to sessile oak population only (g_m: average basal area; d_{mg}: diameter at average basal area; h_{mg}: height at average basal area diameter; Hd: dominant height (average height of the tallest 100 trees); V: standing volume).

	N Trees		G (Basal Area)		Stand Characteristics	
	N/ha	%	m²/ha	%		
Quercus petraea	1125	95.62%	26.969	95.31%	g_m (m²/ha)	0.02
Tilia cordata	20	1.67%	0.260	0.92%	d_{mg} (cm)	17.50
Castanea sativa	2	0.21%	0.005	0.02%	H_{mg} (m)	13.37
Ostrya carpinifolia	12	1.04%	0.147	0.52%	Hd (m)	16.40
Pinus sylvestris	15	1.25%	0.786	2.78%	V (m³/ha)	215.47
Acer spp.	2	0.21%	0.001	0.01%		
Total	1177		28.168			

The forest lies on an acid brown soil originating from porphyritic quartz rock. Average annual temperature at the site is 11.4 °C, whereas average annual precipitation is 800 mm [45]. The experimental design consisted of a set of nine circular plots (12 m radius), three for each different treatment: unfertilized control plot, fertilization by adding $NH_4^+NO_3^-$ directly to the ground (i.e., traditional fertilization, N_{BL}) and fertilization by adding $NH_4^+NO_3^-$ to the canopy by aerial mist spraying (i.e., aerial fertilization, N_{AB}; Figure 2). The nine plots were arranged in a completely randomized design and were established close to each other, in an area of about 200 m of length, but with at least 10 m of distance between each plot. This was chosen in order to minimize the variation of stand conditions while avoiding issues related to drifting of the fertilizer during applications. The $NH_4^+NO_3^-$ solution (4.3 g N/L) was applied at ground or canopy level (at 15–18 m height, depending on plot maximum tree height) by aerial mist spraying, resulting in a deposition of 20 kg ha^{-1} yr^{-1}, which was more than three times higher than background atmospheric N deposition [45]. A gasoline powered pump (Officine Carpi S.R.L., Poviglio, Reggio Emilia, Italy) was used to ensure the needed pressure. The pump was then connected, for aerial fertilization, to rotating sprinklers (Rain Bird SNC, Aix-en-Provence, France) mounted on telescopic posts (Fireco S.R.L., Gussago, Brescia, Italy) installed at the center of the N_{ab} plots. These sprinklers provided a spray radius of 12 m, covering the whole plot when operating at a pressure of 2 bars. Tests on the uniformity of the quantity of water provided in the covered area were performed before installation. For N_{BL} plots, the $NH_4^+NO_3^-$ solution was applied manually with a water hose and a spray nozzle, paying attention to uniformly distribute the solution in the treated area. The fertilization was provided in five application dates, monthly from May to September starting from May 2015. The amount of water provided in each plot with fertilization (210 L H$_2$O yr^{-1}) is equivalent to a precipitation of 0.46 mm for the season, which is negligible if compared to the average annual precipitation of the region.

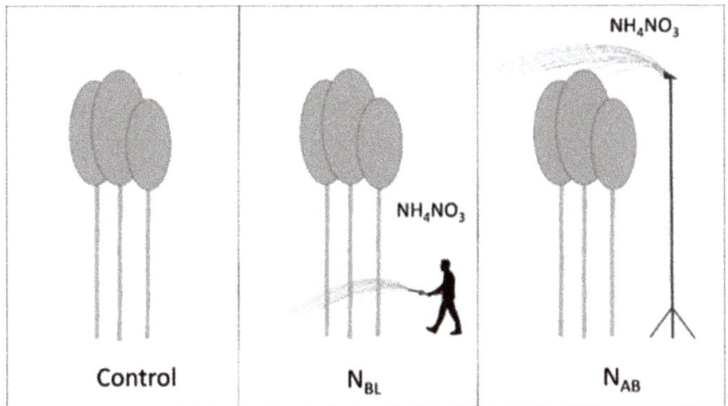

Figure 2. Illustrated scheme of the fertilization system adopted in this study. N_{BL} = below canopy fertilization; N_{AB} = above canopy fertilization; control = no fertilization.

4.2. Tree Growth

At the beginning of the experiment, in 2014, we measured all diameters of each tree within the plots using a tree caliper, with a threshold of 5 cm at breast height (i.e., DBH, 1.30 m). Successively, we placed permanent girth tapes D1 (UMS Gbh, München, DE, Germany) on a subsample of 10 trees per each plot, stratified according to the relative distribution of trees within the DBH classes, in order to obtain measurements as representative as possible of the whole stand growth (see Table S1 for information about number of trees, mean diameter and coefficient of variation for each plot). We then read the girth tapes values every year before the start of the growing season (January or February), from 2015 to 2019. The obtained radial increment values were then transformed in basal area increment (BAI) values, referred both to the plot surface and to the hectare.

4.3. iWUE and Leaves Parameters Data Collection

Discrimination against the heavier C stable isotope occurs during the photosynthetic process. Therefore, plant tissues are generally depleted in ^{13}C if compared to the atmospheric CO_2 and the ^{13}C isotopic signature (δ^{13}C) in leaves reflects changes in the c_i/c_a ratio, which is the ratio between CO_2 concentration in the leaf intercellular spaces (c_i) and in the atmosphere (c_a). These are linked to changes in both A and gs. This relationship is described in the simplified Farquhar equation as follows [46]:

$$\delta^{13}C_p = \delta^{13}C_a - a - (b - a)\,(c_i/c_a), \tag{1}$$

where $\delta^{13}C_a$ is the isotopic signature of atmospheric CO_2, a is the fractionation for $^{13}CO_2$ during diffusion through air (4.4‰) and b is the fractionation occurring during carboxylation (27‰) by the Rubisco enzyme. From this equation, it is possible to derive c_i as follows:

$$c_i = c_a\,((\delta^{13}C_a - \delta^{13}C_p - a)/(b - a)). \tag{2}$$

Therefore, given the values of CO_2 atmospheric concentration and those of atmospheric and leaf δ^{13}C, it is possible to estimate the intrinsic water-use efficiency (iWUE; A/g_s) by the following equation:

$$iWUE = A/g_s = (c_a - c_i)/1.6 = (c_a - (c_a\,(\delta^{13}C_a - \delta^{13}C_p - a)/(b - a)))/1.6. \tag{3}$$

Oak leaves were sampled in July of four consecutive years (2015–2018) for stable isotope analysis and both relative N concentration and leaf mass per area (LMA) determination. We randomly sampled 30 fresh leaves in each plot, taken from several trees and only from the part of the canopies directly exposed to light radiation through telescopic pruning shears. Immediately after the sampling, we weighted each single fresh leaf and measured the respective leaf area through a LI-COR 3000C (LI-COR, Lincoln, NE, USA), in order to determine the LMA. Afterwards, the leaves were oven dried overnight at 60 °C, pooled for each plot and milled. The obtained powder was then used for the $\delta^{13}C$ isotopic analyses. The analysis was done by using an EA elemental analyzer (Flash 2000, Thermo Scientific, Waltham, MA, USA) connected to an isotope ratio mass spectrometer (Delta V Advantage, Thermo Scientific, Waltham, MA, USA) via a continuous flow interface (ConFlo IV, Thermo Scientific, Waltham, MA, USA). Isotope ratios were expressed as permil δ notation ($\delta = (R_{sample}/R_{standard}) - 1 \times 1000$) relative to VPDB international standards for ^{13}C. Moreover, $\delta^{13}C$ values were corrected for the "Suess effect" [47], i.e., the decrease of atmospheric $\delta^{13}C$ due to emissions of ^{13}C depleted carbon dioxide since the onset of industrialization. Data for correction were retrieved from NOAA mean annual CO_2 data (www.esrl.noaa.gov/gmd/ccgg/trends). iWUE values were calculated using Equation (3). Together with the isotope signature of leaves, also N content was measured.

The total leaf area index (LAI) per plot was determined from litterfall quantification, by means of three litter traps with a collecting area of 1590 cm^2 each, placed in every plot and surveyed periodically each year from November to January. Eventually, by multiplying the N concentration by LMA and LAI we obtained the canopy total N content (g N/cm^2).

4.4. Data Analysis

The results obtained in terms of tree growth (BAI), iWUE and leaves functional traits (leaves N content, LMA, LAI and canopy total N) were related through a correlation analysis (IBM SPSS 25, IBM, Armonk, NY, USA) by pooling all the data together and by splitting the data-set by N addition treatment (N_{AB}, N_{BL} and N_{CTRL}), including all sampled years. To understand at what extent the different treatments affected the changes of the considered features over time, we applied a general linear model (GLM) repeated measures procedure, using IBM SPSS 25 software (IBM, Armonk, NY, USA), including all the considered variables, with the sampling years as within-subject factor and the fertilization treatment as between-subject factors, with a significant level of 0.05.

5. Conclusions

After 4 years of additional N supply, both above and below the canopy, we were not able to detect an effect on tree growth, iWUE and some leaves functional traits. On the other hand, we observed a common pattern for almost all the investigated parameters, hinting to a still prevailing background environmental forcing on the studied stand, rather than that of the simulated N deposition. Among the factors responsible for the lack of response, we considered as relevant both the short time-span of the observation and the relatively low rate of N applied, particularly in relation to the forest species and age. However, given the crucial role of N depositions in affecting the C sequestration potential of temperate forests and their response to increased drought, also the long-term effects of the experiment should be investigated in the future, to further understand the interactions of the above-mentioned factors and the overall forest climate change mitigation potential.

Supplementary Materials: The following are available online at http://www.mdpi.com/1999-4907/11/1/47/s1; Table S1: Number of trees, mean diameter and mean diameter coefficient of variation (CV) for each plot at the time of girth tapes placement in 2014.

Author Contributions: P.P., M.V. and G.T. conceived the experiment. F.G., P.P., M.V. and G.T. performed the experiment. F.G., P.P., M.V. and G.T. analyzed the data. F.G. drafted the manuscript, F.G., P.P., M.V. and G.T. reviewed the manuscript. All authors have read and agreed to the published version of the manuscript.

Funding: This research was funded by the Free University of Bolzano, through NITROFOR (grant number 141J12000820005) and DECANITRO (grant number I52F15000170005) research projects.

Acknowledgments: We would like to thank Stefano Minerbi of the Bolzano Province Forest Service for assistance in the selection of the study site as well as Joel Towoua and Lorenzo Panizzon for fieldwork assistance.

Conflicts of Interest: The authors declare no conflict of interest.

References

1. Hyvönen, R.; Agren, G.I.; Linder, S.; Persson, T.; Cotrufo, M.F.; Ekblad, A.; Freeman, M.; Grelle, A.; Janssens, I.A.; Jarvis, P.G.; et al. The likely impact of elevated [CO2], nitrogen deposition, increased temperature and management on carbon sequestration in temperate and boreal forest ecosystems: A literature review. *New Phytol.* **2007**, *173*, 463–480. [CrossRef] [PubMed]

2. Core Writing Team; Pachauri, R.K.; Meyer, L.A. (Eds.) *Climate Change 2014: Synthesis Report. Contribution of Working Groups I, II and III to the Fifth Assessment Report of the Intergovernmental Panel on Climate Change*; IPCC: Geneva, Switzerland, 2014; p. 151.

3. Gagen, M.; Finsinger, W.; Wagner-Cremer, F.; Mccarroll, D.; Loader, N.J.; Robertson, I.; Jalkanen, R.; Young, G.; Kirchhefer, A. Evidence of changing intrinsic water-use efficiency under rising atmospheric CO2 concentrations in Boreal Fennoscandia from subfossil leaves and tree ring δ13C ratios. *Glob. Chang. Biol.* **2011**, *17*, 1064–1072. [CrossRef]

4. Leonardi, S.; Gentilesca, T.; Guerrieri, R.; Ripullone, F.; Magnani, F.; Mencuccini, M.; Noije, T.V.; Borghetti, M. Assessing the effects of nitrogen deposition and climate on carbon isotope discrimination and intrinsic water-use efficiency of angiosperm and conifer trees under rising CO 2 conditions. *Glob. Chang. Biol.* **2012**, *18*, 2925–2944. [CrossRef] [PubMed]

5. Giammarchi, F.; Cherubini, P.; Pretzsch, H.; Tonon, G. The increase of atmospheric CO2affects growth potential and intrinsic water-use efficiency of Norway spruce forests: Insights from a multi-stable isotope analysis in tree rings of two Alpine chronosequences. *Trees-Struct. Funct.* **2017**, *31*, 503–515. [CrossRef]

6. Andreu-Hayles, L.; Planells, O.; Gutiérrez, E.; Muntan, E.; Helle, G.; Anchukaitis, K.J.; Schleser, G.H. Long tree-ring chronologies reveal 20th century increases in water-use efficiency but no enhancement of tree growth at five Iberian pine forests. *Glob. Chang. Biol.* **2011**, *17*, 2095–2112. [CrossRef]

7. Peñuelas, J.; Canadell, J.G.; Ogaya, R. Increased water-use efficiency during the 20th century did not translate into enhanced tree growth. *Glob. Ecol. Biogeogr.* **2011**, *20*, 597–608. [CrossRef]

8. Galloway, J.N. The global nitrogen cycle: Past, present and future. *Sci. China Ser. C Life Sci. Chin. Acad. Sci.* **2005**, *48*, 669–677.

9. Magnani, F.; Mencuccini, M.; Borghetti, M.; Berbigier, P.; Berninger, F.; Delzon, S.; Grelle, A.; Hari, P.; Jarvis, P.G.; Kolari, P.; et al. The human footprint in the carbon cycle of temperate and boreal forests. *Nature* **2007**, *447*, 848–850. [CrossRef]

10. Cleveland, C.C.; Townsend, A.R.; Schimel, D.S.; Fisher, H.; Howarth, R.W.; Hedin, L.O.; Perakis, S.S.; Latty, E.F.; Von Fischer, J.C.; Hlseroad, A.; et al. Global patterns of terrestrial biological nitrogen (N2) fixation in natural ecosystems. *Glob. Biogeochem. Cycles* **1999**, *13*, 623–645. [CrossRef]

11. LeBauer, D.S.; Treseder, K.K. Nitrogen Limitation of Net Primary Productivity in Terrestrial Ecosystems Is Globally Distributed. *Ecology* **2008**. [CrossRef]

12. Aber, J.; McDowell, W.; Nadelhoffer, K.; Magill, A.; Berntson, G.; Kamakea, M.; McNulty, S.; Currie, W.; Rustad, L.; Fernandez, I. Nitrogen Saturation in Temperate Forest Ecosystems. *Bioscience* **1998**, *48*, 921–934. [CrossRef]

13. Townsend, A.R.; Braswell, B.H.; Holland, E.A.; Penner, J.E. Spatial and Temporal Patterns in Terrestrial Carbon Storage Due to Deposition of Fossil Fuel Nitrogen. *Ecol. Appl.* **1996**, *6*, 806–814. [CrossRef]

14. Sutton, M.A.; Simpson, D.; Levy, P.E.; Smith, R.I.; Reis, S.; van Oijen, M.; de Vries, W. Uncertainties in the relationship between atmospheric nitrogen deposition and forest carbon sequestration. *Glob. Chang. Biol.* **2008**, *14*, 2057–2063. [CrossRef]

15. Erisman, J.W.; Van Grinsven, H.; Grizzetti, B.; Bouraoui, F.; Powlson, D.; Sutton, M.A.; Bleeker, A.; Reis, S. The European nitrogen problem in a global perspective Executive summary Major uncertainties/challenges. In *The European Nitrogen Assessment*; Sutton, M.A., Britton, C., Erisman, J.W., Billen, G., Bleeker, A., Greenfelt, P., van Grinsven, H., Grizzetti, B., Eds.; Cambridge University Press: Cambridge, UK, 2011; pp. 9–31.

16. Hyvönen, R.; Persson, T.; Andersson, S.; Olsson, B.; Ågren, G.I.; Linder, S. Impact of long-term nitrogen addition on carbon stocks in trees and soils in northern Europe. *Biogeochemistry* **2008**, *89*, 121–137. [CrossRef]

17. Ewers, B.E.; Oren, R.; Sperry, J.S. Influence of nutrient versus water supply on hydraulic architecture and water balance in Pinus taeda. *Plant Cell Environ.* **2000**, *23*, 1055–1066. [CrossRef]

18. Lambers, H.; Chapin, F.S., III; Pons, T. *Plant Physiological Ecology*; Springer: Berlin/Heidelberg, Germany, 2008.

19. Ripullone, F.; Lauteri, M.; Grassi, G.; Amato, M.; Borghetti, M. Variation in nitrogen supply changes water-use efficiency of Pseudotsuga menziesii and Populus x euroamericana; a comparison of three approaches to determine water-use efficiency. *Tree Physiol.* **2004**, *24*, 671–679. [CrossRef]

20. Guerrieri, R.; Mencuccini, M.; Sheppard, L.J.; Saurer, M.; Perks, M.P.; Levy, P.; Sutton, M.A.; Borghetti, M.; Grace, J. The legacy of enhanced N and S deposition as revealed by the combined analysis of $\delta13C$, $\delta18O$ and $\delta15N$ in tree rings. *Glob. Chang. Biol.* **2011**, *17*, 1946–1962. [CrossRef]

21. Betson, N.R.; Johannisson, C.; Löfvenius, M.O.; Grip, H.; Granström, A.; Högberg, P. Variation in the $\delta13C$ of foliage of *Pinus sylvestris* L. in relation to climate and additions of nitrogen: Analysis of a 32-year chronology. *Glob. Chang. Biol.* **2007**, *13*, 2317–2328. [CrossRef]

22. Zhang, W.; Yan, J.; Zhang, W.; Shen, W.; Zhu, S.; Cai, X.; Liu, Z.; Wang, F.; Rao, X.; Mo, J.; et al. CAN Canopy Addition of Nitrogen Better Illustrate the Effect of Atmospheric Nitrogen Deposition on Forest Ecosystem? *Sci. Rep.* **2015**, *5*, 1–12. [CrossRef]

23. Sievering, H.; Tomaszewski, T.; Torizzo, J. Canopy uptake of atmospheric N deposition at a conifer forest: Part I-canopy N budget, photosynthetic efficiency and net ecosystem exchange. *Tellus Ser. B Chem. Phys. Meteorol.* **2007**, *59*, 483–492. [CrossRef]

24. Gaige, E.; Dail, D.B.; Hollinger, D.Y.; Davidson, E.A.; Fernandez, I.J.; Sievering, H.; White, A.; Halteman, W. Changes in canopy processes following whole-forest canopy nitrogen fertilization of a mature spruce-hemlock forest. *Ecosystems* **2007**, *10*, 1133–1147. [CrossRef]

25. Guerrieri, R.; Vanguelova, E.I.; Michalski, G.; Heaton, T.H.E.; Mencuccini, M. Isotopic evidence for the occurrence of biological nitrification and nitrogen deposition processing in forest canopies. *Glob. Chang. Biol.* **2015**, *21*, 4613–4626. [CrossRef] [PubMed]

26. Houle, D.; Marty, C.; Duchesne, L. Response of canopy nitrogen uptake to a rapid decrease in bulk nitrate deposition in two eastern Canadian boreal forests. *Oecologia* **2015**, *177*, 29–37. [CrossRef] [PubMed]

27. Galloway, J.N.; Dentener, F.J.; Capone, D.G.; Boyer, E.W.; Howarth, R.W.; Seitzinger, S.P.; Asner, G.P.; Cleveland, C.C.; Green, P.A.; Holland, E.A.; et al. Nitrogen Cycles: Past, Present, and Future. *Biogeochemistry* **2004**, *70*, 153–226. [CrossRef]

28. Cape, J.N.; Sheppard, L.J.; Crossley, A.; Van Dijk, N.; Tang, Y.S. Experimental field estimation of organic nitrogen formation in tree canopies. *Environ. Pollut.* **2010**, *158*, 2926–2933. [CrossRef] [PubMed]

29. Gilliam, F.S.; Adams, M.B. Effects of Nitrogen on Temporal and Spatial Patterns of Nitrate in Streams and Soil Solution of a Central Hardwood Forest. *ISRN Ecol.* **2011**, *2011*, 1–9. [CrossRef]

30. Jennings, K.A.; Guerrieri, R.; Vadeboncoeur, M.A.; Asbjornsen, H. Response of Quercus velutina growth and water use efficiency to climate variability and nitrogen fertilization in a temperate deciduous forest in the northeastern USA. *Tree Physiol.* **2016**, *36*, 428–443. [CrossRef]

31. Fleischer, K.; Rebel, K.T.; Van Der Molen, M.K.; Erisman, J.W.; Wassen, M.J.; Van Loon, E.E.; Montagnani, L.; Gough, C.M.; Herbst, M.; Janssens, I.A.; et al. The contribution of nitrogen deposition to the photosynthetic capacity of forests. *Global Biogeochem. Cycles* **2013**, *27*, 187–199. [CrossRef]

32. Bobbink, A.R.; Hicks, K.; Galloway, J.; Spranger, T.; Alkemade, R.; Ashmore, M.; Cinderby, S.; Davidson, E.; Dentener, F.; Emmett, B.; et al. Global assessment of nitrogen deposition effects on terrestrial plant diversity: A synthesis. *Ecol. Appl.* **2010**, *20*, 30–59. [CrossRef]

33. Elhani, S.; Guehl, J.M.; Nys, C.; Picard, J.F.; Dupouey, J.L. Impact of fertilization on tree-ring $\delta15N$ and $\delta13C$ in beech stands: A retrospective analysis. *Tree Physiol.* **2005**, *25*, 1437–1446. [CrossRef]

34. Huang, Z.; Liu, B.; Davis, M.; Sardans, J.; Peñuelas, J.; Billings, S. Long-term nitrogen deposition linked to reduced water use efficiency in forests with low phosphorus availability. *New Phytol.* **2016**, *210*, 431–442. [CrossRef] [PubMed]

35. Kattge, J.; Knorr, W.; Raddatz, T.; Wirth, C. Quantifying photosynthetic capacity and its relationship to leaf nitrogen content for global-scale terrestrial biosphere models. *Glob. Chang. Biol.* **2009**, *15*, 976–991. [CrossRef]

36. Pardo, L.H.; Robin-Abbott, M.J.; Driscoll, C.T. Assessment of nitrogen deposition effects and empirical critical loads of nitrogen for ecoregions of the United States. *Gen. Tech. Rep. NRS* **2011**, *80*, 291.

37. Cornejo-Oviedo, E.H.; Voelker, S.L.; Mainwaring, D.B.; Maguire, D.A.; Meinzer, F.C.; Brooks, J.R. Basal area growth, carbon isotope discrimination, and intrinsic water use efficiency after fertilization of Douglas-fir in the Oregon Coast Range. *For. Ecol. Manag.* **2017**, *389*, 285–295. [CrossRef]
38. Magill, A.H.; Aber, J.D.; Berntson, G.M.; McDowell, W.H.; Nadelhoffer, K.J.; Melillo, J.M.; Steudler, P. Long-term nitrogen additions and nitrogen saturation in two temperate forests. *Ecosystems* **2000**, *3*, 238–253. [CrossRef]
39. Pitcairn, C.E.R.; Leith, I.; Fowler, D.; Hargreaves, K.; Moghaddam, M.; Kennedy, V.; Granat, L. Foliar Nitrogen as an Indicator of Nitrogen Deposition and Critical Loads Exceedance on a European Scale. *Water Air Soil Pollut.* **2001**, *130*, 1037–1042. [CrossRef]
40. Talhelm, A.F.; Pregitzer, K.S.; Burton, A.J. No evidence that chronic nitrogen additions increase photosynthesis in mature sugar maple forests. *Ecol. Appl.* **2011**, *21*, 2413–2424. [CrossRef]
41. Aber, J.D.; Goodale, C.L.; Ollinger, S.V.; Smith, M.-L.; Magill, A.H.; Martin, M.E.; Hallett, R.A.; Stoddard, J.L. Is Nitrogen Deposition Altering the Nitrogen Status of Northeastern Forests? *Bioscience* **2003**, *53*, 375. [CrossRef]
42. Bauer, G.A.; Bazzaz, F.A.; Minocha, R.; Long, S.; Magill, A.; Aber, J.; Berntson, G.M. Effects of chronic N additions on tissue chemistry, photosynthetic capacity, and carbon sequestration potential of a red pine (Pinus resinosa Ait.) stand in the NE United States. *For. Ecol. Manag.* **2004**, *196*, 173–186. [CrossRef]
43. Hu, Y.; Zhao, P.; Zhu, L.; Zhao, X.; Ni, G.; Ouyang, L.; Schäfer, K.V.R.; Shen, W. Responses of sap flux and intrinsic water use efficiency to canopy and understory nitrogen addition in a temperate broadleaved deciduous forest. *Sci. Total Environ.* **2019**, *648*, 325–336. [CrossRef]
44. Sheppard, L.J.; Coward, P.; Skiba, U.; Harvey, F.J.; Crossley, A.; Ingleby, K. Effects of five years of frequent N additions, with or without acidity, on the growth and below-ground dynamics of a young Sitka spruce stand growing on an acid peat: Implications for sustainability. *Hydrol. Earth Syst. Sci.* **2010**, *8*, 377–391. [CrossRef]
45. Marchetti, F.; Tait, D.; Ambrosi, P.; Minerbi, S.; Agrario, I.; Michele, S.; Michele, S.; Tn, A. Atmospheric deposition at four forestry sites in the Alpine Region of Trentino-South Tyrol, Italy. *J. Limnol.* **2002**, *61*, 148–157. [CrossRef]
46. Farquhar, G. Carbon Isotope Discrimination And Photosynthesis. *Annu. Rev. Plant Physiol. Plant Mol. Biol.* **1989**, *40*, 503–537. [CrossRef]
47. Francey, R.J.; Allison, C.E.; Etheridge, D.M.; Trudinger, C.M.; Enting, I.G.; Leuenberger, M.; Langenfelds, R.L.; Michel, E.; Steele, L.P. A 1000-year high precision record of delta13C in atmospheric CO2. *Tellus B* **1999**, *51*, 170–193. [CrossRef]

 *forests*

Article

Atmospheric Nitrogen Dioxide Improves Photosynthesis in Mulberry Leaves via Effective Utilization of Excess Absorbed Light Energy

Yue Wang, Weiwei Jin, Yanhui Che, Dan Huang, Jiechen Wang, Meichun Zhao and Guangyu Sun *

School of Life Sciences, Northeast Forestry University, Harbin 150040, Heilongjiang, China;
wangyue@nefu.edu.cn (Y.W.); jinwei6677066@163.com (W.J.); carcar@nefu.edu.cn (Y.C.);
18804501876@163.com (D.H.); 15771398049@163.com (J.W.); 15846070751@163.com (M.Z.)
* Correspondence: sungy@nefu.edu.cn; Tel.: +86-0451-8219-1507

Received: 10 February 2019; Accepted: 30 March 2019; Published: 5 April 2019

Abstract: Nitrogen dioxide (NO_2) is recognized as a toxic gaseous air pollutant. However, atmospheric NO_2 can be absorbed by plant leaves and subsequently participate in plant nitrogen metabolism. The metabolism of atmospheric NO_2 utilizes and consumes the light energy that leaves absorb. As such, it remains unclear whether the consumption of photosynthetic energy through nitrogen metabolism can decrease the photosynthetic capacity of plant leaves or not. In this study, we fumigated mulberry (*Morus alba* L.) plants with 4 $\mu L \cdot L^{-1}$ NO_2 and analyzed the distribution of light energy absorbed by plants in NO_2 metabolism using gas exchange and chlorophyll a fluorescence technology, as well as biochemical methods. NO_2 fumigation enhanced the nitrogen metabolism of mulberry leaves, improved the photorespiration rate, and consumed excess light energy to protect the photosynthetic apparatus. Additionally, the excess light energy absorbed by the photosystem II reaction center in leaves of mulberry was dissipated in the form of heat dissipation. Thus, light energy was absorbed more efficiently in photosynthetic carbon assimilation in mulberry plants fumigated with 4 $\mu L \cdot L^{-1}$ NO_2, which in turn increased the photosynthetic efficiency of mulberry leaves.

Keywords: nitrogen dioxide; nitrogen metabolism; photorespiration; heat dissipation; excess absorbed light energy; electron transfer; photochemical efficiency

---

1. Introduction

Atmospheric nitrogen oxides (NO_x) mainly include nitric oxide (NO), nitrogen dioxide (NO_2), dinitrogen trioxide (N_2O_3), dinitrogen monoxide (N_2O), and dinitrogen pentoxide (N_2O_5). Nitrogen oxides other than NO_2 are extremely unstable and can be converted to NO_2 in the presence of light, humidity, or heat [1]. NO_2 sources are divided into natural and man-made sources. Natural sources mainly include lightning, stratospheric photochemistry, and microbiological processes in ecosystems. NO_2 formed in nature is generally in ecological balance at a natural point of equilibrium, which is low relative to man-made air pollution [2]. The emission of NO_2 from man-made sources is indeed the main component of atmospheric pollutants, which forms aerosol particles of nitric acid with particulate matter in the air. These aerosols form secondary pollution with pollutants from sources that include fossil fuel and biomass combustion as well as various electroplating, carving, welding, and other industrial emissions [3,4].

NO_2 not only causes acid rain, but also changes the competition and species composition among wetland and terrestrial plant taxa, reduces atmospheric visibility, increases acidification and eutrophication of surface water, and increases the toxin content of fish and other aquatic organisms [5]. Cheng et al [6] found that tiny particles of water in the air act as incubators during hazy conditions and trap NO_2 to interact and form sulfate. Additionally, stationary polluted weather systems accelerate chemical reactions, trapping near-surface NO_2, leading to NO_2 concentrations that are more than three times higher than that found in sunny weather. This increase in aerosol mass concentration leads to an increase in water content, accelerating the accumulation of sulfate and causing severe haze. In addition, NO_2 is also a respiratory system irritant. After being inhaled, NO_2 first affects the respiratory organs, the lungs in particular. The combination of nitrite and nitric acid that occurs when NO_2 encounters mucus membranes is a strong irritant with corrosive effects [7].

NO_2 affects the normal growth of plants. When NO_2 concentrations are higher than the annual average NO_2 concentration limit of 53 ppb in the United States [8], NO_2 can damage the leaves of plants, causing chlorosis in angiosperms, needle burns in conifers [9,10], reduced leaf area [11], and lower stem weights [12]. However, when the concentration of NO_2 is lower than the average annual NO_2 concentration of 53 ppb in the United States, the total leaf area, nutrient intake, and aboveground biomass were more than doubled [13,14]. Similar results have been found in different plant species, including *Arabidopsis thaliana* [15,16], tobacco (*Nicotiana plumbaginifolia* L.) [17], and crops such as lettuce (*Lactuca sativa* L.), sunflower (*Helianthus annuus* L.), cucumber (*Cucumis sativus* L.), and squash (*Cucurbita moschata* L.) [9]. In addition, atmospheric NO_2 can shorten flowering periods in tomato (*Solanum lycopersicum* L. 'Micro-Tom'), increasing the number of flowers and the yield of the fruit [18].

In China, the concentration of NO_2 emission, which caused formation of fine particulate matter (PM2.5), is not enough to injure tree plants. As a result, trees have been used to absorb atmospheric nitrogen dioxide to reduce atmospheric PM2.5 [19].

Photosynthesis is the foundation of plant growth and development, and the primary source of photosynthetic energy is light [20]. When the absorption of light energy is excessive, the excess excitation energy can harm the photosynthetic systems, causing photosynthetic inhibition, and even photooxidation and photodamage. Plants have multiple photoprotective mechanisms that reduce the potential harm of excess light energy to the photosynthetic apparatus under strong light. Reducing excess energy in addition to heat dissipation dissipates energy by other means [21]. Nitrogen metabolism and photorespiration also use and consume light energy or photosynthetic electrons absorbed by leaves [22,23]. However, it is unclear whether this consumption of photosynthetic energy will reduce the photosynthetic capacity of plant leaves and thereby hinder the growth and development of plants.

Mulberry, which has strong adaptability to the environment, has been an important economic tree species in China since ancient times. Nowadays, mulberry can be used for sericulture, new high-protein forage grass, fruit tree, and greening trees in northeast China [24–26]. We have studied the response of mulberry to atmospheric pollutant SO_2 and found that mulberry is very sensitive to it [24]. We also studied the response characteristics of mulberry to nitrogen and obtained significant results in the absorption and metabolism of nitrogen by mulberry trees [27–30]. We think atmospheric NO_2 also affects the nitrogen metabolism of mulberry trees. In the present study, we fumigated mulberry leaves with NO_2 and assessed the impact on nitrogen metabolism, photorespiration, photosynthetic energy distribution, and electron flow distributions. Our assays characterized the response of photosynthetic efficiency to light energy used and consumed by mulberry plants, including the absorption of NO_2 via nitrogen metabolism in vivo.

2. Materials and Methods

2.1. Plant Material and Growth Conditions

Seedlings of *Morus alba* L. were selected as experimental materials. Mulberry seeds were provided by the Heilongjiang Sericulture Institute of Heilongjiang Province in China. Seeds with strong, full, and uniform size were selected, disinfected with 75% ethanol for 3min, rinsed with distilled water for 5–6 times, and soaked with distilled water at 25 °C for 24 h. The seeds were blotted with sterile filler paper and sown into the seedling tray. Two seeds were sown in each hole. After germination and cultivation, the test seedlings were grown to a height of 10 cm and then transplanted into pots with a diameter of 12 cm and a height of 15 cm. Experimental seedlings were cultured in a seedling greenhouse with an average temperature of 28/25 °C (light/dark), light intensity of 400 µmol m^{-2}·s^{-1}, photoperiod of 12 h/12 h (light/dark), and relative humidity of about 75%. The culture substrate was uniformly mixed with peat and vermiculite, and irrigated with 800 mL of tap water every 2 days. In order to ensure relative consistency of the experimental materials, the branches and leaves of the mulberry seedlings were removed at the time of transplantation, such that only 5 cm of the main root and the main stem were preserved. Two plants were planted into each pot, with 80 plants in total cultivated. When the seedlings had grown to a height of 30–40 cm, 12 mulberry seedlings with uniform growth were selected for NO_2 fumigation treatment.

2.2. NO_2 Fumigation Treatment

The gas fumigation box was a custom-made open-top automatic monitoring gas concentration device [31]. The NO_2 cylinder (Dalian Date Gas, Dalian, China) was connected to an electromagnetic valve and axial flow fan as the air supply source. Solenoid valve controls regulated the test gas, while the axial flow of the fan is used to send the test gas into the air chamber. A gas pressure reducer and a gas flowmeter in the middle of the gas delivery system control the gas flow. A solenoid valve associated with the micro-flux switch valve controls the average flow in low-frequency pulse width modulation (PWM) mode. The gas chamber was composed of organic glass in the shape of a six-sided prism, with a cross section diagonal length of 1.16 m and height of 1.85 m. Vent holes were set on the top, with a gas grid plate 0.30 m from the bottom of the chamber. The NO_2 gas first enters the bottom space under the gas grid, and then passes through the gas chamber from the bottom of the 1200 holes, each with a diameter of 2 mm and evenly distributed across the grid plate. A fan was installed at the top of the fume chamber to ensure that the conditions were evenly distributed among the plants. The concentration of NO_2 used in the experiment was 4 $\mu L \cdot L^{-1}$, and the concentration was selected according to our pre-experiment results. We used 0.5, 1, 2, 4, 6, 8 $\mu L \cdot L^{-1}$ NO_2 to fumigate mulberry leaves. The results showed that the net photosynthetic rate in leaves of mulberry was the highest as NO_2 concentration 4 $\mu L \cdot L^{-1}$. The fumigation time is 4 hours/day, from 7:30 to 11:30; the temperature in the fumigation chamber was maintained at 28 °C. Samples were taken before the fumigation began (0 h) and at 4 and 8 h after fumigation. The indicators of N metabolism and photosynthetic gas exchange parameters as well as chlorophyll *a* fluorescence parameters were determined.

2.3. Measurement of NO_3^--N Content

An analysis of nitrate nitrogen was completed by using the salicylic acid method [32]. Standard solutions of 500 mg/L nitrate nitrogen with deionized water to create a 20, 40, 60, 80, and 100 mg/L series of standard solutions. Then, 0.5 g of plant material was placed into each test tube, to which 10 mL of deionized water was added before being placed into a boiling water bath for 30 min. Afterwards, test tubes were cooled with tap water, and extracts were filtered into volumetric flasks. The residues were rinsed, and the samples were allowed to finally settle to 25 mL. Then, 0.1 mL aliquots of the above series of standard solutions and sample extracts were respectively transferred into new test tubes, and the standard solution was replaced with 0.1 mL of distilled water.

Then, 0.4 mL of 5% salicylic acid solution was added prior to each sample being shaken well with a 20 min of room temperature incubation. After incubation, 9.5 mL of 8% NaOH solution was added, and samples were shaken and allowed to cool to room temperature. The total volume of the sample was 10 mL. The absorbance was measured at 410 nm with a blank as a reference. A standard curve was drawn, and a regression equation was fitted to the standard by using the nitrate concentration as the abscissa and the absorbance as the ordinate. The concentration of nitrate nitrogen was calculated using the inferred regression equation, and content was calculated by using the formula NO_3^--N content $= (C \times V/1000)/W$, where C is the regression equation calculated NO_3^--N concentration, V is the extract total mL of sample liquid, and W is the sample fresh weight.

2.4. Measurement of Amino Acid Content

A colorimetric analysis was done on hydrated ninhydrin for amino acids [33]. To assess amino acid content, 200 $\mu g/mL$ amino acid standard solutions were measured into volumes of 0.0, 0.5, 1.0, 1.5, 2.0,

2.5, and 3.0 mL, respectively, and transferred into 25 mL volumetric flasks, to which water was added until a volume of 4.0 mL was reached. Then, 1 mL of ninhydrin solution (20 g/L) and 1 mL of phosphate buffer (pH 8.04) was added to each flask, followed by mixing and incubation in a 90 °C water bath until the color became constant. At this point, samples were removed rapidly and cooled to room temperature, and water was added until a final volume was reached. Samples were then shaken well and allowed to stand for 15 min. The absorbance A of the remaining solutions was then determined relative to a reagent blank as a reference solution at a wavelength of 570 nm. In the standard curve, the micrograms of amino acids represented the abscissa, while absorbance A was the ordinate. The standard curve was drawn based on these data, and the regression equation was then inferred.

Then, 0.5 g plant samples were added to 5 mL of 10% acetic acid and ground in a mortar. These ground samples were then washed into 100 mL volumetric flasks, diluted to a set volume in water, and filtered into triangle bottles for determination of the filtrate. Then, 4 mL of the clarified sample solution was subjected to the standard curve inference procedure outlined above, in which the absorbance A value was measured under the same conditions and the microgram content of amino acids was calculated using a regression equation of the following form: amino acid content (μg/100 g) = $C/(m \times 1000) \times 100$, where C represents the mass number of amino acids and the m represented the mass of the sample.

2.5. Measurement of Nitrate Reductase Activity

Nitrate reductase (NR) activity was determined using the kit produced by Suzhou Keming Company (Suzhou, China). This kit functions by assaying how NR catalyzes the reduction of nitrate to nitrite according to the reaction $NO_3^- + NADH + H^+ \rightarrow NO_2^- + NAD^+ + H_2O$. The resulting nitrite quantitatively produces red azo compounds under acidic conditions with p-aminobenzenesulfonic acid and α-naphthylamine. The generated red azo compound has a maximum absorption peak at 540 nm and was measured by spectrophotometry.

2.6. Measurement of Nitrite Reductase Activity

Nitrite reductase (NiR) activity was determined using the kit produced by Suzhou Keming Company. This kit functions by assaying how nitrite reductase reduces NO_2^- to NO, and the sample is subjected to the diazotization reaction to generate a purple-red compound that indicates a NO_2^- decrease; that is, a change in absorbance at 540 nm reflects the activity of nitrite reductase.

2.7. Measurement of Gas Exchange Parameters

Gas exchange was measured with a portable photosynthesis system (LICOR-6400; LI-COR, Lincoln, NE, USA). All the photosynthetic measurements were taken at a constant airflow rate of 400 μmol·s^{-1}, and the temperature was 28 ± 2 °C. While using the liquefied CO_2 cylinders were used to provide different CO_2 concentrations, across a CO_2 concentration gradient of 50, 100, 200, 300, 400, 600, 800, 1000, and 1200 μmol·mol^{-1}. The photosynthetically active radiation (PAR) was 1000 μmol m^{-2}·s^{-1} in order to avoid light limitation of photosynthesis. The photosynthetic rate (P_n) and stomatal conductance of H_2O (G_s) were measured. Photosynthesis curves plotted against intercellular CO_2 concentrations (P_n-Ci curve) was analyzed to estimate the maximum carboxylation rate (V_{cmax}) of ribulose-1,5-bisphosphate carboxylase/oxygenase (Rubisco) and dark respiration rate (R_d) [34].

2.8. Measurement of Chlorophyll A Fluorescence Transient and Light Absorbance at 820 nm

According to the method developed by Schansker et al. [35], leaves were first dark-adapted for 20 min, and then the rapid chlorophyll fluorescence-induced kinetic curve (OJIP curve) and 820 nm light absorption curve were measured using a plant efficiency analyzer (M-PEA, Hansatech, King's Lynn, UK). The OJIP curve was induced under 3500 μmol·m^{-2}·s^{-1} pulsed light, and the fluorescence signal was recorded from 10 μs to the end of 2 s, with an initial recording rate of 10^5 data per second. The relative value of the difference between the maximum (I_o) and minimum (I_m) absorbance at 820 nm, i.e., $\Delta I/I_o = (I_o - I_m)/I_o$, was used as an index for measuring photosystem

I (PSI) activity. The OJIP fluorescence induction curve was analyzed with reference to the JIP-test used by Strasser et al. [36] to measure the initial fluorescence (F_o), the maximum fluorescence (F_m), the light energy absorbed by the unit reaction center (ABS/RC), the unit reaction center captured for the reducing energy of Q_A (TR_o/RC), energy captured by the unit reaction center for electron transfer (ET_o/RC), and energy dissipated in the unit reaction center (DI_o/RC). The potential activity of photosystem II (PSII) was then calculated as $F_v/F_o = (F_m/F_o) - 1$. The maximum photochemical efficiency of PSII was calculated as $F_v/F_m = 1 - (F_o/F_m)$. Lastly, the number of active reaction centers per unit area was calculated as $RC/CS_m = (F_v/F_m)\,(V_J/M_o)\,(ABS/CS_m)$.

2.9. Measurement of Photochemical Quenching, Electron Transfer Rate, and Absorbed Energy of the PSII Reaction Center

After dark adaptation of the leaves for 30 min, chlorophyll fluorescence quenching analysis was performed using a pulse modulation fluorimeter (FMS-2, Hansatech, King's Lynn, UK). In this procedure, the leaf adapts to the light for the first 30 s, and the light intensity is the same as the ambient light intensity before the measurement of the leaf (1000 $\mu mol \cdot m^{-2} \cdot s^{-1}$). The steady-state fluorescence parameter (F_s) and electron transfer rate (ETR) under the light-adapted condition was measured, and the saturated pulsed light (8000 $\mu mol \cdot m^{-2} \cdot s^{-1}$) was used to determine the photochemical quenching (qP), the maximum fluorescence value (F_m'). Then, the method of Hendrickson et al. [37] was followed to determine the light energy absorbed by the PSII reaction center into each of four parts [38]. That is, the following parameters were inferred: quantum yield (Φ_{PSII}) for the photochemical reaction, quantum yield (Φ_{NPQ}) dependent on the proton gradient lutein cycle on both sides of the thylakoid membrane, quantum yield of fluorescence and heat energy dissipation ($\Phi_{f,D}$), and the thermal dissipation of the quantum yield (Φ_{NF}) of the deactivated PSII reaction center. These parameters were inferred using the following formulae:

$$\Phi_{PSII} = [1 - (F_s/F_m')]\,[(F_v/F_m)/(F_v/F_{mM})]$$
$$\Phi_{NPQ} = [(F_s/F_m') - (F_s/F_m)]\,[(F_v/F_m)/(F_v/F_{mM})]$$
$$\Phi_{f,D} = (F_s/F_m)\,[(F_v/F_m)/(F_v/F_{mM})]$$
$$\Phi_{NF} = 1 - [(F_v/F_m)/(F_v/F_{mM})].$$

The sum of each of these parameters is 1, namely $\Phi_{PSII} + \Phi_{NPQ} + \Phi_{f,D} + \Phi_{NF} = 1$ (Figure 1).

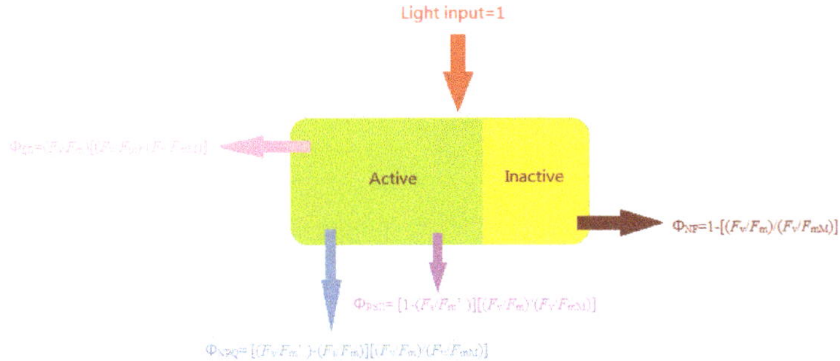

Figure 1. Energy allocation pathways of photosystem II (PSII): proportion of light energy used in the quantum yield of PSII photochemistry (Φ_{PSII}), xanthophyll-mediated thermal dissipation (Φ_{NPQ}), quantum yield used in thermal dissipation in non-functional PSII (Φ_{NF}), and quantum yield of fluorescent and heat energy dissipation ($\Phi_{f,D}$).

2.10. Statistical Analysis

Excel and SPSS software were used to conduct statistical analyses on the measured data. The data in the figure shows means of four plants ± standard deviation. Tukey Multiple Comparisons test was adopted to compare the differences between treatments. Different lowercase letters for the same parameter indicate significant differences among different treatments at $p < 0.05$ levels.

3. Results

3.1. Effects of N Metabolism Indicators

NO_2 fumigation affected nitrogen metabolism in mulberry. The nitrate nitrogen content of mulberry leaves fumigated with NO_2 increased by 56.10% ($p < 0.05$) at 4 h after the start of the treatment, and when the fumigation time lasted for 8 h, it was nearly doubled ($p < 0.05$) compared to the control. The amino acid content in mulberry was not significantly increased ($p > 0.05$) when NO_2 was fumigated for 4 hours; however, it increased by 35.17% ($p < 0.05$) after 8 hours of fumigation.

Nitrate reductase and nitrite reductase activities increased significantly ($p < 0.05$) by 4 h after the start of the treatment, and when the fumigation time lasted for 8 h, the activities enhanced more (Figure 2a–d).

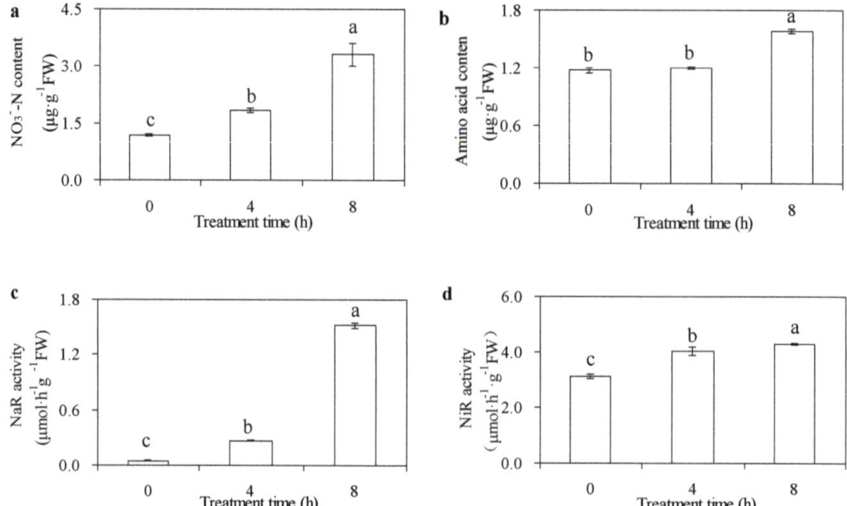

Figure 2. Effects of N Metabolism Indicators. (**a**) $NO_3{}^{-}$-N content, (**b**) amino acid content, (**c**) nitrate reductase activity, and (**d**) nitrite reductase activity in leaves of mulberry seedlings exposure to 4 $\mu L \cdot L^{-1}$ NO_2 for 0 h, 4 h (4 h·d^{-1}, for one day), and 8 h (4 h·d^{-1}, for 2 days). Date represent means of four plants $\pm$ standard deviations. Different lowercase letters for the same parameter indicate significant differences among different treatments at $p < 0.05$ levels.

3.2. Effects of Gas Exchange Parameters

NO_2 fumigation changed the P_n-Ci curve of mulberry leaves, and P_n showed an upward trend with the increase of CO_2 concentration. When the CO_2 concentration was lower than 400 $\mu mol \cdot mol^{-1}$, P_n increases approximately linearly with the increase of CO_2 concentration. Subsequently, as the CO_2 concentration continued to increase, the rate of P_n increase slowed. When the CO_2 concentration reached 1200 $\mu mol \cdot mol^{-1}$, the P_n-Ci gradually flattened (Figure 3a). Under the same CO_2 concentration, the P_n of mulberry leaves with NO_2 fumigation was significantly higher than that of the control, indicating that NO_2 fumigation improved the carbon assimilation capability of mulberry leaves. Stomatal conductance (G_s), the maximum carboxylation rate (V_{cmax}), and dark respiration rate (R_d) of the mulberry leaves increased significantly ($p < 0.05$) after 4 h and 8 h of fumigation compared with the control (Figure 3b–d).

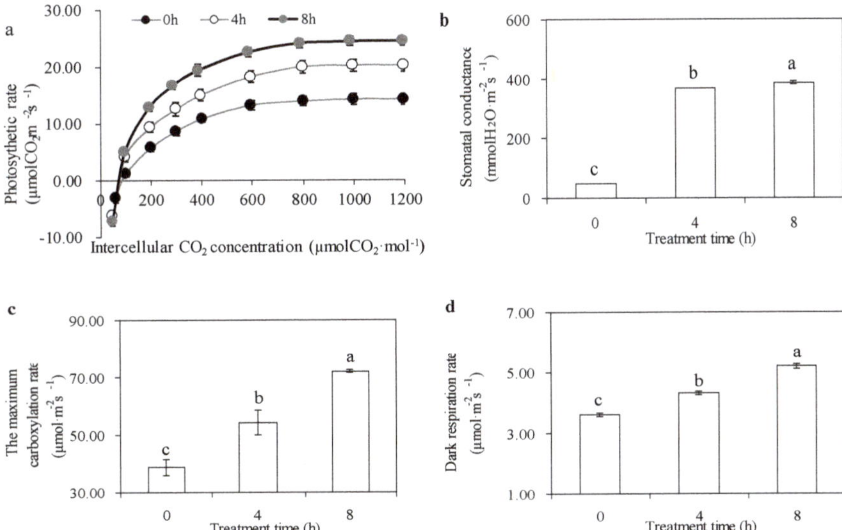

Figure 3. Effects of Gas Exchange Parameters. (**a**) P_n-Ci curve, (**b**) conductance of H_2O, (**c**) the maximum carboxylation rate, and (**d**) dark respiration rate in leaves of mulberry seedlings exposure to 4 $\mu L \cdot L^{-1}$ NO_2 for 0 h, 4 h (4 $h \cdot d^{-1}$, for one day), and 8 h (4 $h \cdot d^{-1}$, for 2 days). Date represent means of four plants ± standard deviations. Different lowercase letters for the same parameter indicate significant differences among different treatments at $p < 0.05$ levels.

3.3. Effects of Distribution of Light Absorbed by PSII

Of the light energy absorbed by PSII, the fraction allocated to photochemical conversion (Φ_{PSII}), increased slightly, while the fraction dissipated nonphotochemically in a manner dependent on the trans-thylakoid proton-gradient and the xanthophyll cycle (Φ_{NPQ}) increased after NO_2 fumigation. By contrast, the proportion of basic fluorescence quantum yields and heat-dissipated quantum yields ($\Phi_{f,D}$) and heat-dissipation quantum yields of inactive PSII reaction centers (Φ_{NF}) declined after NO_2 fumigation. The proportions of Φ_{NPQ} and Φ_{PSII} increased by 2.49% ($p < 0.05$), 36.23% ($p < 0.05$) at 4h and increased by 5.67% ($p < 0.05$), 41.07% ($p < 0.05$) at 8h. By contrast, the proportions of $\Phi_{f,D}$ and Φ_{NF} decreased by 5.01% ($p < 0.05$), 60.41% ($p < 0.05$) at 4h and decreased by 10.34% ($p < 0.05$) and 75.74% ($p < 0.05$) at 8 h (Figure 4). The quantum yield for heat dissipation in mulberry leaves fumigated with 4 $\mu L \cdot L^{-1}$ NO_2 was reduced to increase the photosynthetic energy used for photochemical reaction. As shown in Figure 7, the partial energy from total energy absorbed by mulberry leaves fumigated by NO_2 was used for NO_2 metabolism; whether this affects the activity of PSII and PSI is discussed in the following sections.

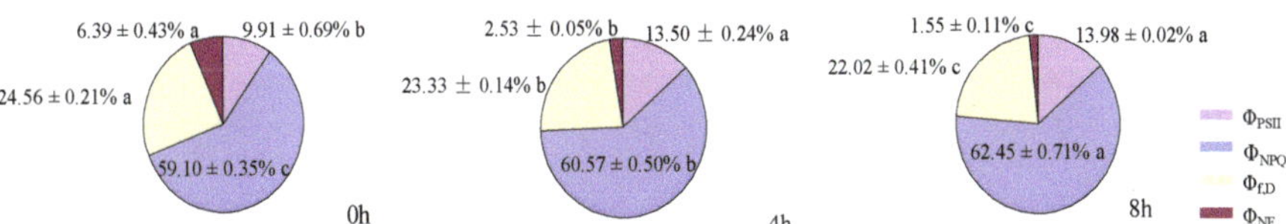

Figure 4. Light energy used for quantum yield of PSII photochemistry (Φ_{PSII}), xanthophyll-mediated thermal dissipation (Φ_{NPQ}), basic fluorescence quantum yield and heat dissipation quantum yield ($\Phi_{f,D}$), and quantum yield used in thermal dissipation in non-functional PSII (Φ_{NF}) in leaves of mulberry seedlings exposure to 4 $\mu L \cdot L^{-1}$ NO_2 for 0 h, 4 h (4 $h \cdot d^{-1}$, for one day), and 8 h (4 $h \cdot d^{-1}$, for 2 days). Date represent means of four plants ± standard deviations. Different lowercase letters for the same parameter indicate significant differences among different treatments at $p < 0.05$ levels.

3.4. Effects of Chlorophyll A Fluorescence Transient

NO_2 fumigation significantly changed the shape of OJIP curves. The fluorescence intensity at points O and J (F_o and F_J, respectively) decreased, and the fluorescence intensity at points I and P (F_I and F_p, respectively) increased significantly. The OJIP curves of the leaves were normalized to the span O-P. The variable fluorescence at point J on the OJIP curve of leaves was significantly increased by NO_2. The difference between the normalized O-P curve and the control was assessed, and the difference between ΔVt and the control at point J was most significant at 2 ms (Figure 5a–d).

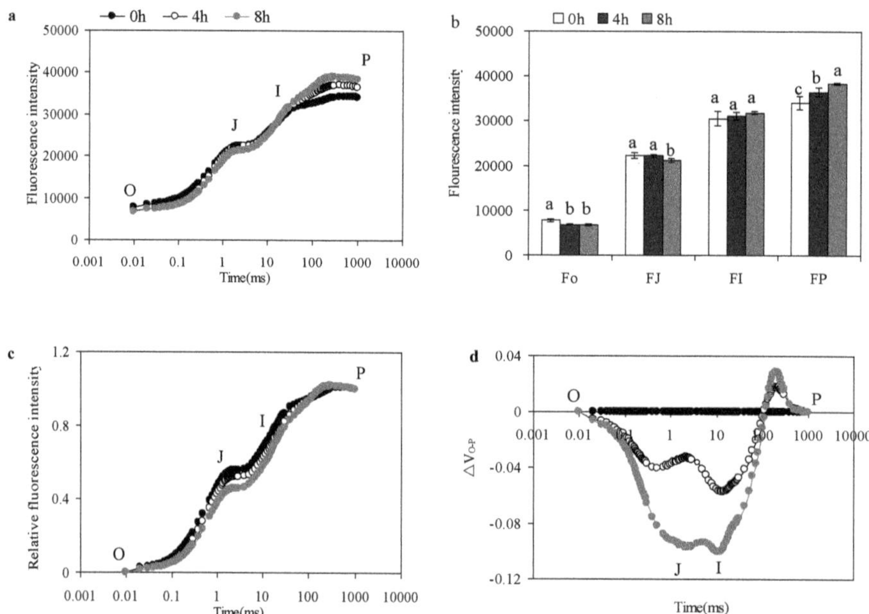

Figure 5. Effects of Chlorophyll A Fluorescence Transient. (**a**) OJIP transmission, (**b**) fluorescence intensity at points O, J, I, and P, (**c**) standardized OJIP transmission, (**d**) standardized OJIP transmission and CK in leaves of mulberry seedlings exposure to 4 $\mu L \cdot L^{-1}$ NO_2 for 0 h, 4 h (4 $h \cdot d^{-1}$, for one day), and 8 h (4 $h \cdot d^{-1}$, for 2 days). Date represent means of four plants $\pm$ standard deviations. Different lowercase letters for the same parameter indicate significant differences among different treatments at $p < 0.05$ levels.

3.5. Effects of PSII activity

NO_2 fumigation had a significant effect on PSII potential photochemical activity (F_v/F_o), PSII maximum photochemical efficiency (F_v/F_m), photochemical quenching (qP), electron transport rate (ETR), and number of active reaction centers per unit area (RC/CS_m), which is shown in Table 1. F_v/F_o, F_v/F_m, qP, ETR, and RC/CS_m increased by 33.44% ($p < 0.05$), 1.22% ($p < 0.05$), 30% ($p < 0.05$), 33.64% ($p < 0.05$), and 27.07% ($p < 0.05$) at 4 h, respectively, and increased by 39.38% ($p < 0.05$), 2.44% ($p < 0.05$), 35% ($p < 0.05$), 37.67% ($p < 0.05$), and 33.37% ($p < 0.05$) at 8 h, respectively.

Table 1. PSII potential photochemical activities (F_v/F_o), PSII photochemical efficiency (F_v/F_m), photochemical quenching (qP), electron transport rate (ETR), and the number of reactive centers per unit area (RC/CS_m) in leaves of mulberry seedlings exposure to 4 $\mu L \cdot L^{-1}$ NO$_2$ for 0 h, 4 h (4 h·d^{-1}, for one day), and 8 h (4 h·d^{-1}, for 2 days). Date represent means of four plants $\pm$ standard deviations. Different lowercase letters for the same parameter indicate significant differences among different treatments at $p < 0.05$ levels.

	0 h	4 h	8 h
F_v/F_o	3.20 ± 0.50b	4.27 ± 0.12a	4.46 ± 0.32a
F_v/F_m	0.82 ± 0.00c	0.83 ± 0.00b	0.84 ± 0.01a
qP	0.20 ± 0.05b	0.26 ± 0.01a	0.27 ± 0.00a
ETR	43.99± 3.32b	58.79 ± 0.92a	60.56 ± 0.17a
RC/CS_m	20501.38 ± 377.35c	26050.96 ± 1194.94b	27341.97 ± 206.688a

3.6. Effects of PSI Activity

The relative drop in 820 nm light signals during far-red light illumination reflects the activity of PSI $\Delta I/Io$. NO$_2$ fumigation increased the drop in the 820 nm optical signal, and the difference increased significantly at 8 h, indicating that NO$_2$ fumigation increased the PSI activity of mulberry leaves (Figure 6a). PSI activity was not significantly raised ($p > 0.05$) for 4 h fumigation, but when fumigated for 8 h, the PSI activity was improved by 20.22% ($p < 0.05$) compared with the control (Figure 6b).

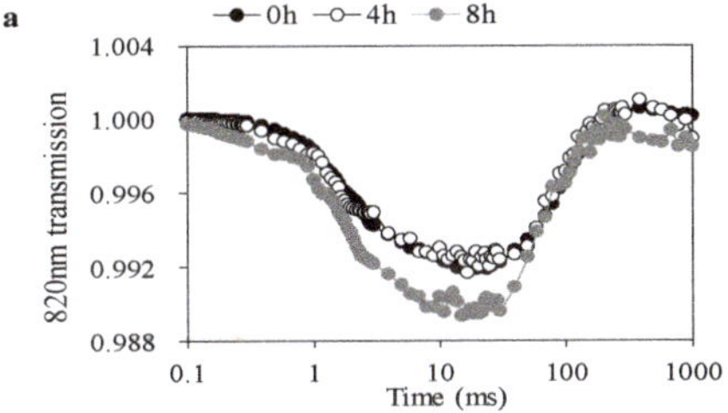
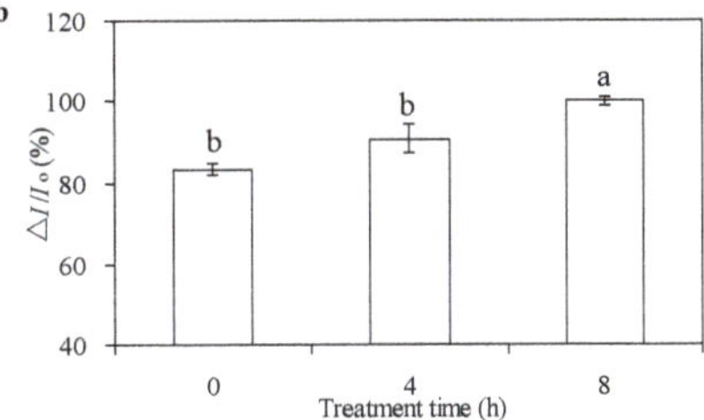

Figure 6. Effects of PSI activity. (**a**) Changes in light transmission at 820 nm, (**b**) Relative values of $\Delta I/I_o$ in leaves of mulberry seedlings exposure to 4 $\mu L \cdot L^{-1}$ NO$_2$ for 0 h, 4 h (4 h·d^{-1}, for one day), and 8 h (4 h·d^{-1}, for 2 days), where the $\Delta I/I_o$ values at 8 h were taken as 100%. Date represent means of four plants $\pm$ standard deviations. Different lowercase letters for the same parameter indicate significant differences among different treatments at $p < 0.05$ levels.

4. Discussion

The present study revealed that the absorption of NO$_2$ by mulberry leaves is not only involved in nitrogen metabolism in vivo, but also increases the leaf content of nitrate nitrogen and amino acids, enhances the activity of nitrate reductase (NR) and nitrite reductase (NiR), and improves the nitrogen metabolism capacity (Figure 2). Zeevaart et al. [39] fumigated pea seedlings with ammonium nitrogen as the only nitrogen source condition and found that NO$_2$ induces NR activity, in perhaps the earliest study on NO$_2$ and nitrogen metabolism. Subsequent studies have since found that atmospheric NO$_2$ has a significant effect on NR. Lower concentrations of NO$_2$ increase NR activity in barley (*Hordeum vulgre*) [40], Scots pine (*Pinus sylvestris*) [41], and red spruce (Picea rubens) [42]. Similarly, when Hisamatsu et al. [43] fumigated squash (*Cucurbita maxima*) seedlings with NO$_2$, NR activity in cotyledon was significantly reduced. In the present study, NR activity had exhibited a small increase at 4 h but increased sharply at 8 h. Enhanced NR activity promotes the improvement of nitrogen metabolism. It not only increases the consumption of excess light energy, but also provides necessary enzymes for carbon metabolism and accelerates carbon metabolism.

The 4 $\mu L \cdot L^{-1}$ NO$_2$ fumigation treatment significantly increased the net photosynthetic rate of mulberry leaves (Figure 3a), accelerating photosynthetic electron transport and enhancing phosphorylation activity in leaves. Additionally, the increase of stomatal conductance promoted the degree of CO_2 acquisition and transportation (Figure 3b), thereby accelerating the production and accumulation of organic matter in plants [44]. At the same time, the maximum carboxylation rate, which is an important parameter for characterizing photosynthetic capacity of plants, also increased (Figure 3b), indicating that NO_2 fumigation increased Rubisco activity of mulberry leaves, thereby enhancing the fixation ability of CO_2 [45]. Moreover, the dark respiration rate was significantly increased (Figure 3d), indicating that its fumigated leaves consumed excess light energy through respiration to protect the photosynthetic system and increase the photosynthetic rate.

Accordingly, the quantitative study of the final destination of light energy absorbed by plant leaves is an important part of research on photosynthesis. Earlier studies on light energy are more directly expressed by actual photochemical efficiency (Φ_{PSII}) and non-photochemical quenching (NPQ) [46,47]. However, in higher plants, NPQ is only determined by the establishment of proton gradients on both sides of the thylakoid membrane and the xanthophyll cycle [48]. Thus, NPQ does not represent all non-photochemical quenching processes. In addition, plant leaves absorb light energy via mechanisms other than photochemical reactions, including physiological processes such as photorespiration, the water–water cycle, and xanthophyll cycle, which can be areas in which light energy is distributed as the final destination [49–51]. The theory developed by Hendrickson et al. [50] clarifies these processes. In this framework, excitation energy can be divided into the light energy absorbed by the PSII reaction centers and used for the quantum yield of photochemical reaction (Φ_{PSII}), the quantum production dependent on the trans-thylakoid proton-gradient for xanthophyll cycle quantum yield (Φ_{NPQ}), fluorescence quantum yield and heat dissipation ($\Phi_{f,D}$), and heat dissipation quantum yield in inactive PSII reaction centers (Φ_{NF}) [51]. In our experiment, mulberry leaves fumigated with NO_2 showed proportional increases in Φ_{PSII} and Φ_{NPQ}, indicating that NO_2 fumigation promoted the establishment of light-induced proton absorption in the chloroplast H^+ concentration gradients on both sides of the thylakoid. In other words, the presence of ΔpH on both sides of the thylakoid membrane increased the power of ATP synthesis in chloroplasts and promoted the photoprotective mechanism based on the xanthophyll cycle. Additionally, $\Phi_{f,D}$ and Φ_{NF} proportionally decreased, indicating that the proportion of heat dissipation and inactivation reaction centers decreased and that the absorbed light energy was mostly used for photosynthetic carbon assimilation, thereby improving the photochemical efficiency of the mulberry leaf (Figure 4).

In this experiment, the changes in PSII of mulberry leaves were analyzed using the JIP test. NO_2 changed the structure and function of PSII and the photosynthetic primary reaction process in mulberry leaves. However, because the OJIP curve is greatly influenced by the environment, its relative fluorescence intensity is affected by various environmental factors. Therefore, the OJIP curve was often normalized to the span O-P. The relative variable fluorescence of V_J and V_I at points J (2 ms) and I point (30 ms) for NO_2-fumigated leaves was significantly decreased (Figure 5), indicating that the PSII reaction center receptor-side electronic primary quinone receptors (Q_A) to the secondary quinone receptor (Q_B) transmission capacity and plastoquinone (PQ) accept electronic ability were enhanced (Figure 7). The potential photochemical activity of PSII (F_v/F_o), PSII maximum photochemical efficiency (F_v/F_m), photochemical quenching (qP), electron transfer rate (ETR), and the number of active reaction centers per unit area (RC/CS_m) were significantly higher under fumigation than in the control treatment (Table 1), indicating that NO_2 fumigation enhances the activity of PSII reaction centers and the degree of openness of reaction centers, which promotes the electron transport of leaves, photosynthetic primary reaction process, and the rate of light photons received by PSII reaction centers. The proportion of light energy used in photochemical reactions increased, thus accelerating the synthesis of NADPH and ATP, as well as the conversion efficiency of light energy and carbon assimilation processes, which increased the photochemical efficiency. However, Hu et al. [31] quantified the photosynthetic responses of hybrid poplar cuttings to 4 $\mu L \cdot L^{-1}$ NO_2. It was found that significant declines in F_v/F_m, indicating inhibition of and even damage to photosynthetic apparatus. The study of the maximum redox capacity ($\Delta I/I_o$) of PSI was also increased (Figure 6) as well as the ability to receive electrons; the ability of PSII to supply electrons in the photosynthetic apparatus was matched by the ability of PSI to receive electrons. Thus, not only can electron transfer be promoted efficiently, but electrons can also be transferred to ferredoxin, which distributes electrons to the nitrogen metabolism, photorespiration, and NO_2 metabolic pathways (Figure 7).

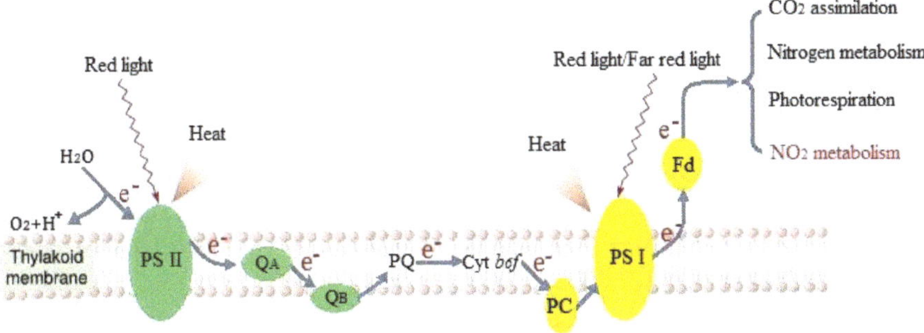

Figure 7. Photosynthetic electron flow distribution. The electrons produced by water splitting in photosynthesis pass through photosystem II (PSII), primary quinone receptor (Q_A), secondary quinone receptor (Q_B), plastoquinone (PQ), cytochrome b_6f (Cyt b_6f), plastid blue pigment (PC), and photosystem I (PSI) to ferredoxin (Fd) were assigned to four pathways: CO_2 assimilation, nitrogen metabolism, photorespiration, and NO_2 metabolic pathways (new pathway).

5. Conclusions

The metabolism of atmospheric NO_2 utilized and consumed light energy and photosynthetic electrons absorbed by leaves of mulberry fumigated with NO_2. However, this part of the photosynthetic energy consumption did not reduce the photosynthetic capacity of the mulberry leaves, but instead increased the photosynthetic efficiency of the plant leaves. This is because the mulberry leaves absorbed NO_2 and conducted nitrogen metabolism and respiration, which consumed excess light energy, and thus protected the photosynthetic apparatus. However, the light energy absorbed by the PSII reaction center in the form of heat dissipation in mulberry plants fumigated with 4 $\mu L \cdot L^{-1}$ NO_2 was also reduced, such that the absorbed light energy was more effectively used in photosynthetic carbon assimilation. Therefore, in the case where the concentration of the atmospheric pollutant NO_2 is lower than the concentration that can damage the mulberry, NO_2 can be absorbed by the mulberry to reduce the haze in the air.

Author Contributions: Y.W. and G.S. conceived and designed the study. Y.W., W.J., Y.C., and D.H. performed the experiments. Y.W., J.W., and M.Z. contributed to the sample measurement and data analysis. Y.W. and G.S. wrote the paper.

Funding: This research was funded by the National Natural Science Foundation of China, grant number 31870373, and the Natural Science Foundation of Heilongjiang Province, grant number ZD201105 and the Applicant and Developmental Project for Agriculture of Heilongjiang Province, grant number GZ13B004.

Acknowledgments: The authors want to appreciate Wah Soon Chow from the Australian National University for revising the manuscript and our colleague Yanbo Hu for his advice and great comments to improve the paper.

Conflicts of Interest: The authors declare no conflict of interest.

References

1. Xia, Y.; Zhao, Y.; Nielsen, C.P. Benefits of China's efforts in gaseous pollutant control indicated by the bottom-up emissions and satellite observations 2000–2014. *Atmos. Environ.* **2016**, *136*, 43–53. [CrossRef]
2. Nouchi, I. Responses of whole plants to air pollutants. In *Air Pollution and Plant Biotechnology*; Springer: Tokyo, Japan, 2002; pp. 3–39.

3. Wang, Y.; Li, L.J.; Liu, Y. Characteristics of atmospheric NO_2 in the Beijing-Tianjin-Hebei region and the Yangtze River delta analyzed by satellite and ground observations. *Chin. J. Environ. Sci.* **2012**, *33*, 3685–3692.

4. Wang, Y.; Zhang, X.L.; Hu, Y.B.; Teng, Z.Y.; Zhang, S.B.; Chi, Q.; Sun, G.Y. Phenotypic response of tobacco leaves to simulated acid rain and its impact on photosynthesis. *Int. J. Agric. Biol.* **2019**, *21*, 391–398.

5. Fang, H.; Mo, J.M. Reactive nitrogen increasing: A threat to our environment. *Ecol. Environ.Sci.* **2006**, *15*, 164–168.

6. Cheng, Y.F.; Zheng, G.J.; Wei, C.; Mu, Q.; Zheng, B.; Wang, Z.B.; Gao, M.; Zhang, Q.; He, K.B.; Carmichael, G.; et al. Reactive nitrogen chemistry in aerosol water as a source of sulfate during haze events in China. *Sci. Adv.* **2016**, *2*, e1601530. [CrossRef] [PubMed]

7. Jian, B.A.; Dai, Q.Q. Investigation report of an acute poisoning of nitrogen dioxide. *Occup. Health* **2004**, *20*, 31.

8. United States Environmental Protection Agency Home Page. Available online: http://www3.epa.gov/ttn/naaqs/criteria.html (accessed on 8 December 2018).

9. Adam, S.E.; Shigeto, J.; Sakamoto, A.; Takahashi, M. Atmospheric nitrogen dioxide at ambient levels stimulates growth and development of horticultural plants. *Botany* **2008**, *86*, 213–217. [CrossRef]

10. Gheorghe, I.F.; Ion, B. The effects of air pollutants on vegetation and the role of vegetation in reducing atmospheric pollution. In *The Impact of Air Pollution on Health, Economy, Environment and Agricultural Sources*; InTech: Timisoara, Romania, 2011; Volume 12, pp. 241–280.

11. Ashenden, T.W. The effects of long-term exposures to SO_2 and NO_2 pollution on the growth of *Dactylis glomerata* L. and *Poa pratensis* L. *Environ. Pollut.* **1979**, *18*, 249–258. [CrossRef]

12. Ashenden, T.W.; Bell, S.A.; Rafarel, C.R. Effects of nitrogen dioxide pollution on the growth of three fern species. *Environ. Pollut.* **1990**, *66*, 301–308. [CrossRef]

13. Takahashi, M.; Nakagawa, M.; Sakamoto, A.; Ohsumi, C.; Matsubara, T.; Morikawa, H. Atmospheric nitrogen dioxide gas is a plant vitalization signal to increase plant size and the contents of cell constituents. *New Phytol.* **2005**, *168*, 149–154. [CrossRef]

14. Takahashi, M.; Morikawa, H. Kinematic evidence that atmospheric nitrogen dioxide increases the rates of cell proliferation and enlargement to stimulate leaf expansion in Arabidopsis. *Plant Signal. Behav.* **2015**, *10*, e1022011. [CrossRef]

15. Takahashi, M.; Furuhashi, T.; Ishikawa, N.; Horiguchi, G.; Sakamoto, A.; Tsukaya, H.; Hiromichi, M. Nitrogen dioxide regulates organ growth by controlling cell proliferation and enlargement in Arabidopsis. *New Phytol.* **2014**, *201*, 1304–1315. [CrossRef] [PubMed]

16. Xu, Q.; Zhou, B.; Ma, C.; Xu, X.; Xu, J.; Jiang, Y.B.; Liu, C.; Li, G.Z.; Herbert, S.J.; Hao, L. Salicylic acid-altering arabidopsis mutants response to NO_2 exposure. *Bull. Environ. Contam. Toxicol.* **2010**, *84*, 106–111. [CrossRef]

17. Morikawa, H.; Takahashi, M.; Kawamura, Y. Metabolism and genetics of atmospheric nitrogen dioxide control using pollutant-philic plants. *Phytoremediat. Transform. Control Contam.* **2004**, 763–786.

18. Takahashi, M.; Sakamoto, A.; Ezura, H.; Morikawa, H. Prolonged exposure to atmospheric nitrogen dioxide increases fruit yield of tomato plants. *Plant Biotechnol.* **2012**, *28*, 485–487. [CrossRef]

19. Wang, Y.; Teng, Z.Y.; Zhang, X.L.; Che, Y.H.; Sun, G.Y. Research progress on the effects of atmospheric nitrogen dioxide on plant growth and metabolism. *Chin. J. Appl. Ecol.* **2019**, *30*, 316–324.

20. Evans, J.R. Improving photosynthesis. *Plant Physiol.* **2013**, *162*, 1780–1793. [CrossRef] [PubMed]

21. Kozaki, A.; Takeba, G. Photorespiration protects C_3 plants from photooxidation. *Nature* **1996**, *384*, 557–560. [CrossRef]

22. Kaiser, W.M.; Spill, D. Rapid modulation of spinach leaf nitrate reductase activity by photosynthesis II. In vitro modulation by ATP and AMP. *Plant Physiol.* **1991**, *96*, 368–375. [CrossRef] [PubMed]

23. Yamori, W.; Shikanai, T. Physiological functions of cyclic electron transport around photosystem I in sustaining photosynthesis and plant growth. *Annu. Rev. Plant Biol.* **2016**, *67*, 81–106. [CrossRef]

24. Wang, Y.; Li, X.P.; Peng, H.X.; Zhu, Y.Y.; Zhang, X.L.; Sun, G.Y. Effect of SO_2 wet deposition on morphology and light energy utilization in mulberry leaves. *Pratacult. Sci.* **2017**, *34*, 2080–2089.

25. Wang, Y.; Xu, B.T.; Teng, Z.Y.; Zhang, S.B.; Zhang, X.L.; Sun, G.Y. Effect of simulated acid rain on the growth and photosystemII in the leaves of mulberry seedlings. *Pratacult. Sci.* **2018**, *35*, 2220–2229.

26. Gu, C.H.; Wang, Y.X.; Bai, S.B.; Wu, J.Q.; Yang, Y. Tolerance and accumulation of four ornamental species seedlings to soil cadmium contamination. *Acta Ecol. Sin.* **2015**, *35*, 2536–2544.

27. Xu, N.; Zhang, X.S.; Zhang, X.L.; Zhang, H.H.; Zhu, W.X.; Li, X.; Yue, B.B.; Sun, G.Y. Effects of different N application levels on yield and physiological characteristics of leaf in mulberry. *Nonwood For. Res.* **2011**, *29*, 45–49.

28. Xu, N.; Zhang, H.H.; Zhu, W.X.; Li, X.; Yue, B.B.; Jin, W.W.; Wang, L.Z.; Sun, G.Y. Effects of nitrogen forms on growth and photosynthetic characteristics. *Pratacult. Sci.* **2012**, *29*, 1574–1580.

29. Xu, N.; Ni, H.W.; Zhong, H.X.; Sha, W.; Wu, Y.N.; Xing, J.H.; Sun, G.Y. Growth and photosynthetic characteristics of forage mulberry in response to different nitrogen application levels. *Jinagsu J. Agric. Sci.* **2015**, *4*, 865–870.

30. Xu, N.; Zhang, H.X.; Sha, W.; Wu, Y.N.; Zhang, H.H.; Sun, G.Y. Different N forms on energy allocation in photosystem of mullberry seedlings leaves. *Chin. Agric. Bull.* **2015**, *31*, 18–25.

31. Hu, Y.; Bellaloui, N.; Sun, G.; Tigabu, M.; Wang, J. Exogenous sodium sulfide improves morphological and physiological responses of a hybrid populus, species to nitrogen dioxide. *J. Plant Physiol.* **2014**, *171*, 868–875. [CrossRef]

32. Gao, J.F. *Guidance for Plant Physiology Experiment*; Beijing Higher Education Press: Beijing, China, 2006; pp. 68–70.

33. Li, H.S. *Principle and Technology of Plant Physiology and Biochemical Experiments*; Beijing Higher Education Press: Beijing, China, 2000; pp. 192–194.

34. Ethier, G.J.; Livingston, N.J. On the need to incorporate sensitivity to CO_2 transfer conductance into the Farquhar-von Caemmerer-Berry leaf photosynthesis model. *Plant Cell Environ.* **2004**, *27*, 137–153. [CrossRef]

35. Schansker, G.; Srivastava, A.; Strasser, R.J. Characterization of the 820-nm transmission signal paralleling the chlorophyll a fluorescence rise (OJIP) in pea leaves. *Funct. Plant Biol.* **2003**, *30*, 785–796. [CrossRef]

36. Strasser, B.J.; Strasser, R.J. Measuring fast fluorescence transients to address environmental questions: The JIP-Test. *Photosynth. Light Biosph.* **1995**, *5*, 977–980.

37. Hendrickson, L.; Furbank, R.T.; Chow, W.S. A simple alternative approach to assessing the fate of absorbed light energy using chlorophyll fluorescence. *Photosynth. Res.* **2004**, *82*, 73–81. [CrossRef]

38. Zhou, Y.; Lam, H.M.; Zhang, J. Inhibition of photosynthesis and energy dissipation induced by water and high light stresses in rice. *J. Exp. Bot.* **2007**, *58*, 1207–1217. [CrossRef] [PubMed]

39. Zeevaart, A.J. Induction of nitrate reductase by NO_2. *Acta Bot. Neerl.* **1974**, *23*, 345–346. [CrossRef]

40. Rowland, A.J.; Drew, M.C.; Wellburn, A.R. Foliar entry and incorporation and incorporation of atmospheric nitrogen dioxide into barley plants of different nitrogen status. *New Phytol.* **1987**, *107*, 357–371. [CrossRef]

41. Wingsle, G.N.; Sholm, T.; Lundmark, T.; Ericsson, A. Induction of nitrate reductase in needles of Scots pine seedlings by NO_X and NO_3^-. *Physiol. Plant.* **2010**, *70*, 399–403. [CrossRef]

42. Norby, R.J.; Weerasuriya, Y.; Hanson, P.J. Induction of nitrate reductase activity in red spruce needles by NO_2. *Can. J. For. Res.* **1989**, *19*, 889–896. [CrossRef]

43. Hisamatsu, S.; Nihira, J.; Takeuchi, Y.; Satoh, S.; Kondo, N. NO_2 suppression of light-induced nitrate reductase in Squash Cotyledons. *Plant Cell Physiol.* **1988**, *29*, 395–401.

44. Wang, Y.H.; He, X.Y.; Zhou, G.S. Characteristics and quantitative simulation of stomatal conductance of Aneurolepidium chinense. *Chin. J. Appl. Ecol.* **2001**, *12*, 517–521.

45. Jiao, N.Y.; Ning, T.Y.; Yang, M.K.; Fu, G.Z.; Ying, F.; Xu, G.W. Effect of maize and peanut intercropping on photosynthetic characters and yield forming of intercropped maize. *Acta Ecol. Sin.* **2013**, *33*, 4324–4330. [CrossRef]

46. Bilger, W.; Schreiber, U.; Bock, M. Determination of the quantum efficiency of photosystem II and of non-photochemical quenching of chlorophyll fluorescence in the field. *Oecologia* **1995**, *102*, 425–432. [CrossRef] [PubMed]

47. Genty, B.; Briantais, J.M.; Baker, N.R. The relationship between the quantum yield of photosynthetic electron transport and quenching of chlorophyll fluorescence. *Biochim. Biophys. Acta* **1989**, *990*, 87–92. [CrossRef]

48. Lavaud, J.; Kroth, P.G. In diatoms, the transthylakoid proton gradient regulates the photoprotective non-photochemical fluorescence quenching beyond its control on the xanthophyll cycle. *Plant Cell Physiol.* **2006**, *47*, 1010–1016. [CrossRef]

49. Demmig-adams, B.; Adams, W.W.; Barker, D.H.; Logan, B.A.; Bowling, D.R.; Verhoeven, A.S. Using chlorophyll fluorescence to assess the fraction of absorbed light allocated to thermal dissipation of excess excitation. *Physiol. Plant.* **1996**, *98*, 253–264. [CrossRef]

50. Kramer, D.M.; Johnson, G.; Kiirats, O.; Edwards, G.E. New fluorescence parameters for the determination of Q_A, redox state and excitation energy fluxes. *Photosynth. Res.* **2004**, *79*, 209–218. [CrossRef] [PubMed]

51. Hendrickson, L.; Förster, B.; Pogson, B.J.; Chow, W.S. A simple chlorophyll fluorescence parameter that correlates with the rate coefficient of photoinactivation of photosystem II. *Photosynth. Res.* **2005**, *84*, 43–49. [CrossRef]

Article

Tree Growth and Wood Quality in Pure Vs. Mixed-Species Stands of European Beech and Calabrian Pine in Mediterranean Mountain Forests

Diego Russo [1], Pasquale A. Marziliano [1,*], Giorgio Macrì [1], Giuseppe Zimbalatti [1], Roberto Tognetti [2,3] and Fabio Lombardi [1]

[1]   Department of AGRARIA, Mediterranean University of Reggio Calabria, 89122 Reggio Calabria, Italy; diego.russo@unirc.it (D.R.); giorgio.macri@unirc.it (G.M.); gzimbalatti@unirc.it (G.Z.); fabio.lombardi@unirc.it (F.L.)

[2]   Department of Agricultural, Environmental and Food Sciences, University of Molise, 86100 Campobasso, Italy; tognetti@unimol.it

[3]   The EFI Project Centre on Mountain Forests (MOUNTFOR), Edmund Mach Foundation, Via Edmund Mach, 38010 San Michele all'Adige, Italy

*   Correspondence: pasquale.marziliano@unirc.it; Tel.: +3909651694256

Received: 2 October 2019; Accepted: 16 December 2019; Published: 18 December 2019

Abstract: Mixed-species forests may deliver more forest functions and services than monocultures, as being considered more resistant to disturbances than pure stands. However, information on wood quality in mixed-species vs. corresponding pure forests is poor. In this study, nine plots grouped into three triplets of pure and mixed-species stands of European beech and Calabrian pine (three dominated by European beech, three dominated by Calabrian pine, and three mixed-species plots) were analysed. We evaluated tree growth and wood quality through dendrochronological approaches and non-destructive technologies (acoustic detection), respectively, hypothesizing that the mixture might improve the fitness of each species and its wood quality. A linear mixed model was applied to test the effects of exogenous influences on the basal area index (BAI) and the dynamic modulus of elasticity (MOEd). The recruitment period (Rp) was studied to verify whether wood quality was independent from stem radial growth patterns. Results showed that the mixture effect influenced both wood quality and BAI. In the mixed-species plots, for each species, MOEd values were significantly higher than in the corresponding pure stands. The mixture effect aligned MOEd values, making wood quality uniform across the different diameter classes. In the mixed-species plots, a significant positive relationship between MOEd and Rp, but also significantly higher BAI values than in the pure plots, were found for European beech, but not for Calabrian pine. The results suggest the promotion of mixing of European beech and Calabrian pine in this harsh environment to potentially improve both tree growth and wood quality.

Keywords: TreeSonic; MOEd; forest productivity; dendrochronology; recruitment period; Aspromonte National Park

---

1. Introduction

Mixed-species forests may deliver forest functions and services more effectively than monocultures [1,2], particularly in threatened mountain environments [3] and in man-made mixed-species forests. Therefore, their spread is an important option to adapt European mountain forests and forestry to future disturbances and extreme events [4]. Mixed-species forests may show less temporal variation in growth and more stable productivity in comparison with pure stands, due to reduced tree species competition for resources [5–13]. Nevertheless, contradictory mixed-species effects on tree growth have been found under conditions favouring drought stress or in dry years [14–16]. Gebauer et al. [17] observed that stand-level canopy transpiration was not higher in mixed broad-leaved forests than in pure European beech stands and that the spatial complementarity in root water uptake of mixed-species stands could be masked by edaphic conditions preventing the vertical stratification of species-specific root systems. An increase in the frequency and severity of drought events, as predicted for the coming decades [18,19], will have dramatic implications for the resilience of Mediterranean mountain ecosystems, particularly in the case of forest stands with a simplified vertical structure [20]. Indeed, species-specific functional traits and allometric relations can be more important for stand water use than tree species diversity *per se*. Therefore, more insight into species mixing, structural diversity and forest dynamics is essential for modeling risk assessment and forest functions aimed at fostering alternative silvicultural practices in harsh environments.

Although mixed-species stands can be more productive in comparison with monocultures [21–24] up to 30% [25], site conditions, stand age, and tree species interactions affect these responses [26,27]. Focusing on the causes of differences in tree growth and stand productivity between mixed-species and pure forests, most research addressed environmental settings and relationships with climatic conditions, species composition, mixture type, and stand age [28–36]. Indeed, few studies exist on the quality and value of wood produced in mixed-species vs. corresponding pure stands [37]. Information on how wood quality can be affected in relation to tree species composition is essential for decision making in adaptive forestry, especially where forest planning and thinning activities favour the occurrence of mixed-species stands [27].

Liu et al. [38] showed that spacing and thinning experiments in pure stands highlighted the strong effect of the surrounding spatial stand structure on tree growth and morphology and, ultimately, wood structure and timber quality. Chomel et al. [39] demonstrated that, in mixed-species plantations, mixing hybrid poplar and white spruce might increase the wood production of poplar in comparison with monocultures of either poplar or white spruce. Battipaglia et al. [40] showed a considerable increase in cumulative basal area and in intrinsic water use efficiency in mixed-species stands of pedunculate oak and Italian alder, largely resulting from an increase in N fixation, which levelled off, when natural mortality or management practices decreased the competitive ability of Italian alder. In addition to stand productivity and water use efficiency, species mixtures with structural stratification may also enhance individual-tree growth rates and stem quality of species in the upper canopies, minimizing the proportion of taller species that reach the highest production [6,41].

Forest management practices oriented to obtain mixed-species stands potentially determine variation in the physical and mechanical properties of the harvested wood, both at stand and tree level [42]. Nevertheless, although mixed-species forestry is gaining popularity in Europe [10,43], a greater understanding of the differences in wood quality between mixed-species vs. corresponding pure stands is needed. In the last decades, the evaluation of wood quality has been preferentially based on non-destructive technologies (NDTs) [44]. More specifically, stress wave-based non-destructive acoustic techniques resulted in very useful methods for predicting the mechanical properties of woody materials [45]. Nowadays, acoustic sensing technology allows for the estimation of wood quality and intrinsic woody properties for standing trees, stems and logs. Among the parameters measurable by acoustic methods, the most important is the dynamic modulus of elasticity (MOEd), being related to wood anatomy and tree physiology. This parameter can be used for the evaluation of wood quality, providing information on the stiffness of material [46,47].

In this study, the wood quality term refers to the use of wood in the field of construction. The wood properties that determine the structural requirements are defined by indicators, such as strength grades, that depend on knottiness, stiffness and density of the wood [48]. These parameters are commonly used worldwide to define strength grades [49]. With reference to stiffness, the dynamic modulus of elasticity (MOEd) is considered a good predictor, so as to be used as a proxy variable for assessing the wood quality and, in particular, its stiffness [50]. As a matter of fact, profound implications for wood properties and utilization derive from the growth and ecophysiological responses of trees to environmental conditions.

We focused on two important tree species, European beech (*Fagus sylvatica* L.) and Calabrian pine (*Pinus nigra* Arnold subsp. calabrica), widespread in mountainous forest ecosystems of Southern Italy; European beech is a late-successional broadleaved tree species, here occurring in its southernmost ecological limit, whereas Calabrian pine is an early-successional conifer, providing timber of high quality. In the Calabrian Apennines, these species occur in both pure and corresponding mixed-species stands, which may help test the effects of mixture on tree growth and wood quality in a drought-prone environment. Our study comprises nine plots grouped into three triplets of pure and mixed-species stands of European beech and Calabrian pine (three dominated by European beech, BP, three dominated by Calabrian pine, PP, and three mixed-species plots, MBP), in the Aspromonte National Park. The three stands grow in the same environmental conditions and show similar structural traits. We hypothesized that the mixture effect may improve the physiological fitness of each species and, thus, the wood quality. The specific objectives of the study were to: (Q1) evaluate the wood quality (MOEd), (Q2) analyse tree growth patterns through the basal area index (BAI), and (Q3) highlight potential relationships between wood quality and tree growth, in mixture vs. monocultures. In particular, we tested whether (i) the mean wood quality in mixed-species and corresponding pure stands are equal, (ii) the mean BAI in pure and mixed-species stands are equal, and if (iii) the wood quality and tree growth patterns are independent.

2. Materials and Methods

2.1. Study Area

The study area is located in the Aspromonte National Park, Calabria (Southern Italy), near the village of Bagaladi (RC). The climate is temperate, with an annual mean temperature of around 8 °C and minimum and maximum monthly means of 0.5 °C (coldest month) and 16 °C (warmest month), respectively. Annual precipitation is 1611 mm unevenly distributed over the year (Meteorological station of Gambarie d'Aspromonte—1187 m above sea level (a.s.l.).

The study is based on nine plots (each with an extension of 5000 m^2 and square shape), grouped into three triplets of pure and mixed-species stands of European beech and Calabrian pine, using a randomized block design. Plots within each block were randomly assigned to the units. Each triplet, extended over 1.5 ha, contains two monospecific plots, dominated by European beech (BP) or Calabrian pine (PP), and one mixed-species European beech—Calabrian pine stand (MBP). The pure stands were selected if the target species represented ~90% of the stand basal area. The mixed-species stand was defined as the stand in which the two species together represented at least ~80% of the total stand basal area and the sum of the basal area of other species was lower than that of each of the two studied species. In order to quantify the mixing effects, the pure stands were used as reference for the mixed-species stand.

Plots are located in similar environmental conditions in terms of environmental settings (topography, slope, substrate) and climatic conditions. Soils developed from igneous and metamorphic rocks and are classified as Umbrisols, Cambisols and Leptosols [51], with an udic soil regime moisture. Further details on the main environmental conditions characterizing the study site are reported in Table 1.

These forest stands are of natural origin, with a dominant age ranging from 70 to 120 years old. They evolved naturally in the last 70 years. However, these forests (approximately until the Second World War) were cut following a management approach based on the "selection cutting" criteria, in which few trees were cut approximately every 20 years in the same forest section [52]. This type of management has preserved the typical forest landscape by maintaining a continuous forest cover, determining a stand structure consisting of clusters of trees in different age classes. These clusters are the result of the natural regeneration occurring in the gaps opened by the selection cutting previously applied [52].

Table 1. The main geographical features of the investigated triplets located nearby the village of Bagaladi, RC (Aspromonte National Park).

	Triplet 1	Triplet 2	Triplet 3
Latitude (N)	38°06′50.53″	38°05′54.59″	38°06′15.69″
Longitude (E)	15°51′53.25″	15°50′37.72″	15°51′23.37″
Exposure (°)	59	59	59
Altitude (m a.s.l.)	1521	1511	1497
Slope (°)	28	30	34

2.2. Tree Sampling and BAI Analyses

In 2017, in order to derive the dendrometric variables at the stand level, the nine plots were fully inventoried. The diameters at breast height (DBH) and the total heights (Ht) of all trees were measured. Tree volume was calculated using the equations derived for the prediction of the aboveground tree volume [53]. Canopy dominant/co-dominant trees were targeted for sampling (coring and measuring MOEd) and were selected through a stratified random sampling, in relation to the frequency of the diameter classes at the plot level. In detail, in each monospecific stand (BP and PP), 30 dominant trees per species were cored, while 60 dominant trees were cored (30 for European beech and 30 for Calabrian pine) in MBP. For each selected tree, two increment cores were collected at breast height and at an angle of 120° to each other. Particular care was taken to select trees with canopies well separated from each other. Moreover, to avoid the effect of wood alteration and exogenous disturbances on tree ring growth, only trees without abrasion scars or other visible evidence of injury were selected. Cores were then mounted on channeled wood, seasoned in a fresh-air dry store and sanded a few weeks later. Ring widths were measured with a resolution of 0.01 mm using the LINTAB measurement equipment (Frank Rinn, Heidelberg, Germany) fitted with a Leica MS5 stereoscope (Leica Microsystems, Wetzlar, Germany). Tree ring widths were then statistically cross-dated and verified with the software TSAP software package (version 4.81c) and COFECHA (version 6.06) [54]. Once all measurement series were validated, tree-ring chronologies were developed for each species. Subsequently, tree ring widths were converted into tree basal area increment (BAI), according to the following standard formula:

$$BAI \ = \ \pi\left(r_n^2 - r_{(n-1)}^2\right)$$

where r is the radius of the stem at breast height and n is the year of tree-ring formation. In order to examine the mean growth trend for the sampled trees, BAI for each year was averaged over all the individuals for removing the variation in radial growth attributable to the increasing tree circumference.

The latest fully built tree ring valid for our analysis was related to 2016. We considered only the tree-ring chronologies starting from 1900, because few trees had rings formed before the year 1900.

2.3. Wood Quality

In order to evaluate the wood quality, acoustic wave tests were conducted at breast height on the same 120 trees from which the woody cores were collected. In order to reduce the effect of air humidity [55], the tests were carried out in the summer of 2017, from June to July.

TreeSonic (FAKOPP TreeSonic™ Fakopp Enterprise, Agfalva, Hungary) [44,56] was used for measuring the acoustic velocity. It is characterized by a hand-held hammer and two probes, a transmitting accelerometer and a receiving accelerometer. The system consists in the insertion of two sensor probes (a transmitting probe and a receiving one) into the sapwood, applying acoustic energy into the tree stem through hammer impact. The probes were aligned within a vertical plane on the same face. In our study, a 1.00-m testing span was roughly centered at breast height. The lowest probe was placed about 60–70 cm above the forest floor. Three measurements were realized for each selected tree and the average of the three recordings was used as final transit time. In order to measure the acoustic velocity wave, the start and stop sensors were driven at a 45° angle through the bark into the wood of the standing tree [57]. Acoustic velocity was determined on the upstream side of the tree (according to standard procedure), the same side where the stem diameter was measured. The acoustic velocity (CT) was then calculated from the span between the two sensor probes and the time-of-flight (TOF) data, using the following formula:

$$CT = S/TOF \tag{1}$$

where CT = tree acoustic velocity (m/s), S = distance between the two probes (sensors) (m), TOF = time-of-flight (s).

The wood density of each measured tree was also determined on the woody cores. The fresh weight and volume of the tree cores were measured in the laboratory. Woody samples were weighed to the nearest 0.01 g with an electronic scale. After ring width measurements, the woody cores were dried in oven at 105 °C, to constant weight. Density ($kg\ m^{-3}$) was calculated by dividing the dry weight by the fresh volume.

Afterwards, it was possible to calculate the modulus of elasticity (MOE), according to the following equation:

$$MOEd = WD_i \times CT^2; \tag{2}$$

where WD_i = tree wood density i ($kg\ m^{-3}$) and CT = velocity ($km\ s^{-1}$).

When MOE is estimated using the above formula, it is termed dynamic modulus of elasticity (MOEd) [58], as the stress wave propagation in wood is a dynamic process, internally related to the physical and mechanical properties of wood [59].

2.4. Relationship between Wood Quality and Tree Growth

In standing trees, the stress wave propagation is affected by tree stem diameter, wood travel distance and internal wood conditions [60]. Furthermore, the mean tree values of timber grade-determining properties (elastic modulus, bending strength, and wood density) are related to the acoustic velocity and tree slenderness. Signals are transferred through the whole stem, depending on stem diameter: when smaller is the diameter, then the signal is more transferred inside the stem [61]. In this study, we used the tree final growth time to evaluate the relationship between wood quality and tree growth, in order to verify if the wood quality was independent from tree growth. In detail, the final growth time is given by the number of radial rings included in the outermost 2.5 cm of the sampled trees. This parameter is called "recruitment period" (RP), which indicates the number of years needed to pass from a diameter class x_i to the largest class $x_i + 5$, assuming classes of 5 cm [62]. Therefore, for each sampled tree, we counted the number of rings contained in the last 2.5 cm.

2.5. Statistical Analysis

The data collected were characterized by a nested structure and they were not independent from each other: in detail, an inter-correlation among samples occurred, since they were referred to the same triplets, and the single tree ring chronologies belonged to one single tree. Therefore, an analysis based on a linear mixed model was used, including both fixed and random effects. We then set up the model functions to test the effect of mixing on tree ring width (expressed as BAI) and on the MOEd. The linear mixed models incorporated the triplet identifier as a random effect, thus considering both the mixing effect (expressed as percentage of basal area of the examined species of beech or pine, compared to the total area) and the nested structure of the data. The initial model included the following independent variables: basal area, stem DBH, tree height, tree age. A stepwise procedure (in each step, a variable was considered for addition to or subtraction from the set of explanatory variables based on *F*-tests) eliminated all variables, except for the basal area, the only variable that had a significant effect on the dependent variable. Therefore, the below model, based only on mixing effect and the nested design of data, was applied to evaluate if significant differences on BAI between mixed-species and corresponding pure stands occurred:

$$BAI_{ij} = \beta_0 + \beta_1 \times Mix_{ij} + u_j + \varepsilon_{ij}$$

where BAI_{ij} is the mean of the basal area increment per tree i on triplet j, β_0 is the intercept, Mix_{ij} is the fixed effect describing the quantification of the mixing effect (expressed as % basal area) on tree BAI for tree i on triplet j, u_j is the random effect for triplet j and ε_{ij} is the error term. Moreover, the following model was used to analyse the effect of both mixture and tree ring width (expressed as recruitment period) on wood quality (MOEd values), incorporating the triplet identifier as a random effect:

$$MOEd_{ij} = \beta_2 + \beta_3 \times Mix_{ij} + \beta_4 \times RP_{ij} + u_j + \varepsilon_{ij}$$

where $MOEd_{ij}$ is the wood quality of tree i on triplet j, RP_{ij} is the recruitment period of tree i on triplet j, examined as the main effect. For the linear mixed-effect model analysis, the nlme package for the R programming language [63] was used. In addition to the linear mixed-effect models, the Mann-Whitney Test for independent-sample (*U*-Test) was also used to compare the average MOEd values and the average BAI values between mixed-species and pure stands. More specifically, the *U*-test is a non-parametric test and requires no assumptions.

3. Results

Table 2 shows the main structural and dendrometric characteristics of the studied stands, while in Table 3, the age, the wood density, the acoustic velocity and the MOEd for the sampled trees are reported. Forest structure did not differ markedly among the studied stands. Analysis of variance (ANOVA) showed no significant differences among plots in stand density ($F_{3/6} = 1.020$; $p = 0.447$), tree DBH ($F_{3/6} = 0.329$; $p = 0.805$), basal area ($F_{3/6} = 2.104$; $p = 0.201$) and stand volume ($F_{3/6} = 2.208$; $p = 0.188$). The number of trees per hectare varied from 743 in the pure European beech plots to 873 in the mixed-species plots. The tree volume ranged between 773 m^3 ha^{-1} in the pure European beech plots and 876 m^3 ha^{-1} in the mixed-species plots.

Table 2. Structural traits obtained for the investigated triplets. Mean values and the related standard deviation (in bracket) are reported.

	Stand Density (N. Trees ha^{-1})	Tree DBH (cm)	Tree Height (m)	Basal Area (m^2 h^{-1})	Stand Volume (m^3 ha^{-1})
Beech pure	743 (155.7)	33.2 (2.8)	24.2 (2.2)	64.3 (4.6)	772.7 (41.5)
Pine pure	754 (96.3)	35.6 (4.4)	23.1 (0.8)	75.1 (10.4)	857.6 (160.9)
MBP-B	470 (75.9)	28.3 (2.3)	18.8 (1.3)	29.6 (6.3)	277.0 (31.1)
MBP-P	403 (74.1)	40.3 (9.6)	23.6 (5.1)	51.4 (14.1)	599.1 (90.7)

Table 3. Tree age, wood density, acoustic velocity, and MOEd values obtained for the two studied species in the pure and mixed-species stands (SD: standard deviation; MOEd: dynamic modulus of elasticity).

Stands	Tree Age (Years)		Wood Density (kg m^{-3})		Wave Velocity (m s^{-1})		MOEd (MPa)	
	Mean	SD	Mean	SD	Mean	SD	Mean	SD
Beech pure	69	41	658	28.4	3841.7	382.4	9523.2	1738.1
Pine pure	97	38	562	27.6	3823.4	427.5	8056.6	1358.2
MBP-B	127	33	653	30.1	4253.0	349.3	11583.0	2212.6
MBP-P	103	21	555	25.2	4066.4	370.6	8999.8	2132.6

Tree age differed between the pure European beech plots (on average 69 years) and the European beech trees sampled in the mixed-species plots (on average 127 years) (Table 3). By contrast, Calabrian pines occurring in the pure and mixed-species plots revealed no significant differences in tree age (on average 100 years).

For both tree species, MOEd values were significantly higher in MBP than in BP and PP, as confirmed by the Mann-Whitney test applied for independent-samples (Figure 1). MOEd was 21.6% and 10.7% higher in MBP than in BP and PP, respectively. Further, considering the different diameter classes investigated, MOEd values were always higher in MBP than in BP and PP.

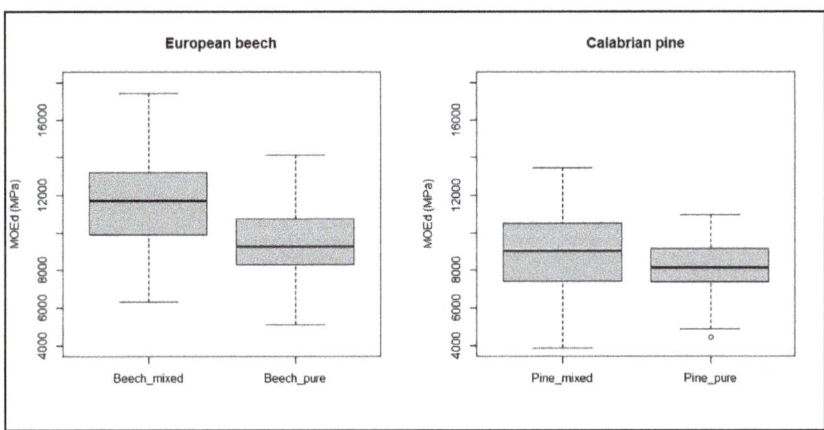

Figure 1. MOEd values in the pure and mixed-species stands for European beech (Mann-Whitney Test, $Z = -6.491$, $p < 0.001$) and Calabrian Pine (Mann-Whitney Test, $Z = -3.134$, $p < 0.010$).

Analysis of variance (ANOVA) showed a significant effect of the diameter classes on MOEd for BP and PP. As the diameter classes are not independent, the Kruskal–Wallis test was also used [64], which confirmed the significant effect of the diameter classes on MOEd. MOEd decreased as tree DBH increased, with significant differences observed between the smallest and the largest diameter classes. On the contrary, no significant differences were observed in MBP (Figure 2). The mixture of species had the effect of aligning MOEd values, both for European beech and Calabrian pine, making wood quality more uniform for each diameter class. In contrast, in the pure stands, MOEd values were significantly lower than in MBP, also varying significantly as tree DBH increased.

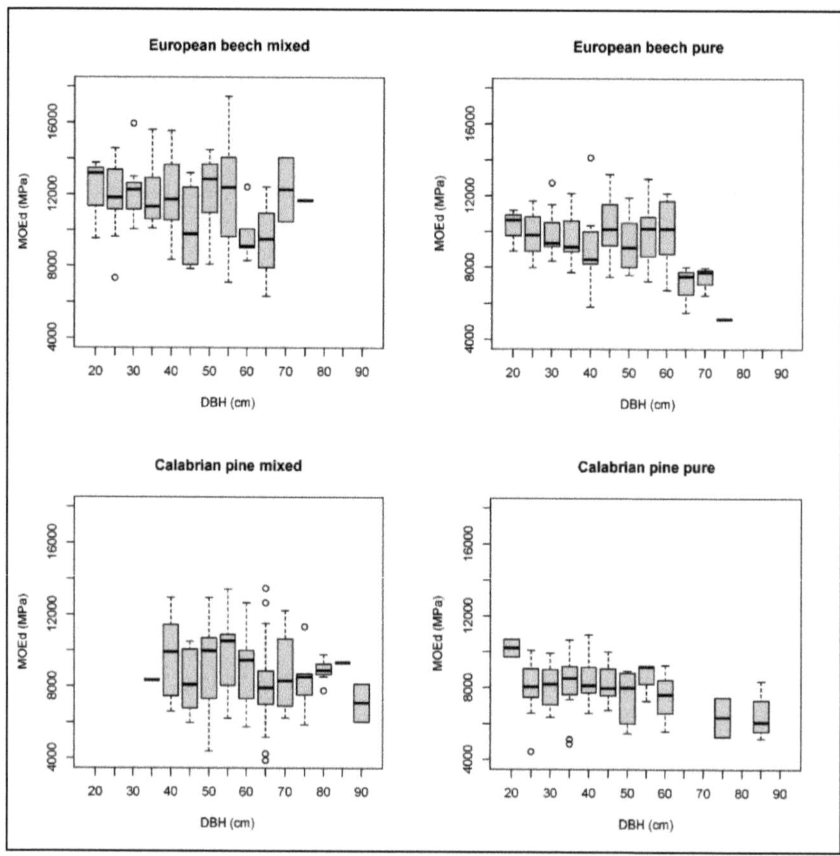

Figure 2. Dynamic modulus of elasticity (MOEd) in relation to the diameter classes for European beech and Calabrian pine growing in the pure and mixed-species stands. European beech in the mixed-species stand: $F_{12;18} = 1.255$, $p < 0.322$; for European beech in the pure stand: $F_{12;18} = 2.461$, $p < 0.050$; Calabrian pine in the mixed-species stand: $F_{12;18} = 0.774$, $p < 0.669$; Calabrian pine in the pure stand: $F_{11;19} = 2.621$, $p < 0.050$.

Figure 3 shows the BAI values obtained for European beech and Calabrian pine. In general, BAI increased over time (at least up to a determined year) in both species. Overall, trees revealed higher BAI in MBP than in BP and PP. However, until 1980, BAI values of European beech growing in mixture were always higher than in BP. Later, a similar trend was observed, until 1990. In the last period (1990–2016), BAI values were again higher in MBP. In the period 1900–2016, BAI always increased in MBP, while it gradually decreased in BP, from 1990. For Calabrian pine, BAI values were higher in MBP than in PP, considering the period 1900–1960. Later, for about 20 years (1960–1980), an opposite trend was observed, with a higher BAI in PP. Furthermore, for 1980–2016, a consistent decrease of BAI values was observed for Calabrian pine, both in MBP and PP.

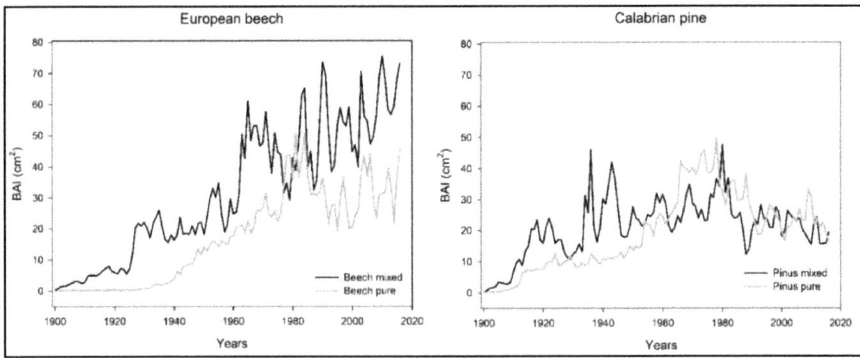

Figure 3. BAI values for European beech and Calabria pine growing in the pure and mixed-species stands. Values were considered starting from 1900.

Figure 4 reports both the variability and the average of BAI values for European beech and Calabrian pine growing in the pure and mixed-species stands. For European beech, the BAI variability was higher in MBP, while for Calabrian pine in PP. European beech growing in mixture had a significantly higher BAI values than in BP (+65%, 30.6 vs. 18.5 cm^2, on average, p-value < 0.001). Yet, Calabrian pine growing in mixture revealed higher BAI values than in PP, although not significantly (+11%, 21.1 vs. 19.1 cm^2, on average, p-value $= 0.0563$).

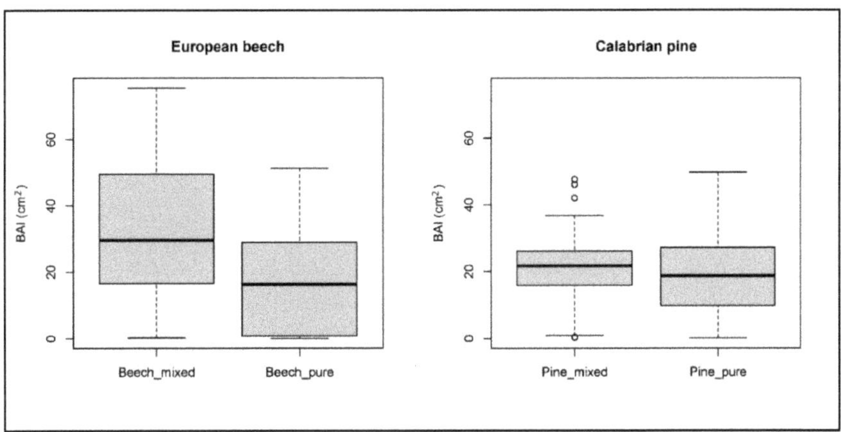

Figure 4. BAI values (mean values) in the pure and mixed-species stands for European beech (linear mixed-effect, $p < 0.05$, $R^2 = 0.437$; Mann-Whitney Test, $Z = -4.345$, $p < 0.0001$) and Calabrian pine (linear mixed-effect, $p = 0.271$, $R^2 = 0.078$; Mann-Whitney Test, $Z = -1.928$, $p = 0.0563$).

In Table 4, results obtained through the application of the linear mixed model are shown. The site identifier was considered as a random effect. The mixture significantly influenced European beech ($p = 0.015$), but not Calabrian pine ($p = 0.601$).

Table 4. Values achieved with the application of the linear mixed-effect model, considering the "triplet" as random effect. The influences of the fixed effects (BA%, RP) on the response variables "basal area increment" (BAI) and "dynamic modulus of elasticity" (MOEd) were examined. Significant variables are reported in bold.

Model	Value	t-Value	p-Value	R^2
European beech (BAI)				0.502
Intercept	17.623	7.720	<0.0001	
Mix (BA)	**0.358**	**4.047**	**0.0155**	
Calabrian pine (BAI)				0.015
Intercept	19.299	8.126	<0.0001	
Mix (BA)	0.028	0.567	0.6009	
European beech (MOEd)				0.278
Intercept	5114.401	4.191	0.0000	
RP	**210.252**	**4.119**	**0.0001**	
Mix (BA)	**76.434**	**3.353**	**0.0285**	
Calabrian pine (MOEd)				0.045
Intercept	6487.034	6.291	<0.0001	
RP	83.52	1.833	0.0684	
Mix (BA)	23.23	2.079	0.1061	

Figure 5 shows the values of the recruitment period (RP) for European beech and Calabrian pine, which invariably decreased as DBH increased. For European beech, RP was lower in MBP than in BP, though with lessening differences as DBH increased. Calabrian pine growing in PP had, for the first diameter classes, a slightly lower RP than in MBP, at least for the first diameter classes. Then, the trend was the opposite and the differences tended to increase, though not for the DBH of about 90 cm.

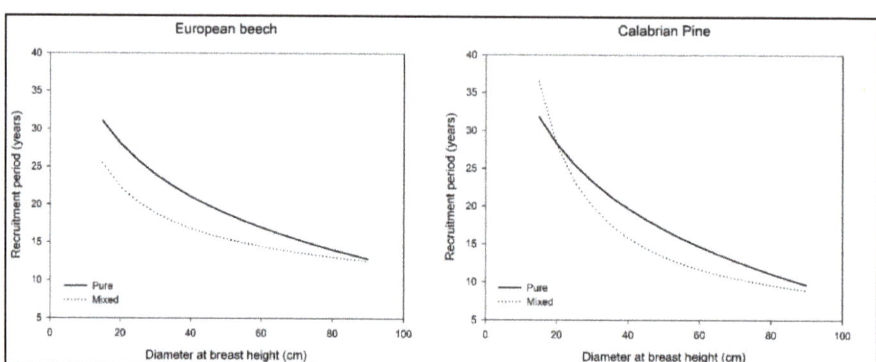

Figure 5. Recruitment period obtained for European beech and Calabrian pine growing in the pure and mixed-species stands. It indicates the number of years needed to pass from a diameter class xi to the largest class xi + 5 cm.

Finally, in Figure 6, variations of MOEd as RP changes are shown. In each stand, MOEd tended to rise as RP increased, even if it revealed higher values in MBP than in BP and PP. In addition, the relationship between MOEd and RP was stronger in MBP, for both species. However, the analyses of the linear mixed-effects model underlined that only for European beech, RP ($p < 0.0001$) and the mixing effect ($p = 0.0285$) had a significant role, explaining the largest proportion of MOEd variation (Table 3). On the contrary, for Calabrian pine, RP ($p = 0.0684$) and the mixing effect ($p = 0.1061$) had marginal effects on MOEd.

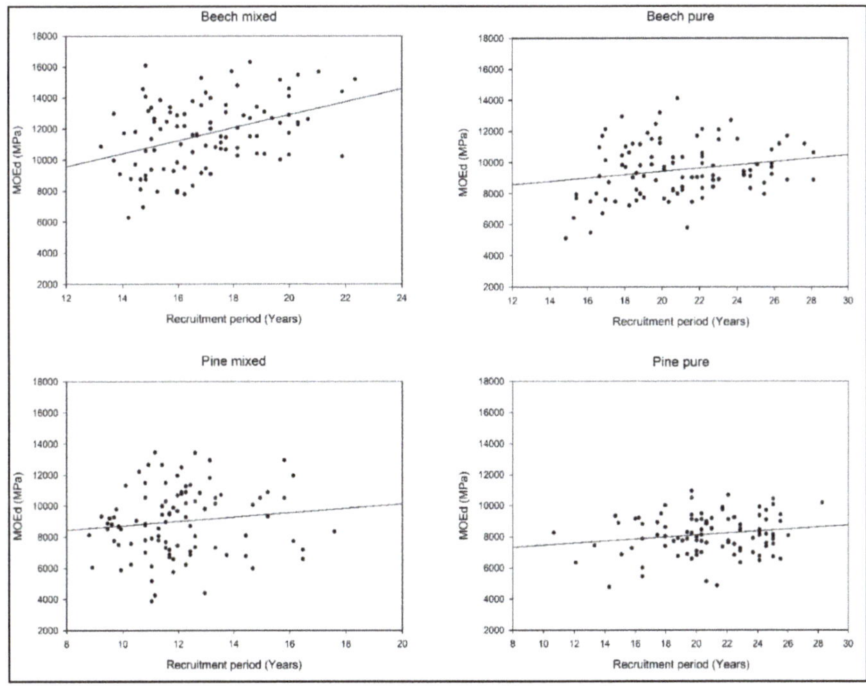

Figure 6. Dynamic modulus of elasticity (MOEd) in relation to the recruitment period for European beech and Calabrian pine growing in the pure and mixed-species stands.

4. Discussion

Results demonstrated how the mixture effect influenced both wood quality and tree ring widths. In both the pure and mixed-species stands, MOEd values were above the minimum quality threshold (MOEd > 9000 MPa) [65]. However, in the mixed-species plots, MOEd values were always higher than in the corresponding pure plots. This difference was more evident for European beech (11583 vs. 9523) than for Calabrian pine (8999 vs. 8056). Tree age differed between European beech trees sampled in the pure and mixed species plots, which might have played a role in determining differences in MOEd values. In the case of Calabrian pine, the differences in MOEd values are less evident, though significant. As a matter of fact, tree age did not differ between trees sampled in the pure and mixed-species plots.

Eventually, effects on MOEd exerted by the other parameters related to the stand structure (stand density, basal area, stand volume) could be excluded, since the parameters did not differ significantly between the pure and mixed-species stands. Several studies showed that, considering other variables being equal (stand density, stand volume, etc.), the mixture effect could have a role in improving the resistance and resilience of trees in comparison with the corresponding monocultures [21–24]. Yet, the mixed-species stands could also be more productive than the corresponding monocultures [26].

Therefore, even with some limitations, a mixture of European beech and Calabrian pine might have a positive effect on MOEd, not only at the stand level, but also considering the different diameter classes. In fact, in the mixed stands, the MOEd values were similar for all the diameter classes, in both species. On the contrary, in the pure stands, we observed a decrease in wood quality as the tree DBH increased. A combination of a light demanding tree species with one more shade tolerant was probably the foundation of these results. In other studies, mixed-species stands were found to show higher light interception and light-use efficiency in comparison with monocultures [10,66–68]. Promoting the coexistence of these two species in the present environmental conditions might improve the quality of wood materials, regardless of the size of trees from which the assortments are obtained.

The MOEd values, strongly related to wood density and acoustic velocity, indicated an overall higher mechanical stability attained in the mixed-species stands. In fact, wood density is correlated with timber strength [69], hardness and abrasiveness [48]. Therefore, mixing these species might provide higher mechanical resistance to natural disturbances, such as windstorms, and higher carbon sequestration potential in these Mediterranean mountain forest ecosystems. It must be pointed out that Pretzsch and Rais [49] observed that the wood strength and stiffness could be lower in complex forests than in homogeneous monocultures, where tree size and shape development progress more continuously. Yet, Torquato et al. [70], in black spruce forests in Canada, detected lower strength and stiffness properties in complex stands.

Growing in mixtures induced an increase in tree biomass production, particularly for European beech, while differences between the pure and mixed-species stands were not significant for Calabrian pine. In European beech, although environmental conditions did not differ between stands, BAI was significantly higher in mixture than in monoculture. However, differences between the pure and mixed-species stands in tree age might also have affected BAI trends. The identification of growth trends and attribution of their drivers are not easy tasks, and require long-term and spatially-representative experiments to control the effect of tree size and age, avoiding sampling biases at the various scales. Indeed, heterogeneous crown structures and the tree size of mixed-species stands might facilitate the penetration of light into the lower canopy layers [10]. In the studied stands, European beech could take advantage of growing underneath the relatively transparent canopy of Calabrian pine. On the other hand, European beech in pure and dense stands develops homogeneous canopy cover, having high shade tolerance and large canopy expansion. Indeed, European beech growing in monocultures showed low self-tolerance [71,72], resulting in severe intraspecific competition due to high lateral canopy expansion [73]. Moreover, other studies reported beneficial effects on European beech productivity from growing in mixture [74,75]. Condes et al. [76] reported that the productivity of pine-beech admixtures was generally greater than in the corresponding pure stands. In light of this, admixture of European beech with Calabrian pine might, therefore, reduce competition or even increase facilitation processes in this Mediterranean mountain environment. Nevertheless, Conte et al. [16] observed higher productivity in the pure than mixed-species stands of European beech and Scots pine, in Alpine environmental conditions. These discrepancies suggest caution in generalizing the positive effects on European beech productivity from growing admixed with pine species.

However, the complexity of the species interactions, which depends on stand development stage, stand density and site conditions [26], suggests caution in generalizing these results, since other studies reported opposite patterns [77]. Equal productivity at the stand level might not necessarily indicate a neutral behaviour of the two species co-occurring in the mixed-species stands. In fact, species-specific reactions, at individual or stand level, might counteract and cancel each other with respect to the stand level productivity [74]. Behind overyielding or underyielding of mixed-species in comparison with nearby pure stands, as revealed in this study for Calabrian pine and European beech, there is always a modified supply and uptake, or different use-efficiency, of available resources [24,26,78]. Zhang et al. [21] reported that beneficial effects of admixing provided an overall 25% increase in productivity across forest types and a 12% increase at European scale in Scots pine—European beech mixtures [10]. Nevertheless, the mechanisms that might promote complementarity effects, leading to increased productivity in pine-beech mixtures, are poorly understood, despite the frequent occurrence and economic importance of these forest types [10].

Values of BAI in Calabrian pine and European beech mixtures, of both species, were generally greater than in the corresponding pure stands, although significant only in the case of European beech. European beech, especially in the mixed-species stands, had a constantly increasing BAI in the investigated time-span, although alternating years of little stem radial growth and years of great stem radial growth. On the contrary, Calabrian pine revealed a decrease in BAI values in the last 30–40 years.

Several studies reported that the inter-annual tree growth variability and its dependence on climatic conditions might be modulated by species composition [4,79–81]. A positive effect of species diversity on stand productivity, due to complementarity effects, might result in a greater size heterogeneity, with unknown consequences on the provision of goods and services in pure vs. mixed-species forests. A reduction in size asymmetry and growth inequality for European beech in mixture with Scots pine was related to the complementary light ecology of the associated species [10].

At the stand level, light absorption of admixed light demanding and shade tolerant species might benefit from light-related interactions, including canopy structure, stand density, tree size, crown architecture, and allometric relationships [82,83]. Nevertheless, in the present Mediterranean environment, although light-related interactions might contribute to the mixing effect on tree growth for Calabrian pine, water- and/or nutrient-related interactions might dominate in the mixing effect on tree growth for European beech [84]. Especially for European beech, we showed that the variability of BAI was higher in the mixed-species vs. the pure stands. Although this variability might have a minor role in determining the strength and stiffness of structural timber [49], it might still affect wood properties, such as the dimensional stability. Nevertheless, we may confirm that (i) wood quality was higher in the mixed-species than pure stands, (ii) wood quality was dependent on tree growth patterns and that (iii) the longer a tree took to pass from one diameter class to the next, the better was the wood quality. However, Genet et al. [85] stated that the way in which tree ring widths and ring wood density are related to each other would depend on whether the tree is a conifer, ring-porous hardwood, or diffuse-porous hardwood species.

We found a positive relationship between final growth time and MOEd, in both species. However, while for European beech MOEd was significantly dependent on tree growth (expressed as recruitment period), this relationship was not significant for Calabrian pine. In these relationships, MOEd was higher in the mixed-species than pure stands, for both species. Again, this might depend on species-specific growth patterns, as well as on technological wood features. However, it must be pointed out that wood quality of coniferous trees, in comparison with broadleaved species, might be strongly influenced by forest management practices. In fact, several studies on conifers [59,86–90], including Calabrian pine [44], showed a reduction in wood quality with increasing spacing and thinning.

Admixture of Calabrian pine and European beech might improve adaptation to rainy winters and dry summers of the Mediterranean climate, increasing both tree growth and wood quality, especially for European beech. Several studies showed that mixed-species forests might be more productive and resistant to disturbances, including drought, than corresponding pure stands [4,79], though contrasting evidence was also reported [14,16]. In particular, European beech populations from marginal ecological conditions were found more resistant to drought than those from the core of the distribution range [91–93], and their admixture with co-occurring drought-tolerant conifers might serve as benchmark for selecting resilient tree species combinations. Additional insight into the effect of tree species mixture on wood quality is required to support the entire wood supply chain, from the integration of wood quality in forest inventories to silvicultural guidelines for decision support tools. In fact, although structure and function of mixed-species stands are relevant for their role as supporting services in Mediterranean mountain forests, such as biodiversity conservation and soil protection, the consequence of structural heterogeneity in admixture on wood quality requires greater consideration in the future.

5. Conclusions

In the present study, the admixture of Calabrian pine and European beech increased stand productivity, particularly for European beech, and improved wood quality. Nevertheless, caution is still needed in drawing general conclusions on the mixture effect on wood quality and stand productivity, in marginal environmental conditions. Nevertheless, European beech and Calabrian pine co-occurring in forest stands of Southern Italy should be managed with the aim of improving the overall quality of wood, which could be promoted by mixing these species. Although any generalization could be misleading, the admixture of tree species in these harsh environments might better maintain forest stability, ensuring also woody products of better quality. These results, though related to marginal environmental conditions, could be significant to implement and support climate-smart measures for managing Mediterranean mountain forests.

Author Contributions: Conceptualization, D.R., P.A.M., G.M. and F.L.; Data curation, D.R. and G.M.; Formal analysis, D.R., P.A.M., G.M., G.Z., R.T. and F.L.; Investigation, D.R. and G.M.; Methodology, D.R., P.A.M., R.T., and F.L.; Resources, R.T. and G.Z.; Writing—original draft, D.R., P.A.M., R.T. and F.L.; Writing—review & editing, P.A.M., R.T. and F.L. All authors have read and agreed to the published version of the manuscript.

Funding: This research was funded by the Aspromonte National Park.

Acknowledgments: We are grateful to the Aspromonte National Park for the support given to realize this study. The research is linked to activities conducted within the COST (European Cooperation in Science and Technology) Action CLIMO (Climate-Smart Forestry in Mountain Regions-CA15226) financially supported by the EU Framework Programme for Research and Innovation HORIZON 2020, and the *Progetto bilaterale di Grande Rilevanza* Italy-Sweden "Natural hazards in future forests: how to inform climate change adaptation".

Conflicts of Interest: The authors declare no conflict of interest.

References

1. Gamfeldt, L.; Snäll, T.; Bagchi, R.; Jonsson, M.; Gustafsson, L.; Kjellander, P.; Ruiz-Jaen, M.C.; Fröberg, M.; Stendahl, J.; Philipson, C.D.; et al. Higher levels of multiple ecosystem services are found in forests with more tree species. *Nat. Commun.* **2013**, *4*, 1340. [CrossRef] [PubMed]

2. van der Plas, F.; Manning, P.; Soliveres, S.; Allan, E.; Scherer-Lorenzen, M.; Verheyen, K.; Wirth, C.; Zavala, M.A.; Ampoorter, E.; Baeten, L.; et al. Biotic homogenization can decrease landscape-scale forest multifunctionality. *Proc. Natl. Acad. Sci. USA* **2016**, *113*, 3557–3562. [CrossRef] [PubMed]

3. Cudlín, P.; Klopčič, M.; Tognetti, R.; Máliš, F.; Alados, C.L.; Bebi, P.; Grunewald, K.; Zhiyanski, M.; Andonowski, V.; La Porta, N.; et al. Frans Emil Wielgolaski Drivers of treeline shift in different European mountains. *Clim. Res.* **2017**, *73*, 135–150. [CrossRef]

4. Pretzsch, H.; Bielak, K.; Bruchwald, A.; Dieler, J.; Dudzinska, M.; Erhart, H.P.; Jensen, A.M.; Johannsen, V.K.; Kohnle, U.; Nagel, J.; et al. Mischung und Produktivität von Waldbeständen. Ergebnisse langfristiger ertragskundlicher Versuche. *Allgemeine Forst-und Jagdzeitung* **2013**, *184*, 177–196.

5. Yachi, S.; Loreau, M. Does complementary resource use enhance ecosystem functioning? A model of light competition in plant communities. *Ecol. Lett.* **2007**, *10*, 54–62. [CrossRef]

6. Piotto, D. A meta-analysis comparing tree growth in monocultures and mixed plantations. *For. Ecol. Manag.* **2008**, *255*, 781–786. [CrossRef]

7. Morin, X.; Fahse, L.; Scherer-Lorenzen, M.; Bugmann, H. Tree species richness promotes productivity in temperate forests through strong complementarity between species. *Ecol. Lett.* **2011**, *14*, 1211–1219. [CrossRef]

8. Paquette, A.; Messier, C. The effect of biodiversity on tree productivity: From temperate to boreal forests. *Glob. Ecol. Biogeogr.* **2011**, *20*, 170–180. [CrossRef]

9. Pretzsch, H.; Schütze, G.; Uhl, E. Resistance of European tree species to drought stress in mixed versus pure forests: Evidence of stress release by inter-specific facilitation. *Plant Biol.* **2013**, *15*, 483–495. [CrossRef]

10. Pretzsch, H.; Schütze, G. Effect of tree species mixing on the size structure, density and yield forest stands. *Eur. J. For. Res.* **2015**, *135*, 1–22. [CrossRef]

11. Jucker, T.; Bouriaud, O.; Avacaritei, D.; Coomes, D.A. Stabilizing effects of diversity on aboveground wood production in forest ecosystems: Linking patterns and processes. *Ecol. Lett.* **2014**, *17*, 1560–1569. [CrossRef] [PubMed]

12. Metz, J.; Annighofer, P.; Schall, P.; Zimmermann, J.; Kahl, T.; Schulze, E.D. Site-adapted admixed tree species reduce drought susceptibility of mature European beech. *Glob. Chang. Biol.* **2016**, *22*, 903–920. [CrossRef] [PubMed]

13. Liang, J.; Crowther, T.W.; Picard, N.; Wiser, S.; Zhou, M.; Alberti, G.; Schulze, E.D.; McGuire, A.D.; Bozzato, F.; Pretzsch, H.; et al. Positive biodiversity–productivity relationship predominant in global forests. *Science* **2016**, *354*, 196. [CrossRef] [PubMed]

14. Grossiord, C.; Granier, A.; Gessler, A.; Jucker, T.; Bonal, D. Does drought influence the relationship between biodiversity and ecosystem functioning in boreal forests? *Ecosystems* **2014**, *17*, 394–404. [CrossRef]

15. Merlin, M.; Perot, T.; Perret, S.; Korboulewsky, N.; Vallet, P. Effects of stand composition and tree size on resistance and resilience to drought in sessile oak and Scots pine. *For. Ecol. Manag.* **2015**, *339*, 22–33. [CrossRef]

16. Conte, E.; Lombardi, F.; Battipaglia, G.; Palombo, C.; Altieri, S.; La Porta, N.; Marchetti, M.; Tognetti, R. Growth dynamics, climate sensitivity and water use efficiency in pure vs. mixed pine and beech stands in Trentino (Italy). *For. Ecol. Manag.* **2018**, *409*, 707–718. [CrossRef]

17. Gebauer, T.; Horna, V.; Leuschner, C. Canopy transpiration of pure and mixed forest stands with variable abundance of European beech. *J. Hydrol.* **2012**, *442*, 2–14. [CrossRef]

18. IPCC. Meeting report of the intergovernmental panel on climate change expert meeting on mitigation, sustainability and climate stabilization scenarios. In *IPCC Working Group III Technical Support Unit*; Shukla, P.R.J., Skea, R., van Diemen, K., Calvin, Ø., Christophersen, F., Creutzig, J., Eds.; Imperial College London: London, UK, 2017.

19. Hanewinkel, M.; Cullmann, D.; Schelhaas, M.J.; Nabuurs, G.J.; Zimmermann, N.E. Climate change may cause severe loss in economic value of European forestland. *Nat. Clim. Chang.* **2013**, *3*, 204–207. [CrossRef]

20. Altieri, V.; De Franco, S.; Lombardi, F.; Marziliano, P.A.; Menguzzato, G.; Porto, P. The role of silvicultural systems and forest types in preventing soil erosion processes in mountain forests: A methodological approach using cesium-137 measurements. *J. Soils Sediments* **2018**, *18*, 1–10. [CrossRef]

21. Zhang, Y.; Chen, H.Y.H.; Reich, P.B. Forest productivity increases with evenness, species richness and trait variation: A global meta-analysis. *J. Ecol.* **2012**, *100*, 742–749. [CrossRef]

22. Kanowski, J.; Catterall, C.P.; Wardell-Johnson, G.W. Consequences of broadscale timber plantations for biodiversity in cleared rainforest landscapes of tropical and subtropical Australia. *For. Ecol. Manag.* **2005**, *208*, 359–372. [CrossRef]

23. Petit, B.; Montagnini, F. Growth in pure and mixed plantations of tree species used in reforesting rural areas of the humid region of Costa Rica, Central America. *For. Ecol. Manag.* **2006**, *233*, 338–343. [CrossRef]

24. Richards, A.E.; Forrester, D.I.; Bauhus, J.; Scherer-Lorenzen, M. The influence of mixed tree plantations on the nutrition of individual species: A review. *Tree Physiol.* **2010**, *30*, 1192–1208. [CrossRef]

25. Bielak, K.; Dudzinska, M.; Pretzsch, H. Mixed stands of Scots pine (*Pinus sylvestris* L.) and Norway spruce [*Picea abies* (L.) Karst] can be more productive than monocultures. Evidence from over 100 years of observation of long-term experiments. *For. Syst.* **2014**, *23*, 573. [CrossRef]

26. Forrester, D.I. The spatial and temporal dynamics of species interactions in mixed-species forests: From pattern to process. *For. Ecol. Manag.* **2014**, *312*, 282–292. [CrossRef]

27. Petráš, R.; Mecko, J.; Bošelâ, M.; Šebeň, V. Wood quality and value production in mixed fir-spruce-beech stands: Long-term research in the Western Carpathians. *For. J.* **2016**, *62*, 98–104. [CrossRef]

28. Magin, R. Ertragskundliche Untersuchungen in montanen Mischwäldern. *Forstwissenschaftliches Centralblatt* **1954**, *73*, 103–113. [CrossRef]

29. Kennel, R. Untersuchungen über die Leistung von Fichte und Buche im Reinund Mischbestand. *Sauerländer* **1965**, *136*, 149–161.

30. Kennel, R. Soziale Stellung, Nachbarschaft und Zuwachs. *Forstwissenschaftliches Centralblatt* **1966**, *85*, 193–204. [CrossRef]

31. Hausser, K.; Troeger, R. Beitrag zur Frage der Massen und Wertleistung gepflanzter Weisstannen und Fichtenbestände auf gleichen Standorten. *Allgemeine Forst-und Jagdzeitung* **1967**, *138*, 150–157.

32. Mitscherlich, G. Ertragskundlich-ökologische Untersuchungen im Rein-und Mischbestand. *Mitteilungen der Forstlichen Bundes-Versuchsanstalt* **1967**, *77*, 9–35.

33. Hink, V. Das Wachstum von Fichte und Tanne auf den wichtigsten Standortseinheiten des Einzelwuchsbezirks "Flächenschwarzwald" (Südwürtemberg-Hohenzollern). *Allgemeine Forst-und Jagdzeitung* **1972**, *143*, 80–85.

34. Mettin, C. Betriebswirtschaftliche und ökologische Zusammenhänge zwischen Standortskraft und Leistung in Fichtenreinbeständen und Fichten Buchen-Mischbeständen. *Allgemeine Forst-und Jagdzeitung* **1985**, *40*, 803–810.

35. Kramer, H. *Waldwachstumslehre (Forest Growth and Yield Science)*; Paul Parey: Hamburg, Germany, 1988; p. 374. (In German)

36. Pretzsch, H. *Forest Dynamics, Growth and Yield: From Measurement to Model*; Springer: Berlin, Germany, 2010.

37. Saha, S.; Kuehne, C.; Kohnle, U.; Brang, P.; Ehring, A.; Geisel, J.; Leder, B.; Muth, M.; Petersen, R.; Peter, J.; et al. Growth and quality of young oaks (*Quercus robur* and *Quercus petraea*) grown in cluster plantings in central Europe: A weighted meta-analysis. *For. Ecol. Manag.* **2012**, *283*, 106–118. [CrossRef]

38. Liu, C.L.C.; Kuchma, O.; Krutovsky, K.V. Mixed secies versus monocultures in plantation forestry: Development, benefits, ecosystem services and perspectives for the future. *Glob. Ecol. Conserv.* **2018**, *15*, e00419. [CrossRef]

39. Chomel, M.; DesRochers, A.; Baldy, V.; Larchevêque, M.; Gauquelin, T. Non-additive effects of mixing hybrid poplar and white spruce on aboveground and soil carbon storage in boreal plantations. *For. Ecol. Manag.* **2014**, *328*, 292–299. [CrossRef]

40. Battipaglia, G.; Pelleri, F.; Lombardi, F.; Altieri, S.; Vitone, A.; Conte, E.; Tognetti, R. Effects of associating *Quercus robur* L. and *Alnus cordata* Loisel. on plantation productivity and water use efficiency. *For. Ecol. Manag.* **2017**, *391*, 106–114. [CrossRef]

41. Kelty, M.J. The role of species mixtures in plantation forestry. *For. Ecol. Manag.* **2006**, *233*, 195–204. [CrossRef]

42. Machado, J.S.; Louzada, J.L.; Santos, A.J.A.; Nunes, L.; Anjos, O.; Rodrigues, J.; Simoes, R.M.S.; Pereira, H. Variation of wood density and mechanical properties of blackwood (*Acacia melanoxylon* r. Br.). *Mater. Des.* **2014**, *56*, 975–980. [CrossRef]

43. Pretzsch, H.; Schütze, G. Transgressive overyielding in mixed compared with pure stands of Norway spruce and European beech in Central Europe: Evidence on stand level and explanation on individual tree level. *Eur. J. For. Res.* **2009**, *128*, 183–204. [CrossRef]

44. Russo, D.; Marziliano, P.A.; Macri, G.; Proto, A.R.; Zimbalatti, G.; Lombardi, F. Does thinning intensity affect wood quality? An analysis of Calabrian Pine in Southern Italy using a non-destructive acoustic method. *Forests* **2019**, *30*, 303. [CrossRef]

45. Guntekin, G.; Emiroglu, Z.G.; Yolmaz, T. Prediction of bending properties for Turkish red pine (*Pinus brutia* Ten.) Lumber using stress wave method. *Bioresources* **2013**, *8*, 231–237. [CrossRef]

46. Teder, M.; Pilt, K.; Miljan, M.; Lainurm, M.; Kruuda, R. Overview of some non-destructive methods for in-situ assessment of structural timber. In Proceedings of the 3rd International Conference Civil Engineering, Jelgava, Latvia, 12–13 May 2011.

47. Wessels, C.B.; Malan, F.S.; Rypstra, T. A review of measurement methods used on standing trees for the prediction of some mechanical properties of timber. *Eur. J. For. Res.* **2011**, *130*, 881–893. [CrossRef]

48. Bacher, M.; Krosek, S. Bending and tension strength classes in European standards. Annals of Warsaw University of Life Sciences—SGGW. *For. Wood Tech.* **2014**, *88*, 14–22.

49. Pretzsch, H.; Rais, A. Wood quality in complex forests versus even-aged monocultures: Review and perspectives. *Wood Sci. Technol.* **2016**, *50*, 845–880. [CrossRef]

50. Rais, A.; Van de Kuilen, J.W.G. Critical section effect during derivation of settings for grading machines based on dynamic modulus of elasticity. *Wood Mater. Sci. Eng.* **2015**, *4*, 189–196. [CrossRef]

51. FAO. World Reference Base for Soil Resources. In *International Soil Classification System for Naming Soils and Creating Legends for Soil Maps*; World Soil Resources Reports, No. 106; FAO: Roma, Italy, 2014.

52. Ciancio, O.; Iovino, F.; Menguzzato, G.; Nicolaci, A.; Nocentini, S. Structure and growth of a small group selection forest of Calabrian pine in Southern Italy: A hypothesis for continuous cover forestry based on traditional silviculture. *For. Ecol. Manag.* **2006**, *224*, 229–234. [CrossRef]

53. Tabacchi, G.; Di Cosmo, L.; Gasparini, P. Aboveground tree volume and phytomass prediction equations for forest species in Italy. *Eur. J. For. Res.* **2011**, *130*, 911–934. [CrossRef]

54. Holmes, R.L. Computer-assisted quality control in tree-ring dating and measurement. *Tree-Ring Bull.* **1983**, *43*, 69–78.

55. Wang, X. Acoustic measurements on trees and logs: A review and analysis. *Wood Sci. Technol.* **2013**, *47*, 965–975. [CrossRef]

56. Vanninen, P.; Ylitalo, H.; Sievanen, R.; Makela, A. Effect of age and site quality on the distribution of biomass in Scots pine (*Pinus sylvestris* L.). *Trees* **1996**, *10*, 231–238. [CrossRef]

57. Vargas-Hernandez, J.; Adams, W.T. Genetic relationships between wood density components and cambial growth rhythm in young coastal Douglas-fir. *Can. J. For. Res.* **1994**, *24*, 1871–1876. [CrossRef]

58. Zhembo, L.; Yixing, L.; Haipeng, Y.; Junqi, Y. Measurement of the dynamic modulus of elasticity of wood panels. *Front. For. China* **2006**, *1*, 425–430.

59. Todaro, L.; Macchioni, N. Wood properties of young Douglas-fir in Southern Italy: Results over a 12-year post-thinning period. *Eur. J. For. Res.* **2011**, *130*, 251–261. [CrossRef]

60. Zhang, H.; Wang, X.; Su, J. Experimental Investigation of Stress Wave Propagation in Standing Trees. *Holzforschung* **2011**, *65*, 743–748. [CrossRef]

61. Krajnc, L.; Farrelly, N.; Harte, A.M. Evaluating Timber Quality in Larger-Diameter Standing Trees: Rethinking the Use of Acoustic Velocity. *Holzforschung* **2019**, *73*, 797–806. [CrossRef]

62. Marziliano, P.A.; Menguzzato, G.; Scuderi, A.; Corona, P. Simplified methods to inventory the current annual increment of forest standing volume. *iForest* **2012**, *5*, 276–282. [CrossRef]

63. Team, R.C. *R: A Language and Environment for Statistical Computing*; R Foundation for Statistical Computing: Vienna, Austria, 2016.

64. Zar, J.H. *Biostatistical Analysis*, 4th ed.; Prentice Hall: Upper Saddle River, NJ, USA, 1999.

65. Detters, A.; Cowell, C.; McKeown, L.; Howard, P. *Evaluation of Current Rigging and Dismantling Practices Used in Arboriculture*; The Health and Safety Executive: London, UK; The Forestry Commission: Edinburgh, UK, 2008; p. 355.

66. Binkley, D.; Dunkin, K.A.; DeBell, D.; Ryan, M.G. Production and nutrient cycling in mixed plantations of eucalyptus and albizia in Hawaii. *For. Sci.* **1992**, *38*, 393–408.

67. Kelty, M.J. Comparative productivity of monocultures and mixed-species stands. In *The Ecology and Silviculture of Mixed-Species Forests*; Springer: Dordrecht, The Netherlands, 1992; pp. 125–141.

68. Forrester, D.I.; Lancaster, K.; Collopy, J.J.; Warren, C.R.; Tausz, M. Photosynthetic capacity of *Eucalyptus globulus* is higher when grown in mixture with *Acacia mearnsii*. *Trees* **2012**, *26*, 1203–1213. [CrossRef]

69. Saranpää, P. Wood density and growth. In *Wood Quality and Its Biological Basis Blackwell*; Biological Sciences Series; Barnett, J.R., Jeronimidis, G., Eds.; CRC Press: Oxford, FL, USA, 2003; pp. 87–113.

70. Torquato, L.P.; Auty, D.; Herândez, R.E.; Duchesne, I.; Pothier, D.; Achim, A. Black spruce trees from fire-origin stands have higher wood mechanical properties than those from older, irregular stands. *Can. J. For. Res.* **2014**, *44*, 118–127. [CrossRef]

71. Metz, J.Ô.; Seidel, D.; Schall, P.; Scheffer, D.; Schulze, E.D.; Ammer, C. Crown modeling by terrestrial laser scanning as an approach to assess the effect of aboveground intra- and interspecific competition on tree growth. *For. Ecol. Manag.* **2013**, *310*, 275–288. [CrossRef]

72. Pretzsch, H.; Biber, P. A re-evaluation of Reineke's rule and stand density index. *For. Sci.* **2005**, *51*, 304–320.

73. Pretzsch, H. Species-specific allometric scaling under self-thinning: Evidence from long-term plots in forest stands. *Oecologia* **2006**, *146*, 572–583. [CrossRef] [PubMed]

74. Pretzsch, H.; Block, J.; Dieler, J.; Dong, P.H.; Kohnle, U.; Nagel, J.; Spellmann, H.; Zingg, A. Comparison between the productivity of pure and mixed stands of Norway spruce and European beech along an ecological gradient. *Ann. For. Sci.* **2010**, *67*, 712. [CrossRef]

75. Pretzsch, H.; Bielak, K.; Block, J.; Bruchwald, A.; Dieler, J.; Ehrhart, H.P.; Kohnle, U.; Nagel, J.; Spellmann, H.; Zasada, M.; et al. Productivity of mixed versus pure stands of oak (*Quercus petraea* (Matt.) Liebl. and *Quercus robur* L.) and European beech (*Fagus sylvatica* L.) along an ecological gradient. *Eur. J. For. Res.* **2013**, *132*, 263–280. [CrossRef]

76. Condés, S.; del Río, M.; Sterba, H. Mixing effect on volume growth of *Fagus sylvatica* and *Pinus sylvestris* is modulated by stand density. *For. Ecol. Manag.* **2013**, *292*, 86–95. [CrossRef]

77. Zeller, L.; Ammer, C.; Annighöfer, P.; Biber, P.; Marshall, J.; Schütze, G.; del Río Gaztelurrutia, M.; Pretzsch, H. Tree ring wood density of Scots pine and European beech lower in mixed-species stands compared with monocultures. *For. Ecol. Manag.* **2017**, *400*, 363–374. [CrossRef]

78. Binkley, D.; Stape, J.L.; Ryan, M.G. Thinking about efficiency of resource use in forests. *For. Ecol. Manag.* **2004**, *193*, 5–16. [CrossRef]

79. Lebourgeois, F.; Gomez, N.; Pinto, P.; Mérian, P. Mixed stands reduce *Abies alba* tree-ring sensitivity to summer drought in the Vosges mountains, western Europe. *For. Ecol. Manag.* **2013**, *303*, 61–71. [CrossRef]

80. Del Río, M.D.; Schütze, G.; Pretzsch, H. Temporal variation of competition and facilitation in mixed species forests in Central Europe. *Plant Biol.* **2014**, *16*, 166–176. [CrossRef]

81. Del Río, M.; Pretzsch, H.; Ruíz-Peinado, R.; Ampoorter, E.; Annighöfer, P.; Barbeito, I.; Bielak, K.; Brazaitis, G.; Coll, L.; Drössler, L. Species interactions increase the temporal stability of community productivity in *Pinus sylvestris-Fagus sylvatica* mixtures across Europe. *J. Ecol.* **2017**, *105*, 1032–1043. [CrossRef]

82. Ellenberg, H.; Leuschner, C. *Vegetation Mitteleuropas Mit den Alpen in Ökologischer, Dynamischer und Historischer Sicht*; Eugen Ulmer Stuttgart: Nürtingen, Germany, 2010.

83. Pretzsch, H. The Effect of Tree Crown Allometry on Community Dynamics in Mixed-Species Stands versus Monocultures. A Review and Perspectives for Modeling and Silvicultural Regulation. *Forests* **2019**, *10*, 810. [CrossRef]

84. Forrester, D.I.; Ammer, C.; Annighöfer, P.J.; Barbeito, I.; Bielak, K.; Bravo-Oviedo, A.; Coll, L.; del Río, M.; Drössler, L.; Heym, M.; et al. Effects of crown architecture and stand structure on light absorption in mixed and monospecific *Fagus sylvatica* and *Pinus sylvestris* forests along a productivity and climate gradient through Europe. *J. Ecol.* **2018**, *106*, 746–760. [CrossRef]

85. Genet, A.; Auty, D.; Achim, A.; Bernier, M.; Pothier, D.; Cogliastro, A. Consequences of faster growth for wood density in northern red oak (*Quercus rubra* Liebl.). *Forestry* **2012**, *86*, 99–110. [CrossRef]

86. Grammel, R. Zusammenhänge zwischen Wachstumsbedingungen und holztechnologischen Eigenschaften der Fichte. *For. Cent.* **1990**, *109*, 119–129. [CrossRef]

87. Brazier, J.D.; Mobbs, I.D. The influence of planting distance on structural wood yields of unthinned Sitka spruce. *Forestry* **1993**, *66*, 333–352. [CrossRef]

88. Larocque, G.R.; Marshall, P. Wood relative density development in red pine (*Pinus resinosa* Ait.) stands as affected by different initial spacings. *For. Sci.* **1995**, *41*, 709–728.

89. Zhang, S.Y.; Chauret, G.; Swift, D.E.; Duchesne, I. Effects of precommercial thinning on tree growth and lumber quality in a jack pine stand in New Brunswick, Canada. *Can. J. For.* **2006**, *36*, 945–952. [CrossRef]

90. Moore, J.R.; Cown, D.J.; McKinley, R.B.; Sabatia, C.O. Effects of stand density and seedlot on three wood properties of young radiata pine grown at a dry-land site in New Zealand. *N. Z. J. For. Sci.* **2015**, *45*, 209. [CrossRef]

91. Tognetti, R.; Michelozzi, M.; Borghetti, M. The response of European beech (*Fagus sylvatica* L.) seedlings from two Italian populations to drought and recovery. *Trees* **1995**, *9*, 348–354. [CrossRef]

92. Bolte, A.; Czajkowski, T.; Cocozza, C.; Tognetti, R.; de Miguel, M.; Pšidová, E.; Ditmarová, Ľ.; Dinca, L.; Delzon, S.; Cochard, H.; et al. Desiccation and mortality dynamics in seedlings of different European beech (*Fagus sylvatica* L.) populations under extreme drought conditions. *Front. Plant Sci.* **2016**, *7*, 751. [CrossRef] [PubMed]

93. Cocozza, C.; de Miguel, M.; Pšidová, E.; Ditmarová, L.; Marino, S.; Maiuro, L.; Alvino, A.; Czajkowski, T.; Bolte, A.; Tognetti, R. Variation in ecophysiological traits and drought tolerance of beech (*Fagus sylvatica* L.) seedlings from different populations. *Front. Plant Sci.* **2016**, *7*, 886. [CrossRef] [PubMed]

 *forests*

Article

Autotoxicity Hinders the Natural Regeneration of *Cinnamomum migao* H. W. Li in Southwest China

Xiaolong Huang [†]**, Jingzhong Chen** [†]**, Jiming Liu** *****, Jia Li, Mengyao Wu and Bingli Tong**

Department of Ecology, College of Forestry, Guizhou University, Guiyang 550025, China;
guidah365@126.com (X.H.); chenjingzhong-2016@foxmail.com (J.C.); lijia22a@163.com (J.L.);
mmmmengyao@126.com (M.W.); DCB0011@126.com (B.T.)
* Correspondence: karst0623@163.com; Tel.: +86-13985015398
† Equal contribution to the study.

Received: 23 September 2019; Accepted: 16 October 2019; Published: 18 October 2019

Abstract: Autotoxicity is a widespread phenomenon in nature and is considered to be the main factor affecting new natural recruitment of plant populations, which was proven in many natural populations. *Cinnamomum migao* H. W. Li is an endemic medicinal woody plant species mainly distributed in Southwestern China and is defined as an endangered species by the Red Paper of Endangered Plants in China. The lack of seedlings is considered a key reason for population degeneration; however, no studies were conducted to explain its causes. *C. migao* contains substances with high allelopathic potential, such as terpenoids, phenolics, and flavonoids, and has strong allelopathic effects on other species. Therefore, we speculate that one of the reasons for *C. migao* seedling scarcity in the wild is that it exhibits autotoxic allelopathy. In this study, which was performed from the perspective of autotoxicity, we collected leaves, pericarp, seeds, and branches of the same population; we simulated the effects of decomposition and release of litter from these different anatomical parts of *C. migao* in the field; and we conducted 210-day control experiments on seedling growth, with different concentration gradients, using associated aqueous extracts. The results showed that the leaf aqueous extract (leaf$_{AE}$) significantly inhibited growth indicators and increased damage of the lipid structure of the cell membrane of seedlings, suggesting that autotoxicity from *C. migao* is a factor restraining seedling growth. The results of the analyses of soil properties showed that, compared with the other treatments, leaf$_{AE}$ treatment inhibited soil enzyme activity and also had an impact on soil fungi. Although leaf$_{AE}$ could promote soil fertility to some extent, it did not change the effect of autotoxic substances on seedling growth. We conclude that autotoxicity is the main obstacle inhibiting seedling growth and the factor restraining the natural regeneration of *C. migao*.

Keywords: *Cinnamomum migao*; autotoxicity; seedling growth; soil substrate; soil enzyme; soil fungi

---

1. Introduction

Autotoxicity is a special type of intraspecific allelopathy of plants [1]. It is known to be widespread in nature, particularly in artificial agroforestry systems [2], leading to population deterioration and regeneration failure [3]. Autotoxicity refers to the phenomenon in which plants release their metabolites into the surrounding environment by volatilization, rainwater leaching, decomposition, and root excretion [4]; these metabolites then inhibit seed germination or other individuals' or their own seedlings' growth directly or indirectly. Canopy trees' autotoxicity to their seedlings may play an important role in forest species replacement [5], and the existence of autotoxicity was proved in many natural populations [4,6]. Current studies on plant autotoxicity have shown that allelochemicals mainly inhibit plant growth in the following ways: (1) Plant growth is directly inhibited by affecting photosynthesis and altering plant cell membrane structures and plant defense systems [7,8]. (2) Plant growth is indirectly affected by inhibiting nutrient absorption or changing soil enzyme activity [9,10].

(3) The rhizosphere microecosystems are changed through the interaction between plant metabolites and fungi to ultimately affect seedling growth [11] (Figure 1a).

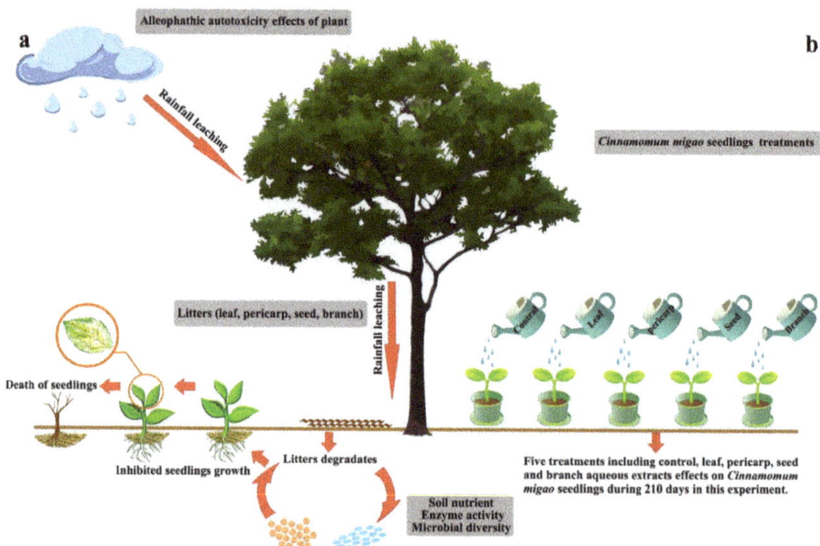

Figure 1. Process of plant autotoxicity (**a**) and our experimental design (**b**).

Autotoxicity is regarded as a negative effect on plant growth, and secondary metabolites produced by maternal plants could hinder the growth of their seedlings [12]. Phenolics and terpenoids released by woody plants play a key role in these interactions by influencing the structure and diversity of plant and soil communities [13–15]. Soil microorganisms can be directly affected by plant phenolics [16], but they can also use these phytochemicals as carbon sources, thereby modifying the chemical plant–plant interactions [17]. Phenolics from litter can inhibit the symbiosis between trees and fungi [18], which may have important consequences on the seedling development of trees. Similarly, phenolics released by plants may also play a key role in influencing the soil microbial community structure [19] and litter decomposition process [20]. For instance, autotoxicity of canopy trees on their own seedlings probably plays a role in forest species turnover along succession in Mediterranean forests [5]. Some autotoxic compounds released by asparagus (*Asparagus officinalis*) probably inhibit its own growth and can also be a reason for "asparagus decline" [21]. Phenolics released by the understory dwarf shrub *Empetrum hermaphroditum* impair the regeneration of the dominant tree *Pinus sylvestris* in boreal forests [22]. In addition, studies by Alías et al. [23] on the soil properties of the invasive rock-rose (*Cistus ladanifer*) population also showed that the compounds released by the plant itself were involved in autotoxicity and regeneration of the rock-rose population. *Cinnamomum migao* contains substances with high allelopathic potential, such as terpenoids, phenolics, and flavonoids [24]; generally, species with high allelopathic potential tend to have strong autotoxicity [25], and medicinal plants are more likely to have autotoxicity than other plants [14]. Therefore, we speculate that one of the reasons for seedling scarcity in the wild population of the species *C. migao* is that this species has autotoxic allelopathy [14].

C. migao H.W. Li is a species of Lauraceae evergreen medicinal woody plant, which is defined as an endangered species by the Red Paper of Endangered Plants in China. It is endemic to China and mainly distributed in Southwestern China. The fruits of *C. migao*, which are effective in treating gastrointestinal and cerebrovascular diseases, are used as traditional medicine in Miao in China [24]. In the last 30 years, researchers have found that the natural population size of *C. migao* is very small, as the majority of populations contain only two or three individuals, and the tree age is relatively high;

the natural regeneration has some obstacles [26]. Moreover, recent investigations of this resource have found that many natural populations have disappeared, and the survival and reproduction of this species are greatly threatened [27]. However, to the best of our knowledge, no research was performed to explain the underlying cause of this phenomenon. Recent studies on the autotoxicity of medicinal plants mainly focused on medicinal herbs, such as *Codonopsis pilosula*, etc. [28]. Few studies have been performed on medicinal woody plants, and the mechanism underlying this autotoxicity has remained unclear [12].

Our previous studies confirmed that *C. migao* has a strong allelopathic effect [29]. To explain the difficulties associated with new recruitment of *C. migao*, the aim of this study was to understand the responses of the plants to autotoxicity and the possible mechanism of autotoxicity that inhibits plant growth. Accordingly, we proposed the following hypotheses: (1) autotoxicity is the main factor affecting seedling growth in *C. migao*, and (2) the main mechanism of autotoxicity is that decomposition and release of autotoxic substances from the litter of *C. migao* can be achieved by altering the soil environment (chemical property, soil enzyme activity, and fungi) to inhibit seedling growth and survival. To test these hypotheses, we adopted a method of irrigating the aqueous extract and simulated the effects of litter decomposition and release from different anatomical parts of *C. migao* on seedling growth under field conditions from the new perspective of plant autotoxicity [30]. The changes in morphology, physiological metabolism, soil substrate, and fungi, which are the four key factors affecting plant growth, were determined during seedling growth.

2. Material and Methods

2.1. Experimental Site

Experiments were conducted at the College of Forestry, Guizhou University, Huaxi District, Guiyang (106°42′E, 26°34′N, 1020 m a.s.l.). The climate in this region is subtropical monsoon, with a mean annual temperature of 15.3 °C and mean annual precipitation of 1129 mm.

2.2. Plant Materials

Fresh mature fruits of *C. migao* were collected from Luodian County, Guizhou Province (25°26′N, 106°31′E) from October to November 2017. They were then transported to the laboratory to remove their flesh immediately, followed by rinsing with water. Before germination, we cleaned the seeds, sterilized them by immersion in 0.5% $KMnO_4$ for 2 h, and then rinsed them in sterile water five times. Seed germination was performed in an artificial climate chamber (RXZ-1500, Ningbo Jiangnan Instrument Factory, Ningbo, China) with a germination box. After germination of the seeds, we transferred them to nutrient soil for planting in January 2018, and seedlings with identical growth were transplanted into nutrient bags for slow seedling treatment in mid-March. The pot-culture experiment began in April after the seedlings (16.25 ± 1.41 cm in height; number of leaves 5 ± 0.58) had been transplanted into plastic plots. The soil in our experiments was selected from the section loess of nongrowing plants and uniformly mixed with humus and perlite (loess: humus: perlite = 7:2:1); each pot contained approximately 2.5 kg of soil. All tested soils were sterilized under high pressure at 121°C twice, with each cycle lasting for 1 h. The litter of *C. migao* used in this experiment was directly collected from the forest floor. In addition, surface soil (0–5 cm) from under the canopy of nine natural populations of *C. migao* in Guizhou, Yunnan, and Guangxi provinces was sampled for sequence analysis to determine fungal diversity (Figure 2).

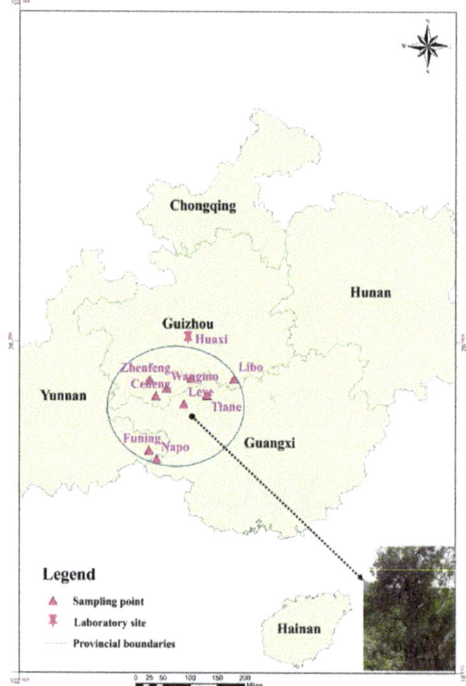

Figure 2. Experimental sites and soil sample collection sites for *C. migao* in Southwest China.

2.3. Experimental Design

Initially, we investigated and statistically analyzed the litter under the canopy of *C. migao* and found that the main components of the litter were leaves, fruits, and branches. Therefore, the materials were divided into four parts: leaf, pericarp, seed, and branch. To simulate natural conditions as much as possible, the litter was collected from under a canopy from November to December 2017 in Luodian County and classified. To prepare aqueous extracts, we cleaned all of the test materials and cut them into pieces (1 cm^2); gathered them into weights of 0, 1, 2, 3, and 4 g; and immersed them in 100 ml of deionized water. Solutions with concentrations of 0.00, 0.01, 0.02, 0.03, and 0.04 g·mL^{-1} were prepared in the dark. In accordance with the above method of aqueous extraction, five types of aqueous extract treatments were set up, namely, the control and treatment with the leaf aqueous extract (leaf$_{AE}$), pericarp aqueous extract (pericarp$_{AE}$), seed aqueous extract (seed$_{AE}$), and branch aqueous extract (branch$_{AE}$). Each treatment had nine replicates, giving a total of $4 \times 5 \times 9 = 180$ pots in this experiment. During the experiment, the aqueous extract was co-irrigated with litter (leaf, pericarp, seed, and branch) every 10 days, to better simulate the decomposition and release the effects of litter, and the positions of the pots were randomly moved. To maintain soil moisture content, quantitative deionized water was added to each pot at different times to supplement water. Through field monitoring, we found that seedlings usually withered or died within 8 months. Therefore, the experiment was designed to determine the growth indices after 210 days of treatment (Figure 1b).

2.4. Chemical Analyses

2.4.1. Analysis of Seedling Growth and Physiology

At the end of the experiment, we determined seedling growth and physiology; the determination methods were as follows. (1) Seedling height: The seedling height for each treatment was measured using Vernier calipers and a tape measure. (2) Leaf area: Three pots of seedlings were selected for each treatment, and the leaf area of the third mature leaf below the apex of the seedlings was measured using a portable leaf-area meter (LI-3000C; LI-COR, Lincoln, NE, USA). (3) Biomass: At least three seedlings from pots undergoing different treatments were carefully removed from the soil to maintain their integrity, after which residual soil and impurities were removed by flushing with running water. After cleaning, the seedlings were oven-dried twice at 65 °C to a constant weight and then weighed.

Fresh plant materials (0.2 g) were collected from each treatment and homogenized with 5 mL of buffer (with 1% PVP), by grinding on ice, and then centrifuged at 4 °C. After centrifugation, the supernatants were used for measuring the levels of malondialdehyde (MDA), soluble protein, peroxidase (POD), and superoxide dismutase (SOD). In brief, the MDA content was estimated using the thiobarbituric acid method reported by Hodges et al. [31], with minor modifications. The soluble protein content was measured using the Folin's phenol reagent method [32]. POD activity was measured following the guaiacol method, with minor modifications [33]. SOD activity was analyzed using the nitroblue tetrazolium chloride method by Lacan and Baccou [34], with minor modifications.

2.4.2. Analysis of Soil Physicochemical Properties and Soil Enzyme Activity

At the end of the experiment, soil samples were collected from each pot immediately after the different treatments and divided into two parts: One was used for soil nutrient analysis, and the other was stored at 4°C for soil enzyme and soil fungal analyses.

Soil samples from the different aqueous extract treatments were naturally air-dried; the samples were sieved using a 2 mm diameter mesh to determine their chemical properties. In brief, the soil total nitrogen (TN) content was analyzed using the Kjeldahl method with a Foss–Kjeltec analyzer [35]. The soil total phosphorus (TP) content was determined by the Mo–Sb colorimetric method after soil was digested with a mixed acid solution of H_2SO_4 and $HCLO_4$ [36]. The soil total potassium (TK) content was measured using an alkali melting-flame photometer. Soil available N (AN) content was determined by the method used by Dorich and Nelson [37]. Soil available P (AP) content was obtained by $NaHCO_3$ extraction and analyzed by the sodium bicarbonate–molybdenum resistance colorimetric method [38]. The soil available K (AK) content was measured by the method of ammonium acetate leaching–flame photometry [38]. Soil enzyme activity, including catalase, urease, phosphatase, and invertase activity, was estimated using a soil enzyme activity kit (Beijing Solebo Biotechnology Co., Ltd.), in accordance with the manufacturer's instructions. The initial soil chemical composition and enzyme activity are detailed in Table 1.

Table 1. Initial chemical composition and soil enzyme activity of the tested soil samples.

Soil Properties	pH	Total N (g·kg^{-1})	Total P (g·kg^{-1})	Total K (g·kg^{-1})	Available N (g·kg^{-1})	Available P (mg·kg^{-1})	Available K (mg·kg^{-1})	Catalase mL·g^{-1}·20 min	Urease mgNH$_3^-$N/g	Phopatase umol·g^{-1}·d^{-1}	Invertase mg·g^{-1}·d^{-1}
	6.06	1.2	0.2	0.29	52.5	3.5	13.03	1.43	0.015	23.53	1.5

2.4.3. Analysis of Soil Fungi

Surface soils (0.04 g·mL^{-1}) from the four aqueous extract treatments and the control from our experiment and surface soils of 0–5 cmfrom nine areas in which *C. migao* is distributed were used to analyze the diversity of soil fungi composition. Each treatment and field sample had three biological replicates. Soil genomic DNA was extracted using an extraction kit (FastDNA®tSPINKit for Soil, MP Biomedicals Co., Ltd., Shanghai, China). PCR amplification was performed using TaKaRa *rTaq* DNA polymerase in a 20 μL reaction system containing the following: 4 μL of 5× Buffer, 2 μL of 2.5 mmol/L dNTPs, 0.8 μL of forward primer ITS1F (5′-CTTGGTCATTTAGAGGAAGTAA-3′) (5 μmol/L), 0.8 μL of reverse primer ITS2R (5′-GCTGCGTTCTTCATCGATGC-3′) (5 μmol/L), 0.4 μL of rTaq polymerase, 0.2 μL of BSA, 10 ng of template DNA, and ddH$_2$O to 20 μL. ABI GeneAmp 9700 was used for PCR. The PCR reaction parameters were as follows: 95 °C for 3 min; 35 cycles of 95 °C for 30 s, 55 °C for 30 s, and 72 °C for 45 s; 72 °C for 10 min; and, finally, 10 °C until stopping the reaction. The PCR products were sequenced using an Illumina Hiseq high-throughput sequencing platform by Shanghai Ouyi Biomedical Information Co., Ltd.

2.5. Statistical Analysis

The homogeneity of variance was determined before performing ANOVA, and the data were logarithmically transformed when required. Duncan's test was used for assessing significant differences in all of the parameters using the SPSS 21.0 statistics package (Chicago, IL, USA). All presented data are indicated as the means and standard errors (SEs) of at least three replicates. The final results of fungal sequencing were collated, and SR statistics tools were used (http://www.mdts-erver.com/srst/) to analyze soil fungal data. Graphs were constructed with OriginPro 9.0 (Origin Lab, Northampton, MA, USA).

3. Results

3.1. Effects of C. migao Litter Aqueous Extract on Seedling Growth

The seedling height, biomass, and leaf area of *C. migao* were significantly decreased compared with those of the control when the concentration of leaf$_{AE}$ was increased. At 0.04 g·mL^{-1}, the seedling height, biomass, and leaf area decreased by 32.07%, 49.02%, and 42%, respectively. Within the experimental concentration range, pericarp$_{AE}$, seed$_{AE}$, and branch$_{AE}$ significantly promoted seedling growth. Among them, at 0.04 g·mL^{-1} of pericarp$_{AE}$, seedling height, biomass, and leaf area increased by 57.29%, 37.36%, and 50.88%, respectively. Similar increases with seed$_{AE}$ treatment of 44.89%, 25.62%, and 39.97% at 0.04 g·mL^{-1}, respectively, and with branch$_{AE}$ treatment of 44.89%, 30.21%, and 47.31% at 0.04 g·mL^{-1}, respectively, compared with those of the control were also identified (Table 2). The response of seedling growth to different litter aqueous extracts differed in the same concentration range.

Table 2. Effects of aqueous extract of litter from different anatomical parts of *C. migao* on seedling growth.

Group	Concentration (g·mL^{-1})	Height (cm)	Biomass (g)	Leaf Area (cm^2)
	0	162.53 ± 3.75 a	148.30 ± 7.10 a	17.69 ± 4.30 a
	0.01	143.07 ± 7.30 b	110.80 ± 8.70 b	15.57 ± 2.88 bc
Leaf$_{AE}$	0.02	131.23 ± 8.60 bc	99.80 ± 9.10 bc	14.35 ± 1.24 bc
	0.03	129.30 ± 6.80 c	95.30 ± 4.40 c	12.41 ± 2.65 bc
	0.04	110.40 ± 4.96 d	75.60 ± 5.70 d	10.26 ± 3.68 c
	0	162.53 ± 3.75 d	148.30 ± 7.10 c	17.69 ± 4.30 b
	0.01	220.60 ± 6.10 c	176.30 ± 6.40 b	23.46 ± 3.41 ab
Pericarp$_{AE}$	0.02	225.50 ± 7.50 bc	182.40 ± 6.30 b	22.33 ± 3.62 ab
	0.03	235.80 ± 5.20 b	183.80 ± 5.50 b	22.89 ± 3.05 ab
	0.04	255.60 ± 5.70 a	203.70 ± 7.70 a	26.69 ± 3.67 a
	0	162.53 ± 3.75 d	148.30 ± 7.10 b	17.69 ± 4.30
	0.01	168.80 ± 5.30 cd	152.50 ± 7.30 b	19.43 ± 2.78
Seed$_{AE}$	0.02	178.50 ± 6.00 c	157.00 ± 6.80 b	19.59 ± 3.88
	0.03	225.50 ± 6.70 b	176.60 ± 8.20 a	21.47 ± 4.26
	0.04	235.50 ± 4.80 a	186.30 ± 5.80 a	24.76 ± 3.86
	0	162.53 ± 3.75 c	148.30 ± 7.10 c	17.69 ± 4.30 c
	0.01	170.60 ± 5.90 c	150.30 ± 4.80 c	19.82 ± 4.18 ab
Branch$_{AE}$	0.02	192.50 ± 6.70 b	164.50 ± 6.20 b	23.52 ± 4.11 ab
	0.03	232.80 ± 7.50 a	187.50 ± 6.60 a	24.38 ± 3.65 ab
	0.04	235.50 ± 5.10 a	193.10 ± 5.70 a	26.06 ± 3.72 a

The indices of height, biomass, and leaf areas for *C. migao* seedlings after the application of leaf$_{AE}$, pericarp$_{AE}$, seed$_{AE}$, and branch$_{AE}$ treatments for 210 days. Values are expressed as mean ± SE. The values marked with different letters are significant at $p < 0.05$ (Duncan's test).

3.2. Effects of C. migao Litter Aqueous Extract on the Antioxidant Systems of Seedlings

Plants activate their own defense systems to resist toxicity caused by reactive oxygen species (ROS) when they are under stress, such as enhancing their antioxidant enzyme activities. Our results show that the MDA content in the leaves of *C. migao* seedlings increased as the concentration of leaf$_{AE}$ increased (Figure 3a). Among the treatments, the MDA content was the highest when the concentration of leaf$_{AE}$ was 0.04 g·mL^{-1}, which was 3.3 times that of the control. In pericarp$_{AE}$, seed$_{AE}$, and branch$_{AE}$ treatments, the MDA content changed linearly and was higher than that of the control. This illustrates that different aqueous extracts could cause different degrees of peroxidation of the cell membrane lipid structure. The soluble protein content decreased as the concentration of the extract increased under different treatments. It is noteworthy that the soluble protein content of leaf$_{AE}$ at 0.04 g·mL^{-1} was the lowest, which was reduced by 55% compared with that in the control (Figure 3b). Moreover, the antioxidant enzyme activities of the seedlings were also affected by the concentration of aqueous extract. The activities of POD and SOD in the seedlings treated with pericarp$_{AE}$, seed$_{AE}$, and branch$_{AE}$ were higher than those of the control. This indicates that, under the synergistic action of POD and SOD, excessive free radicals in the cells could be effectively removed, thereby maintaining the structural stability of the cell membrane. Conversely, under a low concentration of leaf$_{AE}$ (>0.01 g· mL^{-1}), *C. migao* seedlings were placed under stress, which resulted in membrane lipid peroxidation, a sharp decrease in the soluble protein content, reductions in POD and SOD activity (Figure 3c,d), and significant inhibition of the antioxidant capacity of the seedlings.

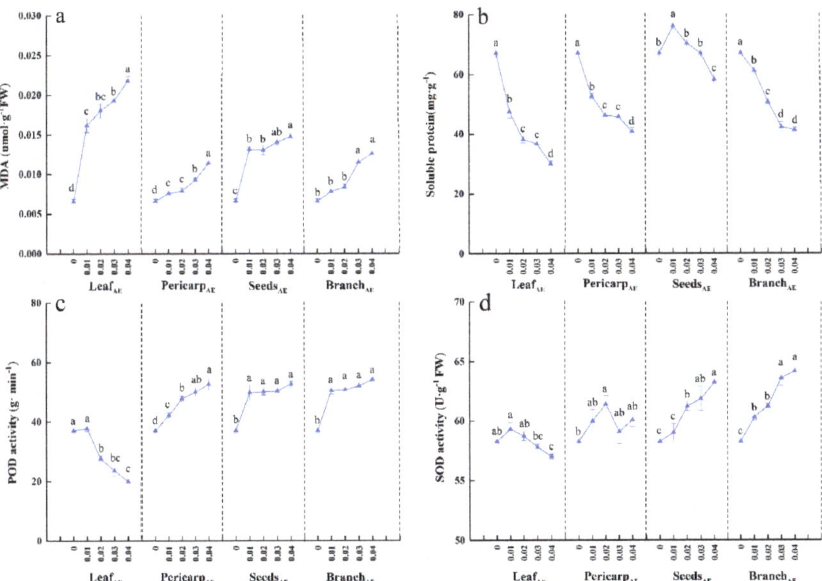

Figure 3. The MDA (**a**), soluble protein (**b**), POD (**c**), and SOD (**d**) contents in *C. migao* seedlings after the application of litter aqueous extract treatments for 210 days. The *x* axis denotes the concentration of the aqueous extract and the *y* axis denotes the content. Values are expressed as mean ± SE. The values marked with different letters are significant at $p < 0.05$ (Duncan's test).

3.3. Effects of C. migao Litter Aqueous Extracts on N, P, and K in Soils of Potted Seedlings

The TN content of the soil was increased with increasing concentration of the aqueous extracts; the TN content increased by 69.68%, 42.26%, and 40.98% with leaf$_{AE}$, pericarp$_{AE}$, and seed$_{AE}$ at 0.04 g·mL^{-1}, respectively, compared with that of the control. The difference was the most significant when the concentration of leaf$_{AE}$ was 0.04 g·mL^{-1} ($p < 0.05$) (Figure 4a). The TP content in soil was higher than that in the control after all treatments. The most significant difference was found when the concentration of pericarp$_{AE}$ was 0.04 g·mL^{-1} ($p < 0.05$) (Figure 4b). The TK content after all treatments significantly increased ($p < 0.05$) compared with that of the control, which increased with increased aqueous extract concentration (Figure 4c). The AN content showed an upward trend under leaf$_{AE}$ and seed$_{AE}$ treatments, and the AN content significantly increased under seed$_{AE}$ ($p < 0.05$). The pericarp$_{AE}$ and branch$_{AE}$ treatments showed downward trends for the AN content, and this variable in soil was the lowest when branch$_{AE}$ was 0.01 g·mL^{-1} (Figure 4d). The soil AP content showed an upward trend with the four kinds of aqueous extracts, with pericarp$_{AE}$ treatment causing the highest increase ($p < 0.05$) (Figure 4e). In addition, with increases of all aqueous extracts, the soil AK content significantly increased compared with that in the control group ($p < 0.05$) (Figure 4f).

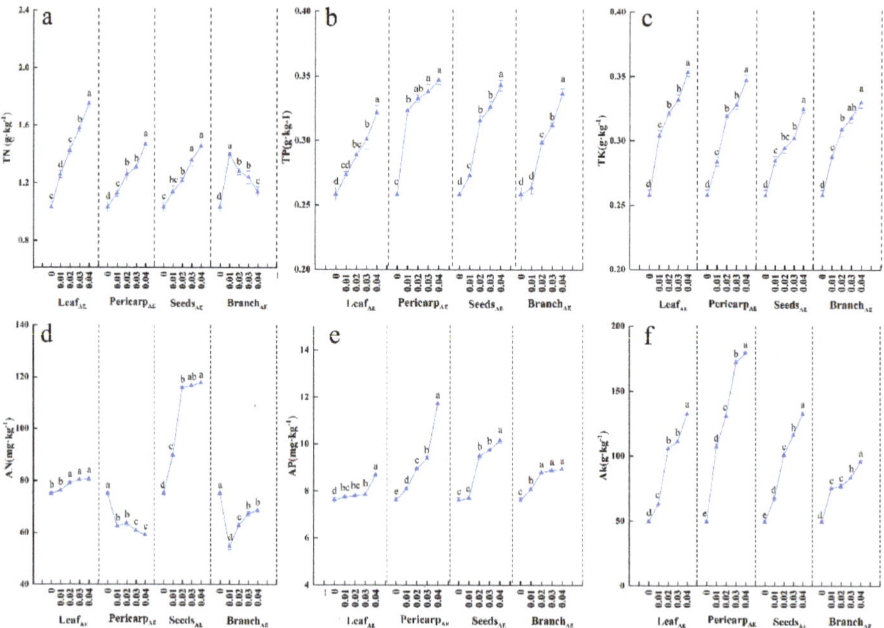

Figure 4. The TN (**a**), TP (**b**), TK (**c**), AN (**d**), AP (**e**), and AK (**f**) contents in soil planted with *C. migao* seedlings after litter aqueous extract treatments for 210 days. The *x* axis denotes the concentration of the aqueous extract, and the *y* axis denotes the content. Values are expressed as mean ± SE. The values marked with different letters are significant at $p < 0.05$ (Duncan's test).

3.4. Effects of C. Migao Litter Aqueous Extract on Soil Enzyme Activities of Potted Seedlings

In the leaf$_{AE}$ and branch$_{AE}$ treatments, the soil catalase activity was significantly decreased compared with that in the control ($p < 0.05$), with increasing extract concentration. The soil catalase activity in the pericarp$_{AE}$ and seed$_{AE}$ treatments first decreased and then increased (Figure 5a). This indicated that the leaf$_{AE}$ and branch$_{AE}$ treatments had significant negative regulatory effects on soil catalase, whereas the pericarp$_{AE}$ and seed$_{AE}$ treatments showed a promoting effect. The soil urease activity showed an increasing trend in all the treatments, being significantly higher than that in the control ($p < 0.05$). All treatments promoted soil urease activity (Figure 5b). The soil phosphatase activity decreased with the increases in the concentrations of leaf$_{AE}$ and pericarp$_{AE}$; the value after the leaf$_{AE}$ treatment was lower than that in the control, while that after the pericarp$_{AE}$ treatment was lower than that in the control at 0.02–0.04 g·mL^{-1}. The seed$_{AE}$ and branch$_{AE}$ treatments could also promote soil phosphatase activity (Figure 5c). Moreover, under all the aqueous extract treatments, the soil sucrase activity was higher than that in the control, and the promoting effect of seed$_{AE}$ at 0.04 g·mL^{-1} was the most pronounced (Figure 5d).

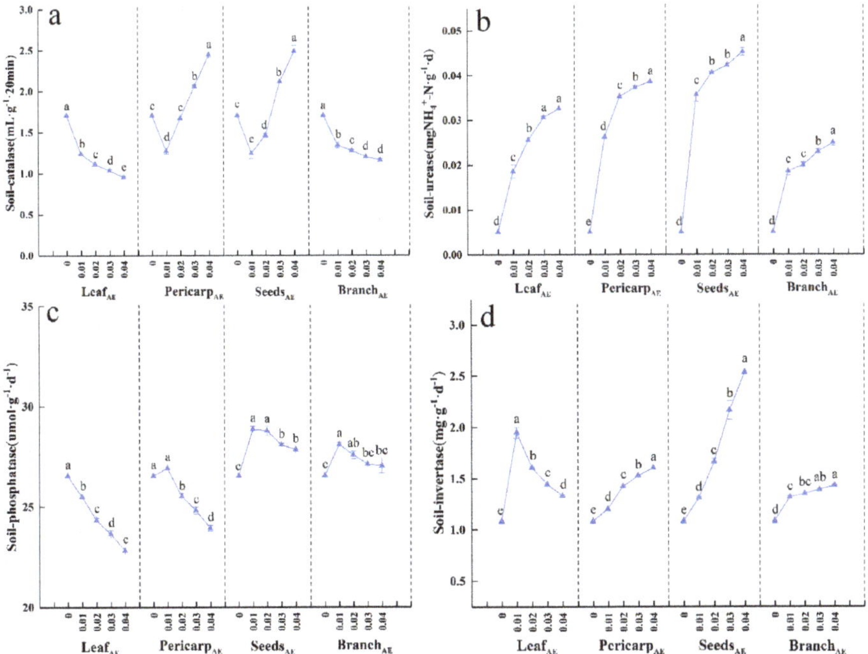

Figure 5. The catalase (**a**), urease (**b**), phosphatase (**c**), and invertase (**d**) contents in soil planted with *C. migao* seedlings after litter aqueous extract treatments for 210 days. The *x* axis denotes the concentration of the aqueous extract and the *y* axis denotes the content. Values are expressed as mean ± SE. The values marked with different letters are significant at $p < 0.05$ (Duncan's test).

3.5. Composition of Fungi in Different Plots and Experimental Treatments

Fungi in the surface soil of nine areas below *C. migao* canopy and upon the treatments with the highest concentration of the extracts were sequenced by high-throughput sequencing technology. As shown in Figure 6a, the common fungi (after the removal of unidentified fungi) in the surface soil below the canopy in the nine counties (Luodian, Ceheng, Libo, Zhenfeng, Wangmo, Napo, Funing, Tian'e, and Leye) were identified as *Archaeorhizomyces sp., Chaetomium nigricolor, Chaetosphaeria vermicularioides, Chloridium sp., Cladophialophora sp., Cryptococcus podzolicus, Cylindrocladiella pseudoparva, Ilyonectria mors-panacis, Lophiostoma sp., Microascaceae sp., Mortierella biramosa, Myxocephala albida, Penicillifer martinii, Penicillium ochrochloron, Rozellomycota sp., Trichoderma koningiopsis,* and *Trichosporon sporotrichoides*. Based on the relative abundance, the most abundant fungal species in Luodian, Ceheng, Libo, Zhenfeng, Wangmo, Funing, and Leye was *Cryptococcus podzolicus*, the most abundant fungal species in Napo was *Mortierella biramosa*, and the most abundant fungal species in Tian'e was *Penicillium ochrochloron*. Comparing the species' composition in terms of the top 30 species under the different treatments (Figure 6b), it could be found that the species overlapping between natural conditions and artificial controlled experimental conditions were *Cladophialophora* and *Lophiostoma*. The average abundance of *Cladophialophora* in each treatment group was in the order pericarp$_{AE}$ > leaf$_{AE}$ > seed$_{AE}$ > control > branch$_{AE}$, whereas that of *Lophiostoma* was branch$_{AE}$ > leaf$_{AE}$ > pericarp$_{AE}$ > seed$_{AE}$ > control.

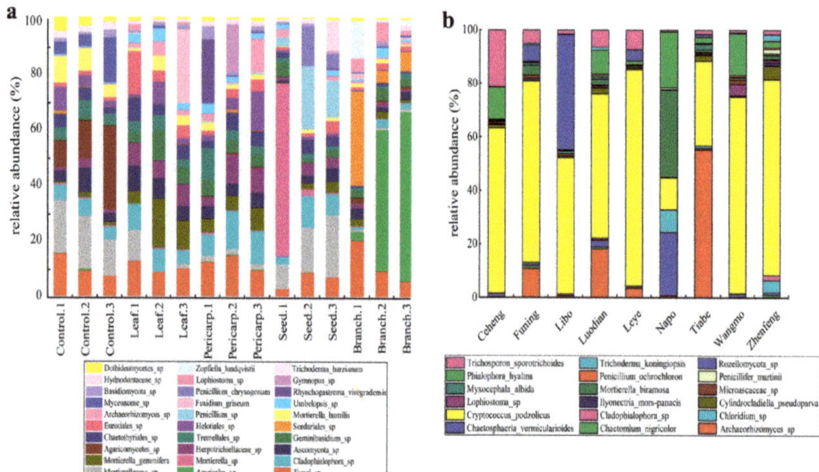

Figure 6. Relative abundance at the species level. Notes: (**a**) Relative abundance of the top 30 species under different treatments. (**b**) The common fungal species in nine areas.

The Adonis analytical method was used for analyzing the differences in the composition of fungi between the different treatment groups (Table 3). The results indicated that the composition of fungi significantly differed among the five treatments. It was also proven that the soil fungal community may have been changed by the extract treatment from different anatomical parts of *C. migao*.

Table 3. Composition of fungi among the four treatments by the Adonis analysis method.

Title	Df	Sum of Squares	Mean Squares	F. Model	R^2	P (>F)
Group_factor	4	1.7057	0.42642	3.4234	0.57795	0.001
Residuals	10	1.2456	0.12456		0.42205	
Total	14	2.9513			1	

$p < 0.05$ indicates that the feasibility of this test is high and that there are significant differences at an intergroup level.

Similarly, UPGMA sample similarity cluster analysis (BinaryJacCard distance) was performed on the three replicates of fungal communities (Figure 7). The results showed that the control, pericarp$_{AE}$, and seed$_{AE}$ (1,2) treatments were grouped into a single branch, whereas the seed$_{AE}$ (3) and leaf$_{AE}$ treatments were grouped together into another single branch. The clustering results showed that there was good similarity within the group, indicating that the change in the fungal composition among the different treatments was closely related to the type of litter.

4. Discussion

Obstacles to the natural regeneration of *C. migao* were always an important factor affecting its population continuity. Early studies also confirmed that *C. migao* has strong allelopathic effects on many plants [29]. Therefore, we hypothesized that *C. migao* is a medicinal plant with autotoxic allelopathy. One of the main methods to affect population regeneration is through the decomposition of litter and release of autotoxic compounds that alter the structure and properties of the ecosystems. Although our laboratory experiments have some limitations, the results indicate that the autotoxicity of *C. migao* is an important factor hindering its seedling growth and survival.

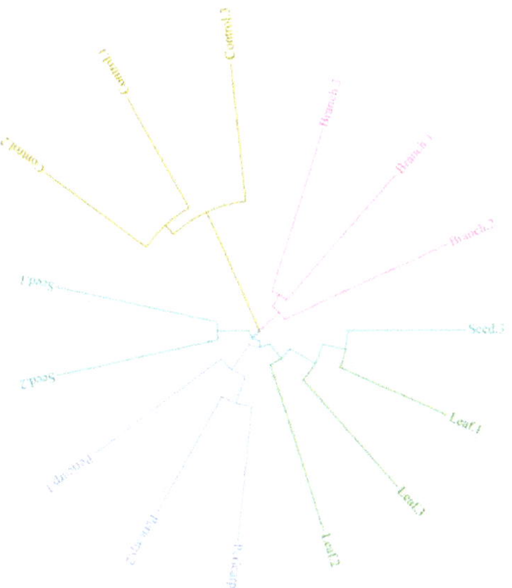

Figure 7. UPGMA sample similarity clustering results.

4.1. Seedling Growth and Antioxidant System Responses to Aqueous Extracts from Different Anatomical Parts of C. migao

Under autotoxicity, plants usually suffer a reduction in biomass, including inhibition of leaf, stem, and root growth [10]. When the concentration of autotoxic substances is low, it may promote plant growth, but if the concentration exceeds a threshold value, it will lead to unhealthy growth or other adverse effects [7]. Our study found that all growth indices of seedlings treated with leaf$_{AE}$ were inhibited (Table 2). The height, biomass, and leaf area of seedlings were significantly decreased with an increase in the concentration of leaf$_{AE}$ ($p < 0.05$), which may be the physio-morphological reaction to the autotoxic compounds contained in leaves on the seedlings [39]. This is similar to the findings in a study on biomass accumulation with the allelopathy of *Eucalyptus* [40]. Contrary to leaf$_{AE}$ treatment, pericarp$_{AE}$, seed$_{AE}$, and branch$_{AE}$ treatments showed different degrees of promotion of seedling growth. The reason for this may be that the types of compounds are different and the levels of autotoxic substances in pericarps, seeds, and branches aqueous extracts at the same concentration are lower than that in leaves, which does not reach the threshold of seedlings' tolerance [7]. For instance, the *Pinus harcus* leave aqueous extract inhibited wheat growth, whereas aqueous extracts from other parts promoted it [41]. Meanwhile, preliminary tests were undertaken to clarify the mechanism of autotoxins in *C. migao* seedlings.

Plants may undergo secondary stress under autotoxicity, for example, oxidative stress, which results in the production of excessive ROS; causes accumulation of osmolytes, such as peroxide anion free radical $O_2{}^-$, hydroxyl radical OH^-, and hydrogen peroxide (H_2O_2) [42]; and leads to peroxidation of the membrane lipid structure in plant cells and, finally, MDA formation. The MDA content can be used as an important index for judging the degree of peroxidation damage in plant cell lipid membranes [43]. Our results showed that the content of MDA in leaf$_{AE}$ was higher than that in other aqueous extracts (Figure 3a). Plants produce protective enzymes to cope with the damage caused by free radicals, including the cooperation of POD and SOD, which can effectively reduce and eliminate the damage caused by free radicals to cell membranes [44]. POD and SOD activities in seedlings were decreased with increasing applications of *C. migao* leaf$_{AE}$ at the end of the experiments (Figure 3c,d). This indicated that the leaf$_{AE}$ treatment had an inhibitory effect on the antioxidant system of seedlings; POD and SOD could not remove free radicals promptly, which made the seedlings undergo membrane lipid peroxidation for a long time. However, the pericarp$_{AE}$, seed$_{AE}$, and branch$_{AE}$ treatments had no significant effect on seedling physiology (Table 2 and Figure 3), whereas the MDA content and antioxidant enzyme activity increased with increasing concentration, and the activity of antioxidant enzymes remained at a high level. This indicated that membrane lipid peroxidation occurs in seedlings after aqueous extract treatment [5] (Figure 3a). POD and SOD antioxidants can remove free radicals in seedlings over time, thereby preventing plant damage [9]. The increases in POD and SOD in our study also supported the notion that seedlings could cause biotic stress under autotoxicity. Similarly, soluble proteins are involved in the metabolism of various enzymes, the content of which could indirectly reflect the ability of plants to perform material synthesis and metabolism [45]. The decrease in the soluble protein content after leaf$_{AE}$ treatment was higher than that after other aqueous extract treatments, indicating that the structure and function of membrane lipids in seedling leaves were damaged for a long time, and this might have led to the decreased activities of various enzymes involved in synthesis and metabolism. Studies have indicated that 30%–50% of the leaf soluble protein is Rubisco, which plays a decisive role in determining the plants' rate of photosynthesis [46]. Therefore, we considered that leaf$_{AE}$ treatment affected the protective enzyme system and photosynthetic capacity of seedlings, which was consistent with the finding of inhibition of seedling growth after leaf$_{AE}$ treatment at the end of the experiment. Although the effects of aqueous extracts from different anatomical parts of *C. migao* on the seedlings differed, the experimental results of leaf$_{AE}$ treatment on the morphological and physiological inhibition of seedlings also validated our first hypothesis, i.e., autotoxicity is the main factor affecting seedling growth.

4.2. Soil Nutrition and Enzyme Activity Responses to Aqueous Extracts of Different Anatomical Parts of C. migao

Soil nutrient elements and enzyme activities play an important role in plant growth. Generally, allelopathic (autotoxic) plant litter has a major impact on soil N, P, and K contents and soil enzymes [47]. Allelopathic substances could alter soil fertility and enzyme activity when they enter the soil, which indirectly leads to the occurrence of autotoxicity [48]. Soil enzymes are catalytically, biologically active substances secreted by fungi, plants, and living animals, which are released after the decomposition of animal and plant residues [49]. They are closely related to the transformation and utilization of soil N, P, and K [50]. Soil catalase mainly decomposes H_2O_2 produced by microorganisms to reduce toxicity; in the present study, the soil catalase content significantly decreased with an increase in the concentration of leaf$_{AE}$ compared with that in the control ($p < 0.05$), indicating that leaf$_{AE}$ treatment significantly inhibited the ability of soil catalase to scavenge H_2O_2, thereby indirectly increasing the toxicity to seedlings [51] (Figures 5a and 3a). Soil urease can catalyze the hydrolysis of urea in soil and improve the soil N supply capacity. Our study found that treatment with aqueous extracts from different anatomical parts of *C. migao* increased soil urease activity, which was similar to the changes in soil urease activity reported in studies of allelopathy on *Allium sativum* bulbs by *Astragalus mongholicus* root aqueous extracts [52]. Generally, invertase activity is positively correlated with soil urease and soil fertility. Compared with the control, the increase in soil fertility had positive effects on soil invertase and urease activities in our study, which illustrates this point [53]. In some autotoxic species, allelochemicals may cause deterioration of the soil substrate, thereby inhibiting the secretion and release of soil phosphatase by plant roots and soil microbial metabolic activities [11] and also affecting the conversion of soil organic P. Our results showed that the TP and AP contents upon different treatments were significantly higher than those in the control, whereas the soil phosphatase activity showed an opposite trend, with an increase in the concentration of the aqueous extract, and leaf$_{AE}$ concentration was always lower than in the control at 0.01–0.04 g·mL^{-1}. One potential reason for this is that the method of co-irrigation of extract and litter was adopted in this experiment, and the later decomposition of litter could supplement soil elements. Furthermore, it is possible that the aqueous extract has a negative effect on the metabolic activity of roots and the composition and activity of soil microorganisms, thus changing the intensity of secretion, release, and modification of soil enzymes [11]. Although aqueous extract treatments supplemented the soil nutrients, they did not change the inhibitory effect of leaf$_{AE}$ on *C. migao* seedling growth, which also confirmed that an improvement in soil fertility could not alleviate the negative effects of autotoxicity on plants [8]. Allelochemicals could affect soil physicochemical properties, microbial composition, and soil enzyme activity after entering the soil, thereby changing the growth of plants [54]. Our study found that litter aqueous extract caused certain changes in soil conditions compared with those in the control, particularly the leaf$_{AE}$ treatment, which altered the soil conditions to indirectly inhibit seedling growth. Therefore, our second hypothesis was partially validated in this experiment. In addition, the inhibition of seedling growth by leaf$_{AE}$ treatment indicates that there are indeed autotoxic substances in *C. migao* leaves, and the types and contents of these substances still need further study.

4.3. Soil Fungal Component with Different Treatments and in Contrast to that in the Field

Soil fungi are essential components of forest ecosystems and are considered to be key factors affecting plant growth. Many autotoxic substances bind in the form of acyl/glycosyl groups in plants. These toxic substances do not directly act on plants and do not exhibit autotoxicity, because of the lack of soil fungi participated [55]. However, when autotoxic compounds enter the soil in the form of litter, they lead to the degradation of acyl/glycosyl groups with the participation of fungi and conversion of inactive to active forms, which finally cause autotoxicity [56]. The soil fungal community is the main part of its microbial population and a decomposer of soil organic compounds [57]. It plays an important role in regulating soil enzyme activities [58] and is directly related to invertase, urease, and phosphatase activities [44]. In this study, soil fungal composition after different aqueous extract treatments were compared with that in the nine areas to explore the differences in fungal diversity in potted soils after different treatments. This could verify the possible effects of autotoxic substances on the soil after entering it. Moreover, the study could determine whether there are common fungi involved in the interaction of litter catabolism and soil substrate change during the process of autotoxicity. Our results showed that there were significant differences in the fungal composition of surface soil between different anatomical parts of aqueous extract treatments. Next, we performed comparisons with 30 common fungi found in the nine areas and the two fungi found in the pot experiments; both of these groups had a high abundance under leaf$_{AE}$ treatment. This indicates that the different secondary metabolites of litter resulted in an interaction between fungi in soil, which resulted in the difference in the fungal composition after different treatments. It also indicates that allelochemicals could affect plant growth by changing soil fungal composition [59]. There are only two kinds of fungi in common between the pot experiment and the natural population. It may be that the pot experiment used high-temperature-sterilized soil, and the species and quantity of fungi carried by the litter itself are limited [13], which cannot fully reflect the fungal composition under natural conditions. In addition, the climate of the experimental site differs from the natural conditions; therefore, the main reason for these differences may be that the in situ environment is more suitable for related fungal growth [12]. From the perspective of revealing the relationship between fungi and autotoxicity, in situ experiments will be indispensable, which will be the focus of our work in the future.

5. Conclusions

In summary, our results demonstrated that the leaf$_{AE}$ treatment could inhibit seedling growth to different degrees, and the inhibitory effects mainly act via decreasing growth targets and damaging cell membrane lipid structures. Our results also revealed that soil enzyme activities were inhibited by leaf$_{AE}$, and the soil fungal composition after leaf$_{AE}$ treatment was significantly different from that after the other treatments (including the control). Although the soil fertility was somewhat improved by leaf$_{AE}$ treatment, it did not promote seedling growth, which indicated that the autotoxic substances contained in leaves are important factors affecting seedling growth. This is consistent with the long-term accumulation of leaf litter in the field and also supports the view that autotoxicity is an important obstacle to the natural regeneration of *C. migao*. From the perspective of species protection, the litter under the canopy should be cleaned up in time to reduce accumulation of autotoxic substances in soils under canopy and conserve the natural population. Meanwhile, measures for ex situ conservation could be considered after artificial seedling cultivation in similar habitats.

Author Contributions: Experimental design, J.L. (Jiming Liu), X.H., and J.C.; field investigation, X.H., J.C., M.W., and B.T.; experiment and date analyzed, X.H., J.C., and J.L. (Jia Li); writing—review and editing, X.H., J.C., and J.L. (Jiming Liu).

Funding: This study was supported by the Guizhou Science and Technology Program "Source Screening and Mycorrhizal Seedling Breeding Technology of *Cinnamomum migao*" (Qiankehe Supporting [2019] 2774).

Acknowledgments: We thank all authors for their contributions to this study. We would like to thank Editage for the English-language revision.

Conflicts of Interest: The authors declare no conflicts of interest.

References

1. Lorenzo, P.; Rodríguezecheverría, S. Influence of soil microorganisms, allelopathy and soil origin on the establishment of the invasive Acacia dealbata. *Plant Ecol. Divers.* **2012**, *5*, 67–73. [CrossRef]
2. Xia, R.; Xiaofeng, H.; Zhongfeng, Z.; Zhiqiang, Y.; Hui, J.; Xiuzhuang, L.; Bo, Q. Isolation, Identification, and Autotoxicity Effect of Allelochemicals from Rhizosphere Soils of Flue-Cured Tobacco. *J. Agri. Food Chem.* **2015**, *63*, 8975.
3. Fuentes-Ramírez, A.; Pauchard, A.; Cavieres, L.A.; García, R.A. Survival and growth of *Acacia dealbata* vs. native trees across an invasion front in south-central Chile. *Forest Ecol. Manag.* **2011**, *261*, 1003–1009. [CrossRef]
4. Blum, U. *Plant—Plant Allelopathic Interactions*; Springer Netherlands: Dordrecht, The Netherlands, 2011; p. 17.
5. Fernandez, C.; Monnier, Y.; Santonja, M.; Gallet, C.; Weston, L.A.; Prévosto, B.; Saunier, A.; Baldy, V.; Bousquet-Mélou, A. The Impact of Competition and Allelopathy on the Trade-Off between Plant Defense and Growth in Two Contrasting Tree Species. *Front Plant Sci.* **2016**, *7*, 594. [CrossRef] [PubMed]
6. Xia, R.; Yan, Z.Q.; He, X.F.; Li, X.Z.; Bo, Q. Allelochemicals from rhizosphere soils of *Glycyrrhiza uralensis* Fisch: Discovery of the autotoxic compounds of a traditional herbal medicine. *Ind. Crops Prod.* **2017**, *97*, 302–307.
7. Ahmed, R. Allelopathic effects of leaf litters of *Eucalyptus camaldulensis* on some forest and agricultural crops. *J. For. Res.* **2008**, *19*, 19–24. [CrossRef]
8. Wardle, D.A.; Bardgett, R.D.; Klironomos, J.N.; Heikki, S.L.; Putten, W.H.V.D.; Wall, D.H. Ecological linkages between aboveground and belowground biota. *Science* **2004**, *304*, 1629–1633. [CrossRef]
9. Mitić, N.; Stanišić, M.; Savić, J.; Ćosić, T.; Stanisavljević, N.; Miljuš-Đukić, J.; Marin, M.; Radović, S.; Ninković, S. Physiological and cell ultrastructure disturbances in wheat seedlings generated by *Chenopodium murale* hairy root exudate. *Protoplasma* **2018**, *255*, 1683–1692. [CrossRef]
10. Yang, L.; Peng, W.; Kong, C. Effect of larch (*Larix gmelini* Rupr.) root exudates on Manchurian walnut (*Juglans mandshurica* Maxim.) growth and soil juglone in a mixed-species plantation. *Plant Soil* **2010**, *329*, 249–258. [CrossRef]
11. Feng, Y.; Hu, Y.; Wu, J.; Chen, J.; Yrjala, K.; Yu, W. Change in microbial communities, soil enzyme and metabolic activity in a *Torreya grandis* plantation in response to root rot disease. *Forest Ecol. Manag.* **2019**, *432*, 932–941. [CrossRef]
12. Chomel, M.; Guittonny-Larchevêque, M.; Fernandez, C.; Gallet, C.; Baldy, V. Plant secondary metabolites: A key driver of litter decomposition and soil nutrient cycling. *J. Ecol.* **2016**, *104*, 1527–1541. [CrossRef]
13. Gavinet, J.; Santonja, M.; Baldy, V.; Hashoum, H.; Bousquet-Mélou, A. Phenolics of the understory shrub *Cotinus coggygria* influence Mediterranean oak forests diversity and dynamics. *Forest Ecol. Manag.* **2019**, *441*, 262–270. [CrossRef]
14. Lobón, N.C.; Cruz, I.F.D.L.; Gallego, J.C.A. Autotoxicity of Diterpenes Present in Leaves of *Cistus ladanifer* L. *Plants* **2019**, *8*, 27. [CrossRef] [PubMed]
15. Margherita, G.; Osborne, B.A. Resource competition in plant invasions: Emerging patterns and research needs. *Front Plant Sci.* **2014**, *5*, 501. [CrossRef]

16. Santonja, M.; Foucault, Q.; Rancon, A.; Gauquelin, T.; Mirleau, P. Contrasting responses of bacterial and fungal communities to plant litter diversity in a Mediterranean oak forest. *Soil Biol. Biochem.* **2018**, *125*, 27–36. [CrossRef]

17. Fernandez, C.; Santonja, M.; Gros, R.; Monnier, Y.; Chomel, M.; Baldy, V.; Bousquet-Mélou, A. Allelochemicals of *Pinus halepensis* as drivers of biodiversity in;mediterranean open mosaic habitats during the colonization stage of secondary Succession. *J. Chem. Ecol.* **2013**, *39*, 298–311. [CrossRef]

18. Rose, S.L.; Perry, D.A.; Pilz, D.; Schoeneberger, M.M. Allelopathic effects of litter on the growth and colonization of mycorrhizal fungi. *J. Chem. Ecol.* **1983**, *9*, 1153–1162. [CrossRef] [PubMed]

19. Chahartaghi, M.; Langel, R.; Scheu, S.; Ruess, L. Feeding guilds in Collembola based on nitrogen stable isotope ratios. *Soil Biol. Biochem.* **2005**, *37*, 1718–1725. [CrossRef]

20. Santonja, M.; Fernandez, C.; Proffit, M.; Gers, C.; Gauquelin, T.; Reiter, I.M.; Cramer, W.; Baldy, V. Plant litter mixture partly mitigates the negative effects of extended drought on soil biota and litter decomposition in a Mediterranean oak forest. *J. Ecol.* **2016**, *105*, 801–815. [CrossRef]

21. Yeasmin, R.; Motoki, S.; Yamamoto, S.; Nishihara, E. Allelochemicals inhibit the growth of subsequently replanted asparagus (*Asparagus officinalis* L.). *Biol. Agric. Hortic.* **2013**, *29*, 165–172. [CrossRef]

22. Nilsson, M.C.A.Z. Characterisation of the differential interference effects of two boreal dwarf shrub species. *Oecologia* **2000**, *123*, 122–128. [CrossRef] [PubMed]

23. Alías, J.C.; Sosa, T.; Escudero, J.C.; Chaves, N. Autotoxicity Against Germination and Seedling Emergence in *Cistus ladanifer* L. *Plant Soil* **2006**, *282*, 327–332. [CrossRef]

24. Li, T.X.; Wang, J.K. Supercritical carbon dioxide extraction of essential oil from *Cinnamomum migao* HW Li. *J. Chin. Med. Mater.* **2003**, *26*, 178–180.

25. Hu, L.; Xue, R.; Xu, C.; Zhang, Z.; Zhang, G.; Zeng, R.; Song, Y. Autotoxicity in the cultivated medicinal herb *Andrographis paniculata*. *Allelopath. J.* **2018**, *45*, 141–151. [CrossRef]

26. Zhou, T.; Yang, Z.N.; Jiang, W.K.; Ai, K.; Guo, W.K. Variation and regularity of volatile oil constituents in fruits of national medicine *Cinnamomum* migao. *China J. Chin. Mater. Med.* **2010**, *35*, 852–856.

27. Zhao, S.; Li, H.Y.; Qiu, D.W.; Liu, Z.R.; Liu, N. *Cinnamomum migao* resources and ecological investigation: Guizhou, Northern Guangxi and the Contiguous Areas of Hunan, Guizhou and Guangxi. *J. Guiyang Univ. Chin. Med.* **1991**, 59–61. [CrossRef]

28. Xie, M.; Yan, Z.Q.; Ren, X.; Li, X.Z.; Qin, B. Autotoxin in cultivated soil of *codonopsis pilosula* (franch.) nannf. *J. Agri. Food Chem.* **2017**, *65*, 2032–2038. [CrossRef]

29. Chen, J.Z.; Liu, J.M.; Xiong, X.; Huang, X.L.; Tong, B.L.; Wen, A.L. Assessing the feasibility of tree-medicinal plant intercropping system of *Cinnamomum migao* and *Aomum villosumt* based on allelopathy. *Chinese J. Ecol.* **2019**, *38*, 1322–1330.

30. Muhl, R.M.W.; Roelke, D.L.; Zohary, T.; Moustaka-Gouni, M.; Sommer, U.; Borics, G.; Gaedke, U.; Withrow, F.G.; Bhattacharyya, J. Resisting annihilation: Relationships between functional trait dissimilarity, assemblage competitive power and allelopathy. *Ecol. Lett.* **2018**, *21*, 1390–1400. [CrossRef]

31. Hodges, D.; Forney, C.; Prange, R.J. Improving the thiobarbituric acid-reactive-substances assay for estimating lipid peroxidation in plant tissues containing anthocyanin and other interfering compounds. *Planta* **1999**, *207*, 604–611. [CrossRef]

32. Lowry, O.H.R.N. Determination of protein by Folin-Phenol reagent. *Food Drug* **2011**, *13*, 147–151.

33. Venisse, J.S.; Malnoy, M.M.; Paulin, J.P.; Brisset, M.N. Modulation of defense responses of Malus spp. during compatible and incompatible interactions with Erwinia amylovora. *Mol. Plant Microbe Interact* **2002**, *15*, 1204–1212. [CrossRef] [PubMed]

34. Lacan, D.; Baccou, J.C. High levels of antioxidant enzymes correlate with delayed senescence in nonnetted muskmelon fruits. *Planta* **1998**, *204*, 377–382. [CrossRef]

35. Wang, Q.; Wang, S.; Huang, Y. Comparisons of litterfall, litter decomposition and nutrient return in a monoculture *Cunninghamia lanceolata* and a mixed stand in southern China. *Forest Ecol. Manag.* **2008**, *255*, 1210–1218. [CrossRef]

36. Olsen, S.R.; Sommers, L.E. Phosphorus. In *Methods of Soil Analysis: Part 2*, 2nd ed.; Page, A.L., Miller, R.H., Keeney, D.R., Eds.; Agronomy Society of America and Soil Science Society of America: Madison, WI, USA, 1982; pp. 403–430.

37. Dorich, R.A.; Nelson, D.W. Evaluation of manual cadmium reduction methods for determination of nitrate in potassium chloride extracts of soils. *Soil Sci. Soc. Am. J.* **1984**, *48*, 72–75. [CrossRef]

38. Bao, S.D. *Soil and Agricultural Chemistry Analysis*, 3rd ed.; China Agricultural Press: Beijing, China, 2013; pp. 56–58.

39. Bughio, F.; Mangrio, S.; Abro, S.; Jahangir, D.T.M.; Bux, H. Physio-morphological responses of native *Acacia nilotica* to *Eucalyptus* allelopathy. *Pak. J. Bot.* **2013**, *45*, 97–105.

40. Santos, F.M.; Balieiro, F.D.C.; Diniz, A.R.; Chaer, G.M. Dynamics of aboveground biomass accumulation in monospecific and mixed-species plantations of *Eucalyptus* and *Acacia* on a Brazilian sandy soil. *Forest Ecol. Manag.* **2016**, *363*, 86–97. [CrossRef]

41. Alrababah, M.A.; Tadros, M.J.; Samarah, N.H.; Ghosheh, H. Allelopathic effects of *Pinus halepensis* and *Quercus coccifera* on the germination of Mediterranean crop seeds. *New For.* **2009**, *38*, 261–272. [CrossRef]

42. An, M.; Pratley, J.E.; Haig, T. Phytotoxicity of vulpia residues: III. Biological activity of identified allelochemicals from *Vulpia myuros*. *J. Chem. Ecol.* **2001**, *27*, 383–394. [CrossRef]

43. Lin, W.X.; Kim, K.U.; Shin, D.H. Rice allelopathic potential and its modes of action on Barnyardgrass (*Echinochloa crus-galli*). *Allelopath. J.* **2000**, *7*, 215–224.

44. Strom, S.L. Microbial ecology of ocean biogeochemistry: A community perspective. *Science* **2008**, *320*, 1043–1045. [CrossRef] [PubMed]

45. Ding, J.; Sun, Y.; Xiao, C.L.; Shi, K.; Zhou, Y.H.; Yu, J.Q. Physiological basis of different allelopathic reactions of cucumber and figleaf gourd plants to cinnamic acid. *J. Exp. Bot.* **2007**, *58*, 3765–3773. [CrossRef] [PubMed]

46. Amit, D.; Portis, A.R.; Henry, D. Enhanced translation of a chloroplast-expressed RbcS gene restores small subunit levels and photosynthesis in nuclear RbcS antisense plants. *Proc. Natl. Acad. Sci. USA* **2004**, *101*, 6315–6320.

47. Haichar, F.E.Z.; Santaella, C.; Heulin, T.; Achouak, W. Root exudates mediated interactions belowground. *Soil Biol. Biochem.* **2014**, *77*, 69–80. [CrossRef]

48. Stefano, M.; Giuliano, B.; Guido, I.; Maria Luisa, C.; Pasquale, T.; Antonio, M.; Mauro, S.; Francesco, G.; Fabrizio, C.; Max, R. Inhibitory and toxic effects of extracellular self-DNA in litter: A mechanism for negative plant-soil feedbacks? *New Phytol.* **2015**, *205*, 1195–1210.

49. Zornoza, R.; Guerrero, C.; Matax-Solera, J.; Arcenegui, V.; García-Orenes, F.M.J. Assessing air-drying and rewetting pre-treatment effect on some soil enzyme activities under Mediterranean conditions. *Soil Biol. Biochem.* **2006**, *38*, 2125–2134. [CrossRef]

50. Freschet, G.T.; Violle, C.; Bourget, M.Y.; Scherer-Lorenzen, M.; Fort, F. Allocation, morphology, physiology, architecture: The multiple facets of plant above- and below-ground responses to resource stress. *New Phytol.* **2018**, *219*. [CrossRef]

51. Chen, C.; Chen, D.; Shu, K.L. Simulation of Nitrous Oxide Emission and Mineralized Nitrogen under Different Straw Retention Conditions Using a Denitrification–Decomposition Model. *CLEAN—Soil Air Water* **2015**, *43*, 577–583. [CrossRef]

52. Mao, J.; Yang, L.; Shi, Y.; Jian, H.U.; Zhe, P.; Mei, L.; Yin, S. Crude extract of *Astragalus mongholicus* root inhibits crop seed germination and soil nitrifying activity. *Soil Biol. Biochem.* **2006**, *38*, 201–208. [CrossRef]

53. Corneo, P.E.; Alberto, P.; Luca, C.; Marco, R.; Marco, C.; Cesare, G.; Ilaria, P. Microbial community structure in vineyard soils across altitudinal gradients and in different seasons. *Fems Microbiol. Ecol.* **2013**, *84*, 588–602. [CrossRef]

54. Zhou, X.; Liu, J.; Wu, F. Soil microbial communities in cucumber monoculture and rotation systems and their feedback effects on cucumber seedling growth. *Plant Soil* **2017**, *415*, 507–520. [CrossRef]

55. van Dam, N.M.; Bouwmeester, H.J. Metabolomics in the Rhizosphere: Tapping into Belowground Chemical Communication. *Trends Plant Sci.* **2016**, *21*, 256–265. [CrossRef] [PubMed]

56. Jackrel, S.L.; Wootton, J.T. Cascading effects of induced terrestrial plant defences on aquatic and terrestrial ecosystem function. *Proc. Biol. Sci.* **2015**, *282*, 264–266. [CrossRef] [PubMed]

57. Frey, S.D.; Ollinger, S.; Nadelhoffer, K.; Bowden, R.; Brzostek, E.; Burton, A.; Caldwell, B.A.; Crow, S.; Goodale, C.L.; Grandy, A.S. Chronic nitrogen additions suppress decomposition and sequester soil carbon in temperate forests. *Biogeochemistry* **2014**, *121*, 305–316. [CrossRef]

58. Saiya-Cork, K.R.; Sinsabaugh, R.L.; Zak, D.R. The effects of long term nitrogen deposition on extracellular enzyme activity in an *Acer saccharum* forest soil. *Soil Biol. Biochem.* **2002**, *34*, 1309–1315. [CrossRef]
59. Hause, B.; Schaarschmidt, S. The role of jasmonates in mutualistic symbioses between plants and soil-born microorganisms. *Phytochemistry* **2009**, *69*, 1589–1599. [CrossRef]

Article

Role of Plant Traits in Photosynthesis and Thermal Damage Avoidance under Warmer and Drier Climates in Boreal Forests

Guiomar Ruiz-Pérez [1,*], Samuli Launiainen [2] and Giulia Vico [1]

[1] Department of Crop Production Ecology, Swedish University of Agricultural Sciences (SLU), Uppsala, 75007, Sweden; giulia.vico@slu.se
[2] Natural Resources Institute Finland (LUKE), Helsinki, 00790, Finland; samuli.launiainen@luke.fi
* Correspondence: guiomar.ruiz.perez@slu.se

Received: 14 March 2019; Accepted: 3 May 2019; Published: 8 May 2019

Abstract: In the future, boreal forests will face warmer and in some cases drier conditions, potentially resulting in extreme leaf temperatures and reduced photosynthesis. One potential and still partially unexplored avenue to prepare boreal forest for future climates is the identification of plant traits that may support photosynthetic rates under a changing climate. However, the interplay among plant traits, soil water depletion and the occurrence of heat stress has been seldom explored in boreal forests. Here, a mechanistic model describing energy and mass exchanges among the soil, plant and atmosphere is employed to identify which combinations of growing conditions and plant traits allow trees to simultaneously keep high photosynthetic rates and prevent thermal damage under current and future growing conditions. Our results show that the simultaneous lack of precipitation and warm temperatures is the main trigger of thermal damage and reduction of photosynthesis. Traits that facilitate the coupling of leaves to the atmosphere are key to avoid thermal damage and guarantee the maintenance of assimilation rates in the future. Nevertheless, the same set of traits may not maximize forest productivity over current growing conditions. As such, an effective trait selection needs to explicitly consider the expected changes in the growing conditions, both in terms of averages and extremes.

Keywords: boreal forest; leaf temperature; photosynthesis; water availability; leaf thermal damage; thermoregulation

1. Introduction

Plant growth and development involve numerous biochemical reactions that are sensitive to temperature [1]. At the leaf level, high temperatures may lead to a decrease in photosynthesis and an increase in respiration, thus reducing ovetrall net carbon dioxide (CO_2) assimilation rate [2]. Under extreme conditions, plant thermotolerance might be exceeded, leading to permanent damage to the photosynthetic machinery [3]. Leaf thermal damage occurs when leaf temperature (T_L) exceeds a critical temperature (T_{CRIT}), but the damage becomes permanent if T_L exceeds an even higher thermal threshold, often termed 'maximum temperature' [3–7].

Leaf temperature is defined by the interplay between plant traits (chiefly, leaf size and shape [8], albedo [9] and stomatal conductance [10]) and environmental conditions (irradiance, air temperature and humidity, plant water availability). When plants are well-watered and the canopy is well-coupled with the surrounding atmosphere, air temperature (T_A) and T_L are generally well correlated, although thermal regulation can result in leaves being cooler than air in warm conditions and warmer otherwise [11]. Low wind speed and plant water shortage can lead to T_L substantially exceeding T_A [12–14], by reducing the leaf evaporative cooling.

Climate change projections suggest a global increase in the frequency of warm days and nights, heat waves and, in some locations, dry spells [15]. Boreal forests are warming up twice as fast as other ecosystems. As a result of higher temperatures and, in some regions, reduced summer precipitation [15], they will be subject to more frequent periods of low water availability. Warming, together with low water, is already negatively affecting forest productivity [16,17] and threatening the survival of sensitive species [18]. This is especially problematic in the boreal region considering the role of boreal forests in the global carbon (C) cycle [19,20] and as a biomass source in climate change mitigation policies [21]. However, while the joint effects of high air temperatures and low water availability and the occurrence of thermal damage and/or limitation of assimilation rates have been extensively studied in arid and semi-arid environments [22,23], the potential occurrence of such conditions has been mostly overlooked in more mesic environments and boreal forests [24]. To better understand the joint effects of high temperatures and low water availability is essential in order to identify the expected frequency of damaging conditions the potential extent of damage, and possible steps towards preparing boreal forests to future growing conditions.

One potential and still partially unexplored avenue for adaptation to climate change is the identification of combinations of plant traits that can support carbon uptake and reduce the risk of thermal damage in a changing climate [25]. In particular, the role of traits leading to high temperature tolerance has been little explored over boreal regions [26]. Similarly, the link between plant traits, water use and water stress has not been fully resolved in boreal forests [27]. Limited information and understanding may be partially due to the fact that measurements of comprehensive sets of traits and long-term monitoring of forest dynamics are expensive and time consuming, constraining the traits considered [27]. Mechanistic models offer a powerful tool to overcome these limitations, allowing the exploration of a wide set of traits and environmental conditions, including the expected warmer and drier conditions.

Here, a mechanistic model describing energy and mass exchanges among the soil, plant and atmosphere [28,29] is applied to explore the role of plant traits and current and future environmental conditions on leaf temperature and net CO_2 assimilation, with a focus on boreal forests. Specifically, we answer three questions, pertaining traits, growing conditions and their joint effects respectively: (i) Which are the dominant plant traits in regulating leaf temperature and CO_2 assimilation? (ii) How do environmental conditions (e.g., air temperature and precipitation timing) interact in determining the occurrence of thermal damage and reduced CO_2 assimilation? (iii) Which are the most suitable trait combinations for sustained CO_2 assimilation and thermal damage avoidance? How do they differ between current and future, warmer and drier conditions? By answering these questions, this study identifies the key plant traits along the thermal safety—C fixation tradeoff, under current and future climates. It also suggests optimal trait combinations for thermal safety and productivity to cope with future climates. Therefore, this study contributes to disentangle the traits—growing conditions nexus, including the joint effect of increasing temperatures and more frequent lack of water in boreal forests.

2. Materials and Methods

2.1. APES-Atmosphere-Plant Exchange Simulator Model

We use the model APES (Atmosphere-Plant Exchange Simulator), previously developed, calibrated and validated in boreal forests. APES is a process-based one-dimensional multilayer, multi-species forest canopy-soil model, designed especially to account for the vertical structure and functional diversity. The model mechanistically solves the coupled energy, water and C cycles in the soil-vegetation-atmosphere system, using physical and physiological constraints. Only the modeling aspects most relevant for the purposes of this work, i.e., those related to the determination of leaf temperature and net CO_2 assimilation rate, are briefly described here and in the Supplementary Materials. A complete description of this model and its validation, as well as some applications, can be found in Launiainen et al. [28,29].

In APES, the leaf temperature, T_L, is calculated separately for sunlit and shaded leaves within each canopy layer, by solving the coupled energy balance and net CO_2 assimilation rate (A_{net}) (see the Supplementary Materials for the equations).

A_{net} is obtained based on the Farquhar model, as the minimum of Rubisco-limited and light-limited rate, which in turn depend on the maximum carboxylation rate V_{CMAX} and the maximum electron transport rate J_{MAX}, respectively [30]. The temperature responses of both V_{CMAX} and J_{MAX} are as in Medlyn et al. [31]. Additionally, leaf respiration increases with T_L, so that A_{net} reaches its maximum at an intermediate temperature (the optimal temperature for net CO_2 assimilation). The parameters V_{CMAX}, J_{MAX} and R_d at reference temperature are affected by the leaf water potential (ψ_L), following Kellomaki and Wang [32]. The stomatal conductance is computed by using the "unified stomatal model" proposed by Medlyn et al. [33].

The microclimatic gradients within the plant canopy are explicitly accounted for. The photosynthetic active (PAR) and near-infrared (NIR) radiation, and the long-wave balance are computed for each canopy layer as in Zhao and Qualls [34,35]. The profiles of air CO_2 and H_2O concentrations, T_A and wind speed (U) are computed using first-order turbulence closure schemes iteratively with solution of leaf energy and CO_2 balance. The above-ground and soil processes are coupled through water and heat fluxes via feedbacks between soil and vegetation (rainfall interception, root water uptake, feedbacks to leaf physiological parameters).

2.2. Metrics Evaluating Thermal Risk and Assimilation Capacity

To explore the role of key plant traits and identify potential trade-offs between photosynthesis and thermoregulation, we focused on four metrics of thermal damage, tolerance and CO_2 fixation.

2.2.1. Maximum T_L ($T_{L,max}$)

$T_{L,max}$ was determined as the maximum leaf temperature within the warmest three-day period during the growing season. For each time step (30 min), we first calculated the average of T_L of each canopy layer in sunlit leaves, which are the ones most likely to experience the warmest temperatures. Using this average T_L series, we then computed the three-day moving average. $T_{L,max}$ is the absolute maximum T_L within the three-day window with the highest moving average. We selected a three-day period, as in O'Sullivan et al. [3], considering that plants can cope with high temperatures for short periods. Knowledge of $T_{L,max}$ allows determining whether conditions may be conducive to thermal damage of the photosynthetic machinery, i.e., if the critical temperature (T_{CRIT}) is exceeded over the growing season.

2.2.2. Cumulated A_{net} ($A_{net,cum}$)

$A_{net,cum}$ is the cumulated A_{net} within the growing season. We first computed the average of A_{net} at each time step (30 min), including both sunlit and shaded leaves, weighted by sunlit and shaded fractions of leaves at each layer and accounting for non-uniform leaf area distribution across canopy layers. Then, $A_{net,cum}$ was calculated as the cumulated sum of the average A_{net} within the growing season.

2.2.3. Maximum A_{net} ($A_{net,max}$)

The $A_{net,max}$ refers to the maximum average A_{net} at each time step (30 min) over the growing season, weighted by sunlit and shaded fractions of leaves at each layer and accounting for non-uniform leaf area distribution across canopy layers.

2.2.4. Thermal Range of High Photosynthesis (ΔT_{90})

The ΔT_{90} is the extent of the temperature range where assimilation rates are $\geq 90\%$ of their seasonal maximum value calculated above, $A_{net,max}$. Considering the previously computed $A_{net,max}$ and, at each time step, the average A_{net} over the whole canopy, we identified the maximum and minimum T_L for which A_{net} is equal to 90% of $A_{net,max}$. ΔT_{90} corresponds to the difference between those maximum and minimum T_L. We interpreted ΔT_{90} as a proxy of leaf thermotolerance: broad ΔT_{90} ranges mean that the CO_2 assimilation rate is close to its maximum under a wide range of thermal conditions, while narrow ΔT_{90} range necessarily implies a higher likelihood to experience sub-optimal thermal conditions for assimilation.

2.3. Plant Traits Selection and Determination of Their Role

2.3.1. Selection of Plant Traits

Only traits expected to have a direct most notable effect on leaf thermoregulation and C-fixation in APES were included in the analyses. Under set environmental conditions, T_L is strongly affected by leaf size and shape [8], albedo [9] and traits related to stomatal conductance and hence regulation of transpiration [10]. These traits also affect the net CO_2 assimilation rate, directly via its maximum rate, and indirectly, via the stomatal opening and the effects of temperature on leaf metabolic rates. Consequently, in the analyses below, we focused on the primary plant traits regulating the CO_2 assimilation rate, the boundary layer and stomatal conductances (g_b and g_s, respectively), the absorption of solar radiation.

A total of six traits were selected: the maximum carboxylation rate at 25 °C, $V_{CMAX,25}$; the two parameters of the stomatal model, i.e., stomatal model slope in well-watered conditions g_1 and a parameter describing the sensitivity of g_1 to soil water potential β [33]; the effective leaf thickness, l_t, and PAR and NIR albedos, α_{PAR} and α_{NIR}. This selection was based on the following considerations. Assuming leaves as flat plates, the effective leaf thickness, l_t, controls the boundary layer conductance g_b [36,37]. Conversely, the stomatal conductance, g_s, is mostly dependent on hydraulic and photosynthetic traits: in the stomatal model employed in APES [33], these aspects are summarized by the parameters g_1 and β. Leaf temperature and photosynthesis also depend on the absorbed radiation, and hence on the leaf albedo for both photosynthetically active and near infrared radiation (α_{PAR} and α_{NIR} respectively). Finally, according to the Farquhar model, the net CO_2 assimilation rate depends on the maximum carboxylation rate, V_{CMAX}, the maximum electron transport rate, J_{MAX}, and the day respiration rates, R_d. These rates are functions of the corresponding rates at 25 °C ($V_{CMAX,25}$, $J_{MAX,25}$ and $R_{d,25}$) and their response to temperature [31]. $J_{MAX,25}$ and $R_{d,25}$ are often well correlated with $V_{CMAX,25}$, following Medlyn et al. [38] and Launiainen et al. [29], respectively, thus allowing to consider only $V_{CMAX,25}$ as a key parameter of photosynthesis. Further details on trait selection and their role in leaf temperature regulation are given in the Supplementary Materials.

For each of these six traits, we considered realistic ranges, based on a literature search. These ranges were maintained intentionally broad, beyond those typical of current dominant coniferous and deciduous tree species in boreal forests (Table 1). Conversely, the understory and forest floor features were fixed in the model. Concretely, the understory consists of seedlings of Norway spruce, Silver birch and other deciduous species (LAI ~0.5 m^2 m^{-2}), as specified in [29].

Table 1. Summary of the traits considered in the General Sensitivity Analysis (GSA), their ranges of variation emerging from the literature globally and for mid-to-high latitudes and corresponding supporting references.

Trait	Range	Range for Mid-to-high Latitude	References	References for Mid-to-high Latitude
$V_{CMAX,25}$ (μmol m^{-2} s^{-1})	20–120	35–70	[39–42]	[29,39,40,43,44]
l_t (m)	0.005–0.15	*	[11,45,46]	[29]
g_1 (kPa$^{0.5}$)	1–15	1.5–7	[33,38,47]	[29,33,38]
β (-)	0.05–3	**	[29,33,38]	[29,33,38]
α_{PAR} (-)	0.02–0.3	0.04–0.12	[48–50]	[29,48–50]
α_{NIR} (-)	0.1–0.7	0.2–0.55		

* No specific values for mid-to-high latitude have been reported. However, boreal forests are dominated by species with needles. ** No specific values for mid-to-high latitude have been reported.

2.3.2. General Sensitivity Analysis (GSA) and Determination of Dominant Traits

A General Sensitivity Analysis (GSA) was run on APES to identify which of the focal traits are most important to maximize both instantaneous and cumulative A_{net} and regulate T_L, reducing the risk of high temperature damage. To this aim, 100 trait sets were randomly generated from uniform distributions of each trait over the ranges specified in Table 1. The traits were sampled independently. The model was run using each trait set for current growing conditions (Section 2.4.1). The obtained simulated time series of A_{net} and T_L allowed the computation of the four metrics described in Section 2.2 for each trait set (maximum leaf temperature, $T_{L,max}$; the thermal range of high photosynthesis, ΔT_{90}, cumulated and maximum net assimilation, $A_{net,cum}$ and $A_{net,max}$).

The model results were then grouped based on whether the metrics exceeded pre-set thresholds. Specifically, a threshold was set for each metric in order to split the 100 trait sets into two groups: one containing the m trait sets for which metric is above the threshold (Group 1) and the others $n = (100 - m)$ (Group 2). Then, for each trait, the cumulative probability distribution for the *m* trait values in Group 1 and *n* trait values in Group 2 were obtained. The higher the distance between the cumulative distribution functions, the more important the role of the trait value in defining the outcome. To measure the distance between the two empirical functions, the Kolmogorov–Smirnov (KS) two-sample test was used. This nonparametric method allows comparing two samples and quantifying the distance between the empirical distributions of the two samples [51]; thus, it can be used to rank the traits according to how dominant they are in determining the final outcome [52–54].

The thresholds for the grouping were selected to be the highest ones that allowed the greatest number of trait sets in the Group 1 while the number of dominant traits remained unchanged [53]. Indeed, choosing an excessively low threshold leads to (almost) no distinction between Group 1 and Group 2 so that the separation between the two distribution functions is negligible for all traits (i.e., no trait emerges as dominant). Conversely, choosing an excessively high threshold reduces substantially the number of trait sets for which the metric is above the threshold: in this case, Group 1 might not be representative except for a minority of trait sets, leading to a substantial separation between the two distribution functions for all traits (i.e., all traits are considered dominant). This approach was followed for each metric except $T_{L,max}$. For the latter, we used T_{CRIT}—i.e., the temperature at which first symptoms of thermal damage appears. Following O'Sullivan et al. [3], based on the latitude of the reference study site (Section 2.4.1), T_{CRIT} was set at 44.0 °C. This threshold also allowed a balanced partitioning of the 100 parameter sets into Group 1 and Group 2, i.e., the approach delineated above would have led to a similar value for the threshold. The resulting thresholds were: $T_{L,max} = 44.0\ °C$, $A_{net,cum} = 1{,}274\ \mu mol\ m^{-2}$, $A_{net,max} = 0.466\ \mu mol\ m^{-2}s^{-1}$ and $\Delta T_{90} = 6.3\ °C$.

2.3.3. Selection of Focal Trait Combinations

Once the dominant traits were identified via Kolmogorov–Smirnov test, four sets of parameters were chosen to represent widely different outcomes in terms of thermotolerance and C fixation (Table 2). Given the focus on boreal forests, the focal trait combinations were chosen among those including trait values within the ranges typically observed in mid-to-high latitude forests (Table 1). We selected four trait combinations that showed contrasting thermal and productivity responses under identical growing conditions (2005 growing season). Specifically, trait combination 1 corresponds to the set of traits with the lowest performance (lowest $A_{net,cum}$ and $T_{L,MAX}$ exceeding T_{CRIT} in 2005): it is representative of those trait combinations with low productivity and high risk of thermal damage, at least under the current conditions (Low Productivity-High Risk; hereafter, LP-HR). Combination 2 was instead the best performing combination, leading to high assimilation rates while preventing thermal damage, i.e., it is representative of individuals with high productivity and low risk of thermal damage (High Productivity-Low Risk; HP-LR). Combination 3 led to high assimilation rates but also to $T_{L,MAX}$ on par with that of combination 1, thus representing trait sets leading to high productivity but also high risk of thermal damage (High Productivity-High Risk; HP-HR). Finally, combination 4 resulted in assimilation rates similar to combination 1 but it did prevent the occurrence of thermal damage; thus, it corresponds to low productivity but also low risk of thermal damage (Low Productivity-Low Risk). These differences in assimilation and thermal metrics are the result of the following differences between the trait values. Combinations 1 (LP-HR) and 4 (LP-LR) have similar and low $V_{CMAX,25}$; combinations 2 (HP-LR) and 3 (HP-HR) have a similar $V_{CMAX,25}$, but these are higher than combinations 1 and 4. Conversely, l_t values in combinations 1 (LP-LR) and 3 (HP-HR) are similar and higher than the ones in combinations 2 (HP-LR) and 4 (LP-LR), the latter of which also have similar values. No patterns emerged relatively to the other traits, which were indeed not dominant as shown by the sensitivity analysis (see results in Section 3.1). These four combinations of traits together with their corresponding assimilation and thermal metrics are listed in Table 2.

Table 2. Combination of traits used to perform the climate scenarios analyses and the resulting values of the assimilation and thermal metrics for the growing season 2005. LP-HR: Low Productivity-High Risk of thermal damage; HP-LR: High Productivity-Low Risk of thermal damage; HP-HR: High Productivity-High Risk of thermal damage; LP-LR: Low Productivity-Low Risk of thermal damage.

Trait Combination	$V_{CMAX,25}$ $\mu mol\ m^{-2}\ s^{-1}$	l_t m	g_1 $kPa^{0.5}$	β (-)	α_{PAR} (-)	α_{NIR} (-)	$A_{net,cum}$ $\mu mol\ m^{-2}$	$T_{L,max}$ °C	$A_{net,max}$ $\mu mol\ m^{-2}s^{-1}$	ΔT_{90} °C
Comb. 1 (LP-HR)	36.16	0.14	6.35	2.41	0.11	0.39	25235	49.50	8.15	6.02
Comb. 2 (HP-LR)	57.36	0.02	2.19	0.70	0.27	0.60	32770	40.08	11.54	5.60
Comb. 3 (HP-HR)	56.93	0.10	6.53	2.34	0.19	0.49	36818	46.18	12.58	5.83
Comb. 4 (LP-LR)	36.37	0.03	6.18	1.92	0.22	0.52	25600	42.14	8.18	6.52

2.4. Growing Conditions

The GSA was performed under current climatic conditions, corresponding to selected years in a study site in Finland (Section 2.4.1). Conversely, the focal combinations of traits were tested under both current and future warmer and/or drier growing conditions, as defined in Section 2.4.2. This approach allowed us to determine whether thermal damage and suboptimal photosynthetic rates would occur given certain temperature increase and intensity of the water stress, thus allowing the quantification of potential effects of climate change and the importance of trait selection in the future.

2.4.1. Current Growing Conditions

Current growing conditions are based on meteorological data observed in the Hyytiälä field station SMEAR II (Station for Measuring Forest Ecosystem-Atmosphere Relations), located in Hyytiälä, Southern Finland (61°51′ N, 24°17′ E). The annual mean temperature is 3 °C and total mean annual precipitation is 700 mm. All the meteorological data necessary for APES are routinely measured at this site. A complete description of the site and its measurements can be found in Kolari et al. [55]; Kulmala et al. [56] and Launiainen et al. [29].

The model was applied over selected growing seasons (2005 and 2006), here defined as the period of May 1st to September 30th [29], with the warmer period falling in the first half of July. The growing season 2005 can be considered as a normal one in the region, while 2006 was a dry year [57], slightly warmer than 2005. The two growing seasons were also characterized by a different precipitation timing: in 2005, the rainiest period occurred in late July and early August and the highest precipitation was recorded at the end of the growing season; in 2006, a dry spell occurred in late July and early August while two of the rainiest peaks were recorded in early July. Therefore, 2005 and 2006 present contrasting plant water availability, allowing us to analyze if these difference might lead to significant changes in thermoregulation and net CO_2 assimilation.

2.4.2. Scenarios of Future Growing Conditions

Boreal regions are experiencing a rapid warming and an increase in the frequency and intensity of periods with low water availability as a consequence of changes in the timing and magnitude of precipitation, as well as increases in evapotranspiration from longer and warmer growing seasons [15]. Therefore, to generate climate change scenarios, we focused on air temperature (T_A) and soil water potential (Ψ_S).

The future scenarios were developed as follows. We considered two single days within the growing season of year 2005 as baseline: (i) the day where T_L reached its maximum value (July 6th) and (ii) an average day with T_A close to the median value for the growing season (May 20th). Then, we built scenarios of future growing conditions by shifting upwards the observed T_A series of 1 to 10 °C combined with decreasing soil water potential ranging from −0.5 to −3 MPa. The soil water potential is considered constant within the day. For each scenario, we computed the instantaneous A_{net} and T_L at midday. While acknowledging their importance, other environmental variables such as wind speed and relative humidity were not modified. Thus, the vapor pressure deficit (VPD) was computed considering the new temperature values and the unchanged relative humidity.

Furthermore, we analyzed how trait combinations might affect soil water availability and how such changes may alter maximum and cumulated net CO_2 assimilation. To this aim, the shift in T_A series by 1 to 10 °C was combined with imposing 5 to 30 day dry spells, with no precipitation. The dry spell was assumed to begin after the last precipitation event before the warmest period registered within the growing season of 2005 (i.e., June 27th). We then computed the corresponding Ψ_s at the end of the 5–30 day dry period, and the $A_{net,cum}$ over the dry down.

In all cases, the length of the growing season was maintained unaltered, even though global warming may lead to longer growing seasons in boreal regions [58,59]. This simplifying assumption bears no consequence of the results discussed below, because the aim was to contrast different trait combinations under the same abiotic conditions.

3. Results

3.1. Dominant Traits Emerging from the General Sensitivity Analysis (GSA)

The General Sensitivity Analysis (GSA) allowed the identification of the traits most relevant for the four metrics: $T_{L,max}$, $A_{net,cum}$, $A_{net,max}$ and ΔT_{90}. The results are presented in Figure 1 and summarized next.

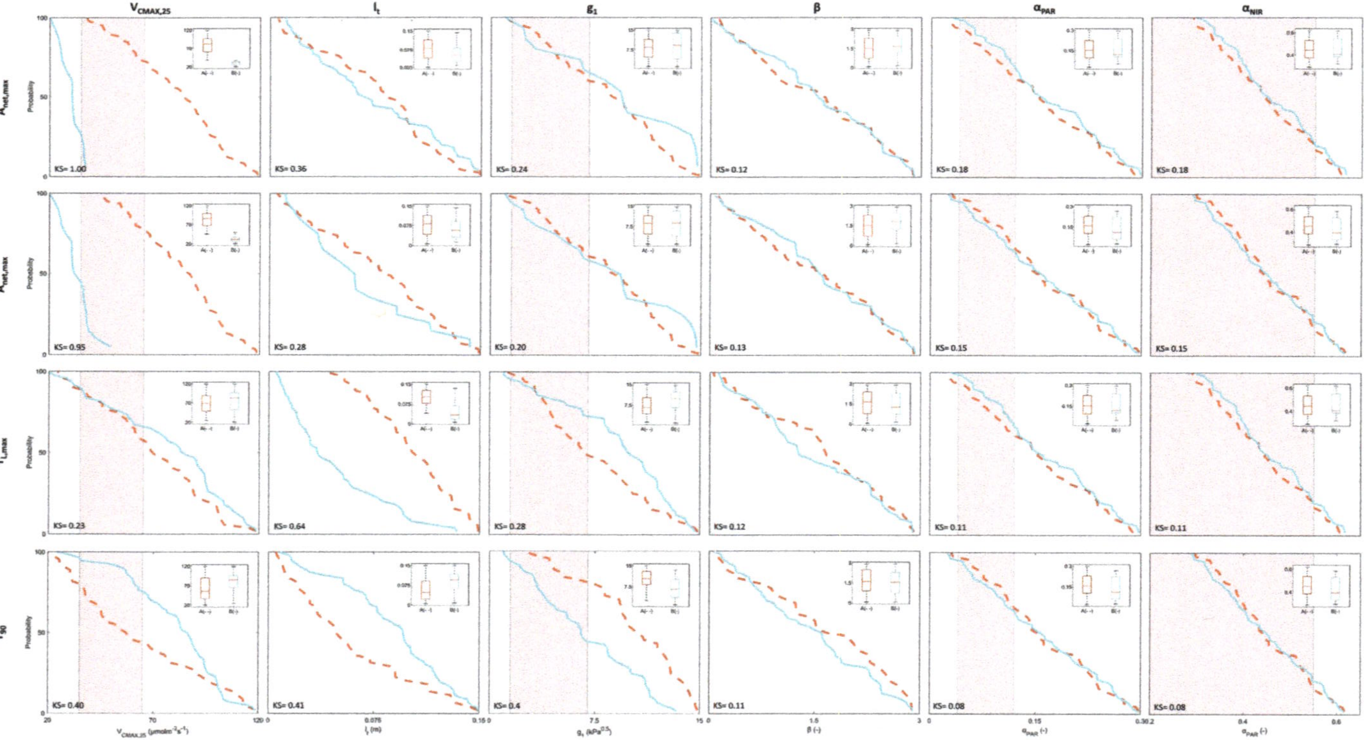

Figure 1. Results of the Global Sensitivity Analysis. Rows refer to the four metrics; columns to the six traits investigated. In each main plot, the lines are the empirical cumulative function of the trait values: blue solid lines refer to the trait values that lead to metric values below the metric-specific threshold; the red dashed lines refer to trait values that lead to metric values above the threshold. Shaded areas identify ranges of trait typically observed in mid-to-high latitudes. In the insets, the box and whisker plots compare and summarize the trait values leading to metric values above (left box; A−) and below (right box; B-) the threshold. The boxes extend from the first to the third quartile, the end of the whiskers are computed as $1.5 \times$ IQR (Interquantile Range). The median values are indicated by the horizontal red lines.

3.1.1. Maximum T_L ($T_{L,max}$)

Ranking the traits according to KS (Figure 1), the most dominant one was l_t, followed by g_1 and $V_{CMAX,25}$. In the case of the leaf albedos (α_{PAR} and α_{NIR}) and β values, the two distribution functions were overlapping and the KS value was negligible, i.e., the traits did not have a large effect on $T_{L,max}$. The group containing the trait sets that did not result in thermal damage (i.e., $T_{L,max} < T_{CRIT}$) had lower l_t, higher g_1 and higher $V_{CMAX,25}$. For values lower than 55 μmol m^{-2}s^{-1} (i.e., those typically observed in trees in mid-to-high latitudes; shaded range), the importance of $V_{CMAX,25}$ decreases as evidenced by the two empirical distribution functions overlapping for such parameter values. The same behavior was shown by g_1, the influence of which diminishes for values higher than 4 kPa$^{0.5}$.

3.1.2. Cumulated A_{net} ($A_{net,cum}$) and Maximum A_{net} ($A_{net,max}$)

$A_{net,cum}$ was dominated by $V_{CMAX,25}$, followed by g_1 and l_t. Higher values of $V_{CMAX,25}$ led to higher $A_{net,cum}$. Similarly, the group with higher $A_{net,cum}$ presented higher values of g_1 and l_t up to values of 11 kPa$^{0.5}$ and 0.1 m, respectively, above which the effect of these parameters on $A_{net,cum}$ shifted. Similar results were obtained regarding $A_{net,max}$ (Figure 1, third row).

3.1.3. Thermal Range of High Photosynthesis (ΔT_{90})

ΔT_{90} was equally dominated by g_1, l_t and $V_{CMAX,25}$. The group with higher thermotolerance (i.e., broader ΔT_{90}) had higher g_1 and lower $V_{CMAX,25}$ and l_t than the less thermotolerant one. Regarding the leaf albedos and β values, the distribution functions were mainly overlapping and the KS value was almost negligible (Figure 1, bottom row). Note that, differently from the case of $T_{L,max}$, $V_{CMAX,25}$ is also dominant for parameter values typical of trees in mid-to-high latitudes.

3.2. Role of the Growing Conditions

3.2.1. Differences between 2005 and 2006

In both the 2005 and 2006 growing seasons, leaves did not reach their maximum temperature when T_A was warmest, but rather after several consecutive days without precipitation (4 and 7 days in 2005 and 2006, respectively). No large differences between the days when the maximum T_L are reached in each year were found in terms of radiation, wind speed and VPD.

Given the same combination of traits, $A_{net,cum}$ was systematically higher in 2006 than in 2005 growing season (Figure 2, top), while $T_{L,max}$ was systematically higher in 2005 than in 2006. Indeed, some combinations of traits led to thermal damage during 2005, while this threshold was not reached by any combination of traits during 2006. Conversely, no marked differences in terms of $A_{net,max}$ and ΔT_{90} were found (Figure 2).

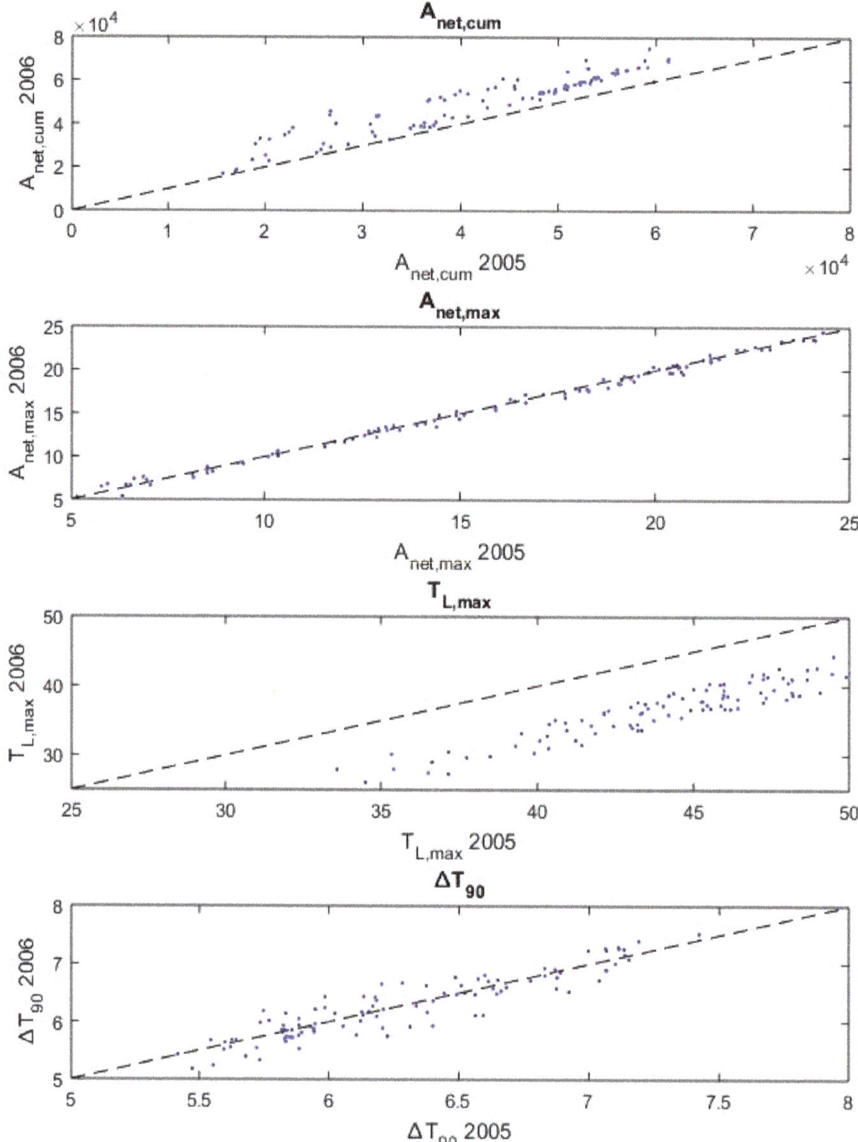

Figure 2. Comparison of model outputs for 2005 and 2006. Value of the four metrics $A_{net,cum}$ (in μmol m^{-2}), $A_{net,max}$ (in μmol m^{-2}s^{-1}), $T_{L,max}$ and ΔT_{90} (in °C) obtained for the 100 combination of traits under the growing conditions 2005 and 2006. The dashed line corresponds to the 1:1 line.

3.2.2. Future Growing Conditions

To investigate the effects of likely future conditions, increases of temperature of 0° C to 10 °C with respect to current conditions were combined with values of soil water potential ranging from −0.5 to −3 MPa. Those changes were applied by considering both a standard day and a warm day of the 2005 growing season as baselines.

Taking the currently normal thermal conditions as baseline, combination 3 (HP-HR) led to the highest assimilation rates under T_A lower than 18°C (Figure 3 bottom row). For T_A below 16°C, T_L reached temperatures similar to combinations 2 (HP-LR) and 4 (LP-LR). Combination 1 (LP-HR) led to the highest difference between T_A and T_L, while combinations 2 (HP-LR), 3 (HP-HR) and 4 (LP-LR) showed similar T_L values over a wide range of T_A conditions (Figure 3 top row). T_L was not sensitive to Ψ_s in any of the four combinations. The A_{net} contour curves showed a parabolic shape with a vertex in T_A equal to 8°C, 10°C, 10°C and 12°C for combination 1, 2, 3 and 4, respectively.

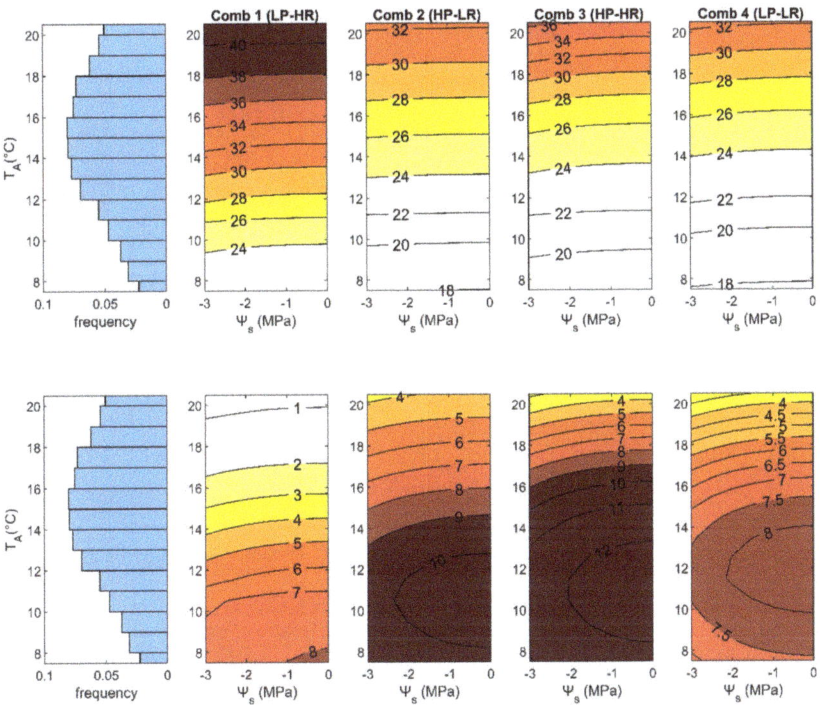

Figure 3. Leaf temperatures (top row; in °C) and net assimilation rates (bottom row; in μmol m^{-2} s^{-1}) for different air temperatures and soil water potentials. Four trait combinations are explored (Table 2). The lowest air temperature corresponds to the median day for the growing season 2005. As a term of comparison, the left column shows the frequency of the air temperature at midday reported in the study site during the growing seasons from 1996 to 2016. LP-HR: Low Productivity-High Risk of thermal damage; HP-LR: High Productivity-Low Risk of thermal damage; HP-HR: High Productivity and High Risk of thermal damage; and LP-LR: Low Productivity and Low Risk of thermal damage.

For the warm day baseline conditions, combinations with high l_t—i.e., combinations 1 (LP-HR) and 3 (HP-HR)—were more likely to result in thermal damage, with T_L exceeding T_{CRIT} even under current thermal conditions (Figure 4 top row). Due to this high T_L, the assimilation rates were also reduced to the point that combination 3 (HP-HR) showed lower A_{net} even though its maximum assimilation rate was higher than the corresponding one in combination 4 (LP-LR) (Figure 4, bottom row). Conversely, combination 2 (HP-LR), and to some extent combination 4 (LP-LR), showed higher thermoregulation capacity (i.e., lower T_L) and higher assimilation rates. When compared, combination 2 (HP-LR) appeared to be more beneficial, resulting in lower T_L and higher assimilation rates. Note that for all four combinations, an increase in T_A with respect to the warm day baseline is detrimental to plant productivity, as A_{net} is reduced and leaves experience thermal damage already under current conditions (combination 1 and 3), or for temperature increases of 1 to 2.5 °C (combination 4 and 2, respectively). T_L was sensitive to Ψ_s under dry conditions, particularly in combination 2 (HP-LR) (Figure 4 top row). Contrarily, in all combinations, A_{net} rates decreased as did water availability (Figure 4 bottom row).

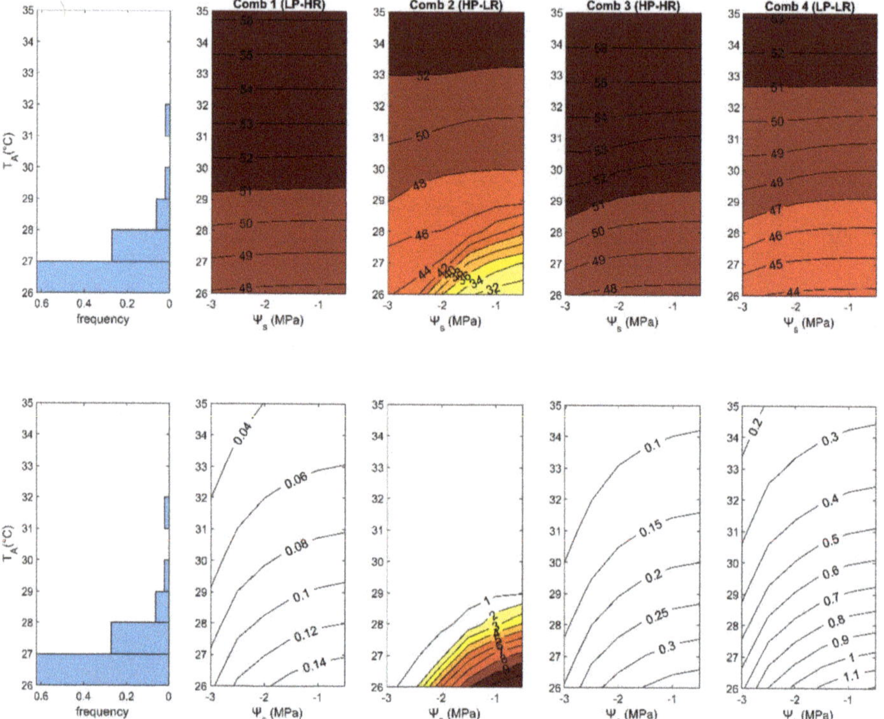

Figure 4. Leaf temperatures (top row; in °C) and net assimilation rates (bottom row; in µmol m^{-2} s^{-1}) for different air temperatures and soil water potentials. The baseline temperature corresponds to the warmest day of the growing season 2005. All the other parameters are as in Figure 3.

Regarding the effects of plant traits on soil water depletion, similar Ψ_s emerged among the four trait combinations after the same number of consecutive days without precipitation. For all four combinations, after intervals shorter than 15 consecutive days without precipitation, soil water potential (i.e., lower Ψ_s) was only slightly affected by increasing temperatures. The lowest Ψ_s—around –0.9 MPa after more than 25 consecutive dry days—was reached by combination 2 (HP-LR) and 4 (LP-LR) (Figure 5).

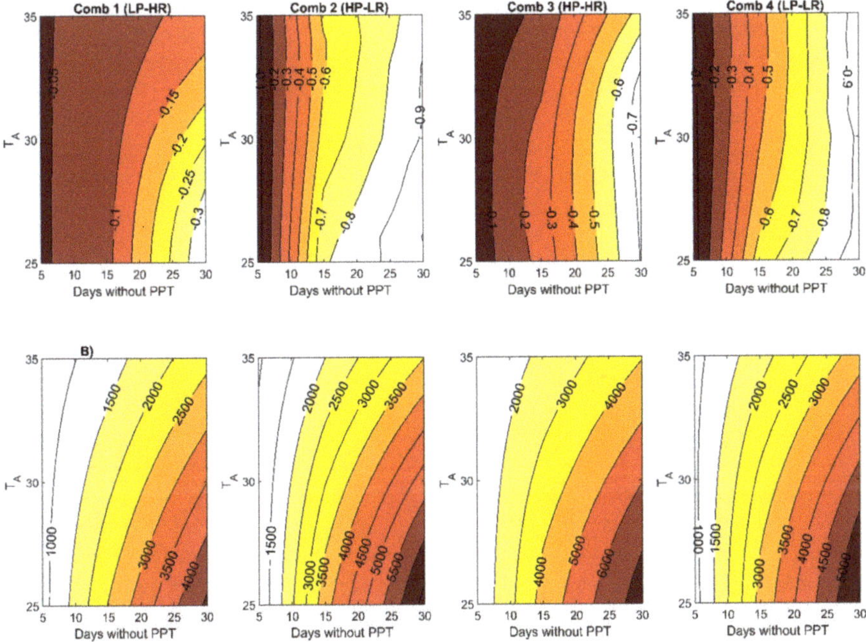

Figure 5. Soil water potential (upper panels; MPa) and cumulated net assimilation (lower panels; μmol m^{-2} s^{-1}) for each focal trait combination (Table 2) for specific T$_A$ (y-axis) and after a certain number of consecutive days without precipitation (x-axis). The baseline temperature corresponds to the warmest day in 2005. The rest of the growing conditions correspond to those observed in the study site during the same period within the growing season of 2005.

4. Discussion

4.1. Role of Traits: Potential Mechanisms Explaining the Dominance of Plant Traits

The GSA allowed identifying the role of six traits affecting net CO$_2$ assimilation, A$_{net}$, leaf temperature and T$_L$, and hence the risk of thermal damage. As expected, both the maximum and cumulated net CO$_2$ assimilation are strongly dependent on the maximum carboxylation rate, V$_{CMAX,25}$. They are also affected by the effective leaf thickness, l$_t$, although less markedly. In particular, higher l$_t$ values were also associated with higher cumulated net CO$_2$ assimilation, A$_{net,cum}$, but only up to l$_t$ ~0.1m, above which this trend reverts. In general, higher l$_t$ leads to higher T$_L$, by reducing the leaf boundary layer conductance g$_b$ [60]. Increase in T$_L$ may have opposite effects on A$_{net}$, depending on the initial temperature and the species: when T$_L$ is below the temperature that maximizes net CO$_2$ assimilation, an increase in T$_L$ stimulates A$_{net}$ [11]. In our results relative to current climates, T$_L$ was generally below such threshold, so that higher l$_t$ values were associated with higher A$_{net,cum}$. Nevertheless, as apparent from our results, there is a maximum l$_t$, above which increases in l$_t$ are no longer beneficial for CO$_2$ assimilation because of an insufficient leaf cooling. Note that most boreal forests have needles, i.e., low l$_t$. It is important to note that plants grown in high light generally have thick leaves to protect them from high-irradiance damage [61,62]. Conversely, in low light available conditions (frequent in boreal forests), leaf thickness reduces to maximize the light capturing area and reduce self-shading. As such, the low l$_t$ typical of many boreal species enhances light use, and hence potentially net CO$_2$ assimilation via light capture as opposed to optimal temperature for photosynthesis. Finally, within the range of values observed in mid-to-high latitudes (shaded area in Figure 1), the two parameters of the photosynthesis model (g_1, β), and both albedos (α_{PAR} and α_{NIR}) did not have any conspicuous influence on the net assimilation rates.

Regarding the thermal metrics, the range of high photosynthesis, $\Delta T_{90}w$, was equally influenced by the traits l_t, g_1 and $V_{CMAX,25}$. Conversely, the maximum temperature $T_{L,max}$, was mostly influenced by l_t, while g_1 and $V_{CMAX,25}$ played secondary roles. In particular, lower l_t values and higher g_1 values at a given $V_{CMAX,25}$ lead to higher boundary layer and stomatal conductances, respectively, preventing thermal damage and resulting in lower T_L. Similarly, higher $V_{CMAX,25}$ were associated to higher g_s, which in turn enhances the cooling effect of transpiration, resulting in lower T_L. This mechanism might explain the effect of $V_{CMAX,25}$ on $T_{L,max}$. However, with focus on values observed in mid-to-high latitudes (shaded area in Figure 1), the role of $V_{CMAX,25}$ in regulating maximum T_L diminished. Therefore, under warm conditions, the cooling effect of the transpiration process is generally neither benefited from nor hampered by the assimilation capacity of plants, but rather regulated by the traits l_t and g_1.

There is no evidence in the literature of how ΔT_{90} changes according to plant traits, to the best of our knowledge. The width of the temperature range that realizes >80% of the maximum photosynthetic rate varies among plant functional types, in particular at low growth temperatures [1,63,64]; however, thus far these differences have not been explained based on specific traits. Our results suggest that $V_{CMAX,25}$, l_t and g_1 are the key traits that determine ΔT_{90}, possibly through the regulation of stomatal and boundary layer conductances. The influence of $V_{CMAX,25}$ might also be due to the dependence of V_{CMAX} on T_L per se, as discussed in the Supplementary Materials. As apparent from Figure 1, leaves that are well coupled with the atmosphere (i.e., high g_b and g_s and, therefore, high g_1 and low l_t) correspond to wider ΔT_{90}, whereas leaves less coupled with the atmosphere present narrower ΔT_{90}. This might mirror two different plant strategies. Plants that are well coupled with the atmosphere respond more easily to changes on the atmospheric conditions so that T_L fluctuates in a wider range than plants that are less coupled with the atmosphere. Therefore, plants that have lower assimilation rates and are well coupled have a wider range of optimal temperatures so that the lower assimilation rates and wider T_L fluctuations are compensated by a wider range of semi-optimal conditions. In contrast, the ones with higher assimilation rates and less coupled with the atmosphere present a narrower range of optimal temperatures.

Briefly, l_t, $V_{CMAX,25}$ and, to some extent, g_1 are key for both assimilation rates and thermoregulation capacities, mainly via the regulation of stomatal and boundary layer conductances. With focus on mid-to-high latitudes, while $V_{CMAX,25}$ has no large influence on the maximum temperature reached by plants under warm conditions, it is key on their thermoregulation capacity under normal conditions (Figure 1). Moreover, over this region, g_1 has no effect on assimilation rates, which are completely dominated by $V_{CMAX,25}$. Finally, the traits β, α_{PAR} and α_{NIR} seem to play secondary roles for all four metrics analyzed here and they will therefore not be further discussed.

It is important to acknowledge that the results of this analysis might slightly change if other growing conditions were to be used. However, this potential limitation is mitigated by the method employed to split the 100 combinations of traits between Group 1 and Group 2 (Section 2.3.2), which identifies suitable threshold irrespective of the model outputs. As such, similar conclusions on the relevance of the different traits would be drawn when considering other realistic growing conditions. A further aspect not accounted for here is thermal acclimation. Hence, assimilation rates and thermoregulation capacities and how they are influenced by the analyzed traits might change along with growth temperature, even in existing leaves, at scales of few days to weeks [1,63]. This might represent a limitation when comparing species exposed to contrasting growing conditions and/or with different acclimation capability. However, this analysis still provides key information by identifying the most dominant traits, their inter-relation and the potential mechanisms that explain plant responses.

4.2. Role of Growing Conditions: the Timing of Precipitatio Affects the Risk of Thermal Damage

In combination with its inherent traits, plant CO_2 assimilation and the risk of thermal damage are also determined by the environmental conditions. The contrasting weather conditions of the growing seasons 2005 and 2006 showed how the same combinations of traits led to different values for the four metrics. While no substantial differences were found in terms of maximum net CO_2 assimilation rate, $A_{net,max}$, and the range of high photosynthesis, ΔT_{90}, substantial differences emerged in terms of maximum temperature, $T_{L,max}$ and cumulated net CO_2 assimilation, $A_{net,cum}$ (Figure 2). Specifically, despite the lower mean air temperature and higher precipitation of 2005 over 2006, warm periods were more damaging in the growing season 2005 than 2006 regardless of the combination of traits (Figure 2). This difference can be ascribed to the timing of rainfall events: in 2005, the longest dry spell occurred during the warmest days, while this was not the case in 2006. As a result, in 2006, plants with sets of traits that reached thermal damage under growing conditions of 2005 could cope with similar or higher air temperatures and even benefit from these slightly warmer conditions, as photosynthetic capacity was enhanced and water was available for evaporative cooling to stave off the risk of thermal damage.

This result also emerges when exploring in more detail the combined effect of temperature and water availability as part of the future scenario analyses. While leaf temperature, T_L, was sensitive to soil water potential, Ψ_s, under warm conditions (Figure 4, top row), T_L did not show any substantial change with Ψ_s under normal ones (Figure 3, top row). Moreover, higher air temperature enhances soil water depletion but only under warm conditions (Figure 5, top row). These conclusions are not affected by the length of the growing season, which was held constant, since the aim is to identify differences in assimilation rates and leaf temperatures for different trait combinations but given the same abiotic conditions. The results highlight the relevance of the timing of high temperature and lower water availability and the importance of considering these events in conjunction [65–68]. The joint effects of high temperatures and low water availability are expected to become even more important in the future since high temperatures and water deficiency during the growing season are likely to become more frequent in boreal regions [69]. Hence, there will be a likely increase in the probability that boreal forests will need to cope with more severe combination of heat waves and droughts [15].

4.3. Interactions of Traits and Growing Conditions: Most Suitable Traits for Enhanced CO$_2$ Assimilation and Reduced Risk of Thermal Damage Under Current and Future Climates

Within the general pattern discussed above, specific trait sets can reduce or enhance thermal risks in specific climatic conditions. Therefore, attempts to identify the set of traits that could maximize productivity while preventing thermal damage under current and future conditions necessarily require the joint analysis of traits and specific growing conditions. Thus, we tested four combination of traits that presented contrasting assimilation and thermal responses during the 2005 growing season—a normal season at the reference site—and analyzed their response under multiple scenarios of future climatic conditions. The values for each trait within each combination are restricted to those observed in boreal tree species—in line with our focus on boreal forests—to account for the fact that trait values per se are related to climatic conditions. For example, global $V_{CMAX,25}$ distribution has been recently proved to be mainly explained by climate [70].

As discussed in Section 4.2, the combination of low water availability and heat stress causes a disproportionate damage compared to each stress component occurring in isolation [65,66,71–73]. Due to the compound effect of low water availability and high temperatures, the focus of the multi-scenarios assessment was on these abiotic stressors, holding constant (i.e., as observed in 2005) the other climatic variables (chiefly, relative humidity, wind speed and atmospheric CO_2 concentration). Nevertheless, reduced wind speed may lead to a further decoupling of leaf and air temperature, with potentially high and damaging temperatures in sunlit leaves. Declines of long-term wind speed (stilling) have been reported in both hemispheres [74,75]. Therefore, all else being equal, even higher leaf temperatures, T_L, are to be expected as the result of the reduced cooling due to stilling, in particular for leaves with large effective leaf thickness l_t. Similarly, the stomatal closure potentially caused by enhanced CO_2 concentration may further increase T_L.

Our model results clearly show that, during the warmest period of the growing season, future warmer and drier conditions may cause reductions in net CO_2 assimilation A_{net} regardless of plant traits (Figure 4 bottom row). Conversely, during periods with lower air temperatures, warming temperatures might be still beneficial (Figure 3 bottom row). However, the analysis of temperature data over the period 1996–2006 (upper row histogram in Figure 3) shows that, for the site of reference, higher temperatures were either not beneficial for productivity or even harmful in 75% to 93% of the days of the growing season depending on the trait combination. Similarly, future warmer and drier conditions may substantially increase the risk of thermal damage regardless of the plant traits. Nonetheless, how damaging these new conditions are depends on the trait set (Figure 4, upper row).

The GSA clearly shows that high photosynthetic capacity (i.e., high maximum carboxylation rate, $V_{CMAX,25}$) is needed to ensure high A_{net}; small l_t is the key trait to prevent thermal damage. Under the growing conditions of 2005, enhancement of cumulated net CO_2 assimilation $A_{net,cum}$ emerges when $V_{CMAX,25}$ increases, as apparent when comparing the $A_{net,cum}$ for combination 1 (Low Productivity-High Risk; LP-HR) and combination 4 (Low Productivity-Low Risk, LP-LR) against combinations 2 (High Productivity-Low Risk; HP-LR) and 3 (High Productivity-High Risk; HP-HR) (Table 2). Indeed, under standard conditions, the combination with the highest $V_{CMAX,25}$ (combination 3; HP-HR) showed the highest A_{net} (Figure 3 bottom row). However, this enhancement does not hold under warm conditions: in this case, assimilation rate is also limited by leaf temperature as the optimal temperature for photosynthesis is exceeded (Figure 4 bottom row). Temperatures above this threshold might constrain assimilation rates to the extent that plants with high $V_{CMAX,25}$ can show lower assimilation rates than plants with lower $V_{CMAX,25}$, as apparent by comparing combination 4 (LP-LR) versus combination 3 (HP-HR) under the warmest conditions within the growing season. This pattern is also partly due to the sensitivity to temperature of V_{CMAX} increasing with $V_{CMAX,25}$. Moreover, lower $V_{CMAX,25}$ was also associated with broader the range of high photosynthesis ΔT_{90}, i.e., a broader range of conditions that allow high assimilation rates. In fact, only those plants with the ability to be well-coupled with the atmosphere (i.e., low l_t) can keep high assimilation rates as shown by combination 2 (HP-LR) and, to a lower extent, combination 4 (LP-LR) (Figure 4 bottom row), even though the former has substantially higher assimilation rates and lower T_L than the latter.

Nonetheless, combination 3 (HP-HR), which has high l_t, was the one with the highest $A_{net,cum}$ within the growing season 2005. This is because, up to air temperature T_A around 18 °C, combination 3 (HP-HR) shows the highest A_{net} (Figure 3 bottom row). Based on the observed T_A at midday in the reference site during the growing seasons in 1996-2016, T_A was lower than 18 °C in the 70.3% of days, making combination 3 (HP-HR) the most appealing one in terms of overall productivity. However, this combination can also lead to thermal damage (i.e., T_L exceeding the critical temperature for leaf damage, T_{CRIT}; Section 2.3.2) for $T_A \geq 26$ °C, particularly under limited water availability (Figure 4 top row). The maximum T_A at midday reported in the reference site exceeded this value in half the years between 1996 and 2016 (Figure 4, top left). Therefore, high l_t might be beneficial for assimilation over the growing season at the expense of increasing the likelihood of thermal damage, particularly under water stress. Indeed, across all plant types, leaf size is distributed geographically according to a combination of the mean temperature of the warmest month of the year and the mean annual precipitation: higher mean temperature in the warmest month and lower mean annual precipitation correspond to smaller leaf sizes [76]. Thus, the selection of l_t is not straightforward since it might represent a trade-off between productivity and prevention of thermal damage. The net result depends largely on the climatic conditions.

As explained in the Methods, the trait combinations were generated randomly because this study aimed to identify the role of each trait when combined with other traits and under specific growing conditions. Thus, the trait combinations do not to represent specific species. However, the combination of traits that leads to the highest thermal damage risk and/or is more likely to experience sub-optimal conditions for assimilation can be used to speculate which species are likely to be more vulnerable to future growing conditions. For example, *Pinus sylvestris* L. and *Picea abies* L.—the two dominant species in Northern Europe - have small and rounded needle leaves (i.e., low effective leaf thickness l_t) in common but *Pinus sylvestris* L. usually exhibits higher photosynthetic capacity (i.e., higher $V_{CMAX,25}$) than *Picea abies* L. e.g., [77]. Although our results are not aimed to be representative of specific boreal species, *Pinus sylvestris* L. can be considered similar to combinations 2 (HP-LR) and *Picea abies* L. to combination 4 (LP-LR). As such, based on our conclusions, one could expect *Picea abies* L. to be more vulnerable to warming climates than *Pinus sylvestris* L., even suffering thermal damage under current conditions. Indeed, negative effects of warming on *Picea abies* L. species have been already reported [78], while *Pinus sylvestris* L. appears less sensitive to warming [79,80].

5. Conclusions

Understanding how specific plant traits, and combination of traits, prevent or enhance heat stress and C uptake is vital to predict how projected increases in frequency of heat waves and droughts in future climate may affect boreal forests. To disentangle the role of plant traits on thermoregulation and C uptake, we focused on six traits: the maximum carboxylation rate, two parameters that regulate the stomatal conductance and its sensibility to water stress, the effective leaf thickness and the PAR and NIR albedos. Four performance metrics related to leaf temperature and assimilation were evaluated, with a focus on the growing season.

Among the analyzed traits, photosynthetic capacity (as represented by maximum carboxylation rate at 25 °C, $V_{CMAX,25}$) and the effective leaf thickness, l_t, were the dominant ones regarding both thermoregulation and assimilation. Higher values of $V_{CMAX,25}$ are needed to enhance assimilation under current and future conditions. To prevent thermal damage, high $V_{CMAX,25}$ should be combined with low l_t. However, the selection between low or high l_t is not straightforward since l_t seems to represent a trade-off between thermal damage prevention and productivity.

Moreover, the climate change scenario analyses highlighted that the projected joint changes in temperature and water availability needs to be considered in our prognosis of future boreal forest wellbeing, because combination of traits that prevent thermal damage under current growing conditions will not be able to limit the occurrence of thermal damage under warmer and/or drier conditions. Likewise, substantial differences were observed when considering currently normal versus warm conditions within the same growing season. This suggests that trait selection should not only rely on the overall productivity over the whole growing season but should also consider specifically the warmest period.

Further analyses exploiting databases of traits and how they are distributed geographically and regional projections of growing conditions, combined with the understanding of the role played by plant traits provided by our results, can support the identification of species and regions most vulnerable to climate change, where appropriate forest management should be focused.

Supplementary Materials: The following are available online at http://www.mdpi.com/1999-4907/10/5/398/s1, Supplementary Materials document.

Author Contributions: This article is the result of the joint effort between G.R.-P., G.V. and S.L. The tasks were divided as follows: conceptualization, G.R.-P., G.V. and S.L.; methodology, G.R.-P. and G.V.; software, S.L.; validation, G.R.-P. and G.V.; formal analysis, G.R.-P.; investigation, G.R.-P. and G.V.; resources, G.V. and S.L.; writing—original draft preparation, G.R.-P.; writing—review and editing, G.R.-P., G.V. and S.L. supervision, G.V. and S.L.; and funding acquisition, G.V.

Funding: This research was supported by the Swedish Research Council Formas through grant 2018-01820 and by the Swedish government through the project Trees and Crops for the Future (TC4F). S.L. was also supported by the Academy of Finland, through the Academy Research Fellow project CLIMOSS (No. 296116 and 307192).

Conflicts of Interest: The authors declare no conflict of interest. The funders had no role in the design of the study; in the collection, analyses, or interpretation of data; in the writing of the manuscript, or in the decision to publish the results.

References

1. Yamori, W.; Hikosaka, K.; Way, D.A. Temperature response of photosynthesis in C-3, C-4, and CAM plants: Temperature acclimation and temperature adaptation. *Photosynth. Res.* **2014**, *119*, 101–117. [CrossRef]
2. Hasanuzzaman, M.; Nahar, K.; Alam, M.M.; Roychowdhury, R.; Fujita, M. Physiological, Biochemical, and Molecular Mechanisms of Heat Stress Tolerance in Plants. *Int. J. Mol. Sci.* **2013**, *14*, 9643–9684. [CrossRef] [PubMed]
3. O'Sullivan, O.S.; Heskel, M.A.; Reich, P.B.; Tjoelker, M.G.; Weerasinghe, L.K.; Penillard, A.; Zhu, L.L.; Egerton, J.J.G.; Bloomfield, K.J.; Creek, D.; et al. Thermal limits of leaf metabolism across biomes. *Glob. Chang. Biol.* **2017**, *23*, 209–223. [CrossRef] [PubMed]
4. Schreiber, U.; Berry, J.A. Heat-induced changes of chlorophyll fluorescence in intact leaves correlated with damage of the photosynthetic apparatus. *Planta* **1977**, *136*, 233–238. [CrossRef]
5. Huve, K.; Bichele, I.; Rasulov, B.; Niinemets, U. When it is too hot for photosynthesis: Heat-induced instability of photosynthesis in relation to respiratory burst, cell permeability changes and H2O2 formation. *Plant Cell Environ.* **2011**, *34*, 113–126. [CrossRef] [PubMed]
6. Knight, C.A.; Ackerly, D.D. Evolution and plasticity of photosynthetic thermal tolerance, specific leaf area and leaf size: Congeneric species from desert and coastal environments. *New Phytol.* **2003**, *160*, 337–347. [CrossRef]
7. O'Sullivan, O.S.; Weerasinghe, K.W.L.K.; Evans, J.R.; Egerton, J.J.G.; Tjoelker, M.G.; Atkin, O.K. High-resolution temperature responses of leaf respiration in snow gum (*Eucalyptus pauciflora*) reveal high-temperature limits to respiratory function. *Plant, Cell* **2013**, *36*, 1268–1284. [CrossRef]
8. Chaves, M.; Costa, J.; Zarrouk, O.; Pinheiro, C.; Lopes, C.; Pereira, J.; Costa, J. Controlling stomatal aperture in semi-arid regions—The dilemma of saving water or being cool? *Plant Sci.* **2016**, *251*, 54–64. [CrossRef]
9. Monteiro, M.V.; Blanuša, T.; Verhoef, A.; Hadley, P.; Cameron, R.W.F. Relative importance of transpiration rate and leaf morphological traits for the regulation of leaf temperature. *Aust. J. Bot.* **2016**, *64*, 32. [CrossRef]
10. Radin, J.W.; Lu, Z.; Percy, R.G.; Zeiger, E. Genetic variability for stomatal conductance in Pima cotton and its relation to improvements of heat adaptation. *Proc. Natl. Acad. Sci. USA* **1994**, *91*, 7217–7221. [CrossRef]
11. Okajima, Y.; Taneda, H.; Noguchi, K.; Terashima, I. Optimum leaf size predicted by a novel leaf energy balance model incorporating dependencies of photosynthesis on light and temperature. *Ecol. Res.* **2012**, *27*, 333–346. [CrossRef]
12. Vogel, S. Leaves in the lowest and highest winds: temperature, force and shape. *New Phytol.* **2009**, *183*, 13–26. [CrossRef]
13. Leigh, A.; Sevanto, S.; Ball, M.C.; Close, J.D.; Ellsworth, D.S.; Knight, C.A.; Nicotra, A.B.; Vogel, S. Do thick leaves avoid thermal damage in critically low wind speeds? *New Phytol.* **2012**, *194*, 477–487. [CrossRef]
14. Luquet, D.; Bégué, A.; Vidal, A.; Clouvel, P.; Dauzat, J.; Olioso, A.; Gu, X.; Tao, Y. Using multidirectional thermography to characterize water status of cotton. *Remote. Sens. Environ.* **2003**, *84*, 411–421. [CrossRef]
15. Intergovernmental Panel on Climate Change. *Climate Change 2014—Impacts, Adaptation and Vulnerability: Part B: Regional Aspects*; Cambridge University Press: Cambridge, UK, 2014.
16. Allen, C.D.; Macalady, A.K.; Chenchouni, H.; Bachelet, D.; McDowell, N.; Vennetier, M.; Kitzberger, T.; Rigling, A.; Breshears, D.D.; Hogg, E. (Ted); et al. A global overview of drought and heat-induced tree mortality reveals emerging climate change risks for forests. *Ecol. Manag.* **2010**, *259*, 660–684. [CrossRef]
17. Boisvenue, C.; Running, S.W. Impacts of climate change on natural forest productivity—Evidence since the middle of the 20th century. *Chang. Boil.* **2006**, *12*, 862–882. [CrossRef]
18. Sastry, A.; Barua, D. Leaf thermotolerance in tropical trees from a seasonally dry climate varies along the slow-fast resource acquisition spectrum. *Sci. Rep.* **2017**, *7*, 11246. [CrossRef]
19. Bonan, G.B. Forests and Climate Change: Forcings, Feedbacks, and the Climate Benefits of Forests. *Science* **2008**, *320*, 1444–1449. [CrossRef]
20. Pan, Y.; Birdsey, R.A.; Fang, J.; Houghton, R.; Kauppi, P.E.; Kurz, W.A.; Phillips, O.L.; Shvidenko, A.; Lewis, S.L.; Canadell, J.G.; et al. A Large and Persistent Carbon Sink in the World's Forests. *Science* **2011**, *333*, 988–993. [CrossRef]
21. Baul, T.K.; Alam, A.; Ikonen, A.; Strandman, H.; Asikainen, A.; Peltola, H.; Kilpeläinen, A. Climate Change Mitigation Potential in Boreal Forests: Impacts of Management, Harvest Intensity and Use of Forest Biomass to Substitute Fossil Resources. *Forests* **2017**, *8*, 455. [CrossRef]

22. Grossiord, C.; Sevanto, S.; Adams, H.D.; Collins, A.D.; Dickman, L.T.; McBranch, N.; Michaletz, S.T.; Stockton, E.A.; Vigil, M.; McDowell, N.G. Precipitation, not air temperature, drives functional responses of trees in semi-arid ecosystems. *J. Ecol.* **2017**, *105*, 163–175. [CrossRef]

23. Helman, D.; Osem, Y.; Yakir, D.; Lensky, I.M. Relationships between climate, topography, water use and productivity in two key Mediterranean forest types with different water-use strategies. *Agric. Meteorol.* **2017**, *232*, 319–330. [CrossRef]

24. Grossiord, C.; Granier, A.; Gessler, A.; Jucker, T.; Bonal, D. Does Drought Influence the Relationship Between Biodiversity and Ecosystem Functioning in Boreal Forests? *Ecosystems* **2014**, *17*, 394–404. [CrossRef]

25. Aubin, I.; Munson, A.; Cardou, F.; Burton, P.; Isabel, N.; Pedlar, J.; Paquette, A.; Taylor, A.; Delagrange, S.; Kebli, H.; et al. Traits to stay, traits to move: a review of functional traits to assess sensitivity and adaptive capacity of temperate and boreal trees to climate change. *Environ. Rev.* **2016**, *24*, 164–186. [CrossRef]

26. Girardin, M.P.; Bouriaud, O.; Hogg, E.H.; Kurz, W.; Zimmermann, N.E.; Metsaranta, J.M.; de Jong, R.; Frank, D.C.; Esper, J.; Buntgen, U.; et al. No growth stimulation of Canada's boreal forest under half-century of combined warming and CO2 fertilization. *Proc. Natl. Acad. Sci. USA* **2016**, *113*, E8406–E8414. [CrossRef]

27. O'Brien, M.J.; Engelbrecht, B.M.J.; Joswig, J.; Pereyra, G.; Schuldt, B.; Jansen, S.; Kattge, J.; Landhäusser, S.M.; Levick, S.R.; Preisler, Y.; et al. A synthesis of tree functional traits related to drought-induced mortality in forests across climatic zones. *J. Appl. Ecol.* **2017**, *54*, 1669–1686. [CrossRef]

28. Launiainen, S.; Katul, G.G.; Kolari, P.; Lindroth, A.; Lohila, A.; Aurela, M.; Varlagin, A.; Grelle, A.; Vesala, T. Do the energy fluxes and surface conductance of boreal coniferous forests in Europe scale with leaf area? *Chang. Boil.* **2016**, *22*, 4096–4113. [CrossRef]

29. Launiainen, S.; Katul, G.G.; Laurén, A.; Kolari, P. Coupling boreal forest CO2, H2O and energy flows by a vertically structured forest canopy – Soil model with separate bryophyte layer. *Ecol. Model.* **2015**, *312*, 385–405. [CrossRef]

30. Farquhar, G.D.; Von Caemmerer, S.; Berry, J.A. A biochemical model of photosynthetic CO2 assimilation in leaves of C3 species. *Planta* **1980**, *149*, 78–90. [CrossRef]

31. Medlyn, B.E.; Dreyer, E.; Ellsworth, D.; Forstreuter, M.; Harley, P.C.; Kirschbaum, M.U.F.; Le Roux, X.; Montpied, P.; Strassemeyer, J.; Walcroft, A.; et al. Temperature response of parameters of a biochemically based model of photosynthesis. II. A review of experimental data. *Plant Cell Environ.* **2002**, *25*, 1167–1179. [CrossRef]

32. Kellomäki, S.; Wang, K.-Y. Photosynthetic responses to needle water potentials in Scots pine after a four-year exposure to elevated CO2 and temperature. *Tree Physiol.* **1996**, *16*, 765–772. [CrossRef] [PubMed]

33. Medlyn, B.E.; Duursma, R.A.; Eamus, D.; Ellsworth, D.; Prentice, I.C.; Barton, C.V.M.; Crous, K.Y.; De Angelis, P.; Freeman, M.; Wingate, L. Reconciling the optimal and empirical approaches to modelling stomatal conductance. *Chang. Boil.* **2011**, *17*, 2134–2144. [CrossRef]

34. Zhao, W.; Qualls, R.J. A multiple-layer canopy scattering model to simulate shortwave radiation distribution within a homogeneous plant canopy. *N.a. Resour.* **2005**, *41*. [CrossRef]

35. Zhao, W.; Qualls, R.J. Modeling of long-wave and net radiation energy distribution within a homogeneous plant canopy via multiple scattering processes. *N.a. Resour.* **2006**, *42*. [CrossRef]

36. Schuepp, P.H. Tansley Review No. 59 Leaf Boundary-Layers. *New Phytol.* **1993**, *125*, 477–507. [CrossRef]

37. Campbell, G.S.; Norman, J.M. *An Introduction to Environmental Biophysics*; Springer Nature: Basingstoke, UK, 1998.

38. Medlyn, B.E.; Badeck, F.W.; De Pury, D.G.G.; Barton, C.V.M.; Broadmeadow, M.; Ceulemans, R.; De Angelis, P.; Forstreuter, M.; Jach, M.E.; Kellomaki, S.; et al. Effects of elevated [CO2] on photosynthesis in European forest species: a meta-analysis of model parameters. *Plant Cell Enivron.* **1999**, *22*, 1475–1495. [CrossRef]

39. Baldocchi, D.; Meyers, T. On using eco-physiological, micrometeorological and biogeochemical theory to evaluate carbon dioxide, water vapor and trace gas fluxes over vegetation: A perspective. *Agric. Meteorol.* **1998**, *90*, 1–25. [CrossRef]

40. Kattge, J.; Knorr, W. Temperature acclimation in a biochemical model of photosynthesis: A reanalysis of data from 36 species. *Plant Cell* **2007**, *30*, 1176–1190. [CrossRef]

41. Beerling, D.; Quick, W. A new technique for estimating rates of carboxylation and electron transport in leaves of C3 plants for use in dynamic global vegetation models. *Chang. Boil.* **1995**, *1*, 289–294. [CrossRef]

42. Aerts, R.; Brovkin, V.; Cavender-Bares, J.; Cavender-Bares, J.; Verheijen, L.M.; Cornelissen, J.H.C.; Kattge, J.; Van Bodegom, P.M. Inclusion of ecologically based trait variation in plant functional types reduces the projected land carbon sink in an earth system model. *Chang. Boil.* **2015**, *21*, 3074–3086.

43. A Arain, M.; A Black, T.; Barr, A.G.; Jarvis, P.G.; Massheder, J.M.; Verseghy, D.L.; Nesic, Z. Effects of seasonal and interannual climate variability on net ecosystem productivity of boreal deciduous and conifer forests. *Can. J.* **2002**, *32*, 878–891. [CrossRef]

44. Medlyn, B.E.; Berbigier, P.; Clement, R.; Grelle, A.; Loustau, D.; Linder, S.; Wingate, L.; Jarvis, P.G.; Sigurdsson, B.D.; McMurtrie, R.E. Carbon balance of coniferous forests growing in contrasting climates: Model-based analysis. *Agric. Meteorol.* **2005**, *131*, 97–124. [CrossRef]

45. Niinemets, Ülo Research review. Components of leaf dry mass per area - thickness and density - alter leaf photosynthetic capacity in reverse directions in woody plants. *New Phytol.* **1999**, *144*, 35–47. [CrossRef]

46. Niinemets, Ülo GLOBAL-SCALE CLIMATIC CONTROLS OF LEAF DRY MASS PER AREA, DENSITY, AND THICKNESS IN TREES AND SHRUBS. *Ecology* **2001**, *82*, 453–469. [CrossRef]

47. Lin, Y.-S.; Medlyn, B.E.; Duursma, R.A.; Prentice, I.C.; Wang, H.; Baig, S.; Eamus, D.; de Dios,, V.R.; Mitchell, P.; Ellsworth, D.S.; et al. Optimal stomatal behaviour around the world. *Nat. Clim. Change* **2015**, *5*, 459. [CrossRef]

48. Breuer, L.; Eckhardt, K.; Frede, H.-G. Plant parameter values for models in temperate climates. *Ecol. Model.* **2003**, *169*, 237–293. [CrossRef]

49. Bartlett, M.K.; Ollinger, S.V.; Wicklein, H.F.; Hollinger, D.Y.; Richardson, A.D. Canopy-scale relationships between foliar nitrogen and albedo are not observed in leaf reflectance and transmittance within temperate deciduous tree species. *Botany* **2011**, *89*, 491–497. [CrossRef]

50. Dickinson, R.E. Land Surface Processes and Climate—Surface Albedos and Energy Balance. *Adv. Geophys.* **1983**, *25*, 305–353.

51. Hornberger, G.M.; Spear, R.C. Eutrophication in Peel Inlet.1. Problem-Defining Behavior and a Mathematical-Model for the Phosphorus Scenario. *Water Res.* **1980**, *14*, 29–42. [CrossRef]

52. McIntyre, N.; Jackson, B.; Wade, A.; Butterfield, D.; Wheater, H.; Wade, A. Sensitivity analysis of a catchment-scale nitrogen model. *J. Hydrol.* **2005**, *315*, 71–92. [CrossRef]

53. Medici, C.; Wade, A.; Frances, F.; Wade, A. Does increased hydrochemical model complexity decrease robustness? *J. Hydrol.* **2012**, *440*, 1–13. [CrossRef]

54. Ruiz-Pérez, G.; Medici, C.; Latron, J.; Llorens, P.; Gallart, F.; Francès, F.; Ruiz-Pérez, G. Investigating the behaviour of a small Mediterranean catchment using three different hydrological models as hypotheses. *Hydrol. Process.* **2016**, *30*, 2050–2062. [CrossRef]

55. Kolari, P.; Kulmala, L.; Pumpanen, J.; Launiainen, S.; Ilvesniem, H.; Hari, P.; Nikinmaa, E. CO2 exchange and component CO2 fluxes of a boreal Scots pine forest. *Boreal Environ. Res.* **2009**, *14*, 761–783.

56. Kulmala, L.; Launiainen, S.; Pumpanen, J.; Lankreijer, H.; Lindroth, A.; Hari, P.; Vesala, T. H2O and CO2 fluxes at the floor of a boreal pine forest. *N.a. B: Chem. Phys. Meteorol.* **2008**, *60*, 167–178.

57. Gao, Y.; Markkanen, T.; Aurela, M.; Mammarella, I.; Thum, T.; Tsuruta, A.; Yang, H.; Aalto, T. Response of water use efficiency to summer drought in a boreal Scots pine forest in Finland. *Biogeosciences* **2017**, *14*, 4409–4422. [CrossRef]

58. Richardson, A.D.; Keenan, T.F.; Migliavacca, M.; Ryu, Y.; Sonnentag, O.; Toomey, M. Climate change, phenology, and phenological control of vegetation feedbacks to the climate system. *Agric. Meteorol.* **2013**, *169*, 156–173. [CrossRef]

59. Gill, A.L.; Gallinat, A.S.; Sanders-DeMott, R.; Rigden, A.J.; Gianotti, D.J.S.; Mantooth, J.A.; Templer, P.H. Changes in autumn senescence in northern hemisphere deciduous trees: A meta-analysis of autumn phenology studies. *Ann. Bot.* **2015**, *116*, 875–888. [CrossRef]

60. Michaletz, S.T.; Weiser, M.D.; McDowell, N.G.; Zhou, J.; Kaspari, M.; Helliker, B.R.; Enquist, B.J. Corrigendum: The energetic and carbon economic origins of leaf thermoregulation. *Nat. Plants* **2016**, *2*, 16147. [CrossRef]

61. Davi, H.; Barbaroux, C.; Dufrene, E.; François, C.; Montpied, P.; Breda, N.; Badeck, F. Modelling leaf mass per area in forest canopy as affected by prevailing radiation conditions. *Ecol. Model.* **2008**, *211*, 339–349. [CrossRef]

62. Xu, F.; Guo, W.; Xu, W.; Wei, Y.; Wang, R. Leaf morphology correlates with water and light availability: What consequences for simple and compound leaves? *Prog. Sci. Mater. Int.* **2009**, *19*, 1789–1798. [CrossRef]

63. Vico, G.; Way, D.A.; Hurry, V.; Manzoni, S. Can leaf net photosynthesis acclimate to rising and more variable temperatures? *Plant Cell Environ.* **2019**. [CrossRef]

64. Way, D.A.; Yamori, W. Thermal acclimation of photosynthesis: On the importance of adjusting our definitions and accounting for thermal acclimation of respiration. *Photosynth. Res.* **2014**, *119*, 89–100. [CrossRef]

65. Barnabas, B.; Jager, K.; Feher, A. The effect of drought and heat stress on reproductive processes in cereals. *Plant Cell Environ.* **2008**, *31*, 11–38. [CrossRef] [PubMed]

66. De Boeck, H.J.; Bassin, S.; Verlinden, M.; Zeiter, M.; Hiltbrunner, E. Simulated heat waves affected alpine grassland only in combination with drought. *New Phytol.* **2016**, *209*, 531–541. [CrossRef]

67. Suzuki, N.; Rivero, R.M.; Shulaev, V.; Blumwald, E.; Mittler, R. Abiotic and biotic stress combinations. *New Phytol.* **2014**, *203*, 32–43. [CrossRef]

68. Zscheischler, J.; Westra, S.; van den Hurk, B.J.J.M.; Seneviratne, S.I.; Ward, P.J.; Pitman, A.; AghaKouchak, A.; Bresch, D.N.; Leonard, M.; Wahl, T.; et al. Future climate risk from compound events. *Nat. Clim. Chang.* **2018**, *8*, 469–477. [CrossRef]

69. Belyazid, S.; Giuliana, Z. Water limitation can negate the effect of higher temperatures on forest carbon sequestration. *Eur. J.* **2019**, *138*, 287–297. [CrossRef]

70. Smith, N.G.; Keenan, T.F.; Prentice, I.C.; Wang, H.; Wright, I.J.; Niinemets, Ü.; Crous, K.Y.; Domingues, T.F.; Guerrieri, R.; Ishida, F.Y.; et al. Global photosynthetic capacity is optimized to the environment. *Ecol. Lett.* **2019**, *22*, 506–517. [CrossRef] [PubMed]

71. Keleş, Y.; Öncel, I. Response of antioxidative defence system to temperature and water stress combinations in wheat seedlings. *N.a. Sci.* **2002**, *163*, 783–790. [CrossRef]

72. Rizhsky, L.; Liang, H.; Mittler, R. The Combined Effect of Drought Stress and Heat Shock on Gene Expression in Tobacco1. *N.a. Physiol.* **2002**, *130*, 1143–1151.

73. Zscheischler, J.; Seneviratne, S.I. Dependence of drivers affects risks associated with compound events. *Sci. Adv.* **2017**, *3*, e1700263. [CrossRef]

74. McVicar, T.R.; Roderick, M.L.; Donohue, R.J.; Van Niel, T.G. Less bluster ahead? Ecohydrological implications of global trends of terrestrial near-surface wind speeds. *Ecohydrol.* **2012**, *5*, 381–388. [CrossRef]

75. Roderick, M.L.; Rotstayn, L.D.; Farquhar, G.D.; Hobbins, M.T. On the attribution of changing pan evaporation. *Geophys. Lett.* **2007**, *34*, L17401–L17403. [CrossRef]

76. Wright, I.J.; Dong, N.; Maire, V.; Prentice, I.C.; Westoby, M.; Díaz, S.; Gallagher, R.V.; Jacobs, B.F.; Kooyman, R.; Law, E.A.; et al. Global climatic drivers of leaf size. *Science* **2017**, *357*, 917–921. [CrossRef]

77. Niinemets, U. Stomatal conductance alone does not explain the decline in foliar photosynthetic rates with increasing tree age and size in Picea abies and Pinus sylvestris. *Tree Physiol.* **2002**, *22*, 515–535. [CrossRef]

78. Zhang, X.W.; Wang, J.R.; Ji, M.F.; Milne, R.I.; Wang, M.H.; Liu, J.-Q.; Shi, S.; Yang, S.-L.; Zhao, C.-M. Higher Thermal Acclimation Potential of Respiration but Not Photosynthesis in Two Alpine Picea Taxa in Contrast to Two Lowland Congeners. *PLoS One* **2015**, *10*, e0123248. [CrossRef]

79. Taeger, S.; Sparks, T.H.; Menzel, A. Effects of temperature and drought manipulations on seedlings of Scots pine provenances. *Plant Biol.* **2015**, *17*, 361–372. [CrossRef]

80. Kurepin, L.V.; Stangl, Z.R.; Ivanov, A.G.; Bui, V.; Mema, M.; Hüner, N.P.; Öquist, G.; Way, D.; Hurry, V. Contrasting acclimation abilities of two dominant boreal conifers to elevated CO2 and temperature. *Plant Cell* **2018**, *41*, 1331–1345. [CrossRef]

Article

No Ontogenetic Shifts in C-, N- and P-Allocation for Two Distinct Tree Species along Elevational Gradients in the Swiss Alps

Jian-Feng Liu [1,2], Ze-Ping Jiang [1,3], Marcus Schaub [2,4], Arthur Gessler [2,4], Yan-Yan Ni [3], Wen-Fa Xiao [3] and Mai-He Li [2,4,5,*]

[1] Key Laboratory of Tree Breeding and Cultivation of State Forestry Administration, Research Institute of Forestry, Chinese Academy of Forestry, Beijing 100091, China; liujf@caf.ac.cn (J.-F.L.); jiangzp@caf.ac.cn (Z.-P.J.)

[2] Swiss Federal Research Institute WSL, CH-8903 Birmensdorf, Switzerland; marcus.schaub@wsl.ch (M.S.); arthur.gessler@wsl.ch (A.G.)

[3] Research Institute of Forest Ecology, Environment and Protection, Chinese Academy of Forestry, Beijing 100091, China; nyy_ecology@126.com (Y.-Y.N.); xiaowenf@caf.ac.cn (W.-F.X.)

[4] SwissForestLab, Zuercherstrasse 111, CH-8903 Birmensdorf, Switzerland

[5] School of Geographical Sciences, Northeast Normal University, Changchun 130024, China

[*] Correspondence: maihe.li@wsl.ch; Tel.: +41-44-739-24-91

Received: 29 March 2019; Accepted: 3 May 2019; Published: 5 May 2019

Abstract: Most of our knowledge about forest responses to global environmental changes is based on experiments with seedlings/saplings grown in artificially controlled conditions. We do not know whether this knowledge will allow us to upscale to larger and mature trees growing in situ. In the present study, we used elevation as a proxy of various environmental factors, to examine whether there are ontogenetic differences in carbon and nutrient allocation of two major treeline species (*Pinus cembra* L. and *Larix decidua* Mill.) along elevational gradients (i.e., environmental gradient) in the Swiss alpine treeline ecotone (~300 m interval). Young and adult trees grown at the same elevation had similar levels of non-structural carbohydrates (NSCs), total nitrogen (TN), and phosphorus (TP), except for August leaf sugars and August leaf TP in *P. cembra* at the treeline. We did not detect any interaction between tree age and elevation on tissue concentration of NSCs, TN, and TP across leaf, shoot, and root tissues for both species, indicating that saplings and mature trees did not differ in their carbon and nutrient responses to elevation (i.e., no ontogenetic differences). With respect to carbon and nutrient allocation strategies, our results show that young and adult trees of both deciduous and evergreen tree species respond similarly to environmental changes, suggesting that knowledge gained from controlled experiments with saplings can be upscaled to adult trees, at least if the light is not limited. This finding advances our understanding of plants' adaptation strategies and has considerable implications for future model-developments.

Keywords: altitude; non-structural carbohydrates; nutrients; ontogeny; *Pinus cembra* L.; *Larix decidua* Mill

1. Introduction

Anthropogenic drivers of global change have been increasingly evident during the last centuries, which include rising atmospheric concentrations of CO_2 and other greenhouse gases and associated changes in the climate, nitrogen deposition, biotic invasions, and land-use change [1]. These environmental pressures greatly challenge performance and persistence of forest species around the world, which provide abundant products and services to support human society [2]. Therefore, understanding forest responses under global change is of fundamental and practical value for predicting the potential and limitations of forest productivity, and for defining mitigation and adaptation policies.

In the field, forest trees are often exposed to a myriad of single and combined stresses with varying strength and duration, the effects of which on plant performance are difficult to predict by solely manipulative experiments. For example, a meta-analysis covering 1634 plant species spanning four continents indicated that warming experiments (average experimental duration of 3.8 years) strongly underestimated advances in flowering and leafing, compared to long-term observations (average duration of 31.0 years) [3]. On the other hand, stress sensitivity, tolerance, and resistance of forest species vary with ontogenetic stage, making it challenging to analyze or identify the key environmental stressors driving plant performance throughout their life histories. Compared to adult trees, early life stages (e.g., seedlings or saplings) of trees are more sensitive to environmental variation or stress [4]. Numerous studies have reported wide ontogenetic variations in carbon assimilation and allocation [5,6], differing resource use strategies [7–9] and stress tolerances [4,10]. For example, Bansal and Germino [11] found that needles in transplanted saplings of two evergreen conifers, *Abies lasiocarpa* (Hook.) Nutt. and *Pseudotsuga menziesii* (Mirb.) Franco, had higher soluble sugar content than established adults at timberline in the Rocky Mountains during late seasons. Hence, to predict the responses of tree species to climate change, disentangling ontogenetic variations along synthesized environmental gradients is indispensable for a more profound understanding of differing adaptation and acclimation strategies between different life stages.

Among environmental gradients, increasing elevation is characterized by decreasing temperature interacting with other factors, such as soil water content, precipitation, season length, atmospheric pressure, and nutrient availability [12,13], and offer a natural laboratory to predict vegetation dynamics under climate change in recent decades [13–16]. To our knowledge, however, only a few studies have investigated the discrepancies between seedlings (saplings or juveniles) and adult trees along field environmental gradients, elevation transects, respectively. Among these are studies on *A. lasiocarpa* and *P. menziesii* [11], *A. faxoniana* Rehder & E.H.Wilson [17], and *Picea crassifolia* Kom. [18] and they generally found differing performances between seedlings and adults. Nevertheless, they are mostly restricted to the leaf level, while a whole-tree approach would be more appropriate to properly address the discrepancies between various life stages along environmental gradients. Furthermore, apart from ontogenetic differences for the same species, evergreen vs. deciduous tree species may vary in their response to environmental gradients in terms of carbon balance. So far, most studies have found higher non-structural carbohydrate concentrations (NSC) in deciduous compared to evergreen tree species [19–21]. However, whether this holds true when integrating ontogenetic variance is still unclear because different plant responses between life stages to climate warming have been found by other studies [22–24].

Pinus cembra L. (evergreen) and *Larix decidua* Mill. (deciduous) are two co-occurring treeline species growing in the Swiss Alps. In the present study, we aimed to answer the questions of whether the responses of carbon and nutrient allocation to environmental changes (i.e., using elevation as a proxy for changes in environmental factors) vary with ontogenetic stages of trees, and with leaf habits (evergreen vs. deciduous). We hypothesize that different life stages of both deciduous and evergreen trees lead differing carbon and nutrient allocation responses to elevational changes, because seedlings and saplings may be more sensitive to environmental changes than adult trees [4,25,26]. Our study could provide field data that reveal ontogenetic variation in response to environmental gradients and fill the gap where most predictions are restricted to manipulative experiments with young trees in early life stages.

2. Materials and Methods

2.1. Description of Sites and Species

Our study was conducted on *Pinus cembra* and *Larix decidua*, two native treeline species growing in the Swiss Alps. One transect was established within the alpine treeline ecotone in Chandolin, LaTzoumaz, Moosalp, and Sievz, respectively, in southern Switzerland (Figure 1, Table S1). In each transect, four sampling plots ($n = 4$) were designed at the alpine treeline (H) and the timberline (L) with ~300 m elevational difference between H and L, respectively (Table S1). The two species were selected because they have different leaf habits (evergreen vs. summergreen) and they co-exist in the Swiss alpine treeline ecotone. The four transects were selected for the present study because where every individual of both adult trees and saplings of the two species co-occurred is isolated, and thus no light effects on any individual sampled need to be considered. Each transect has a relatively homogeneous gentle slope (~30°) from the timberline to the alpine treeline, and hence, we can assume that the growth condition including temperature, water availability, and soil chemical and physical properties changes gradually with increasing elevation. This means that the present study uses elevation as a proxy for the whole of all growth-related factors, to study the ontogenetic responses of C-, N-, and P-allocation to combined environmental changes rather than to changes in any single factor.

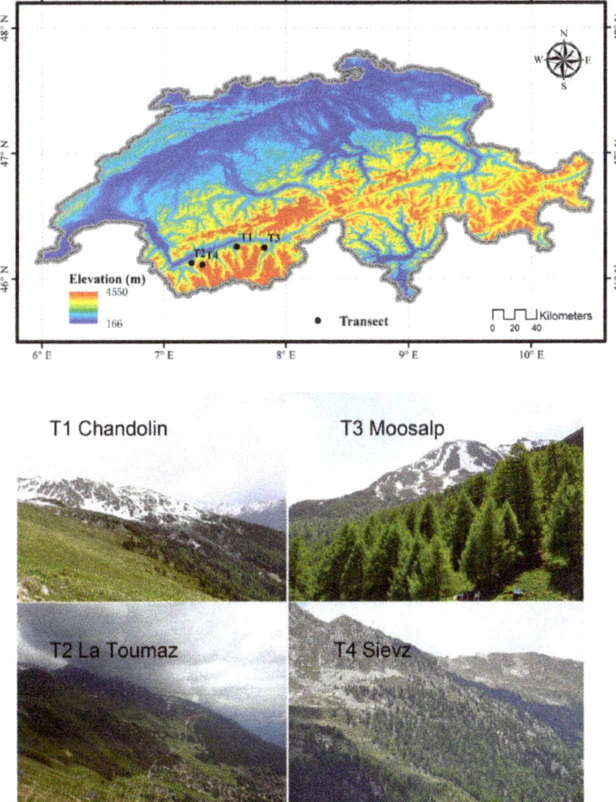

Figure 1. Locations of the sampled sites (T1: Chandolin, T2: La Toumaz, T3: Moosalp, and T4: Sievz) in the southern Swiss Alps.

2.2. Sampling Protocol

In each plot (mixed stand for the two targeted species), two age classes (saplings of <1.5 m in height, and adult trees of >5 m in height) per species were selected in the mid- and late-growing season (20 August and 17 October 2014) (Tables S1 and S2). Twice samplings accounted for seasonal transition which could bring variations in NSCs or nutrient status in tree species [25,27,28]; and the sampling dates selected were the time when the phenological difference among populations or individuals along elevational gradients is negligible [29–31]. For each age class, 3~5 isolated individuals with no signs of browsing or other damage were selected, with a minimum distance of 10 m from each other. The height (m) and diameter (cm) (10 cm aboveground diameter for saplings, and diameter at breast height for adults) of sampled individuals were carefully recorded. The individuals from the same age class from the same plot were mixed as one replicate and we had thus four replicates in total for each age class and each species.

From each targeted individual, we cut three or four upper and outermost sun-exposed branches. From these branches, current-year needles (~50 g) and shoots (~50 g) were collected. Fine roots (<5 mm in diameter, without bark) (~50 g) attached to coarse roots of each sample tree were manually excavated using a mini-spade and carefully collected. All samples were stored in a cool box for transportation after bagging and labeling. All samples were heated in a microwave oven at 600 W for 60 s to minimize the enzymic and physiological activity to reduce the respiration carbon loss of tissues [32,33] and then dried at 65 °C for 72 h and ground to pass a 0.15 mm sieve for further analyses.

2.3. Plant Analysis

2.3.1. Non-Structural Carbohydrates

The powdered material (~0.1 g) was put into a 10 ml centrifuge tube, where 5 mL of 80% ethanol was added. The mixture was incubated at 80 °C in a water bath shaker for 30 min, and then centrifuged at 4000 rpm for 5 min. The pellets were extracted two more times with 80% ethanol. Supernatants were retained, combined and stored at −20 °C for soluble sugar determinations. The ethanol-insoluble pellets were used for starch extraction. Glucose was used as a standard. Soluble sugars were determined using the anthrone method [34]. The starch concentration was measured spectrophotometrically at 620 nm using anthrone reagent and was calculated by multiplying glucose concentrations by the conversion factor of 0.9 [35]. The concentration of sugars and starch was described on a dry matter basis (% DW), and NSC was the sum of total soluble sugar and starch, and NSCs include NSC, soluble sugars, and starch.

2.3.2. Nitrogen and Phosphorus Concentrations

For the determination of tissue nitrogen (N) and phosphorus (P) concentrations (mg g^{-1} DW), finely ground material (~50 mg) was first digested with H_2SO_4 and H_2O_2 for further analysis. The nitrogen concentration was then measured applying the Kjeldahl method (Kjeltec 2200, FOSS, Hoganas, Sweden), while the phosphorus concentration was determined with the molybdenum blue spectrophotometric procedure (6505 UV spectrophotometer, Essex, UK) [36].

2.3.3. Carbon and Nitrogen Isotopic Abundance

The abundance of stable carbon (δ^{13}C) and nitrogen (δ^{15}N) isotopes in current needles was determined as described by Gebauer and Schulze [37]. To avoid tissue age-effects on δ^{13}C and δ^{15}N, we analyzed them in the current-year needles only. Aliquots (1.2~1.5 mg) of ground material were weighed into tin capsules, and then analyzed in an elemental analyzer (Euro EA; Hekatech GmbH, Wegberg, Germany) coupled inline with an isotope ratio mass spectrometer (DELTA V Advantage; Thermo Scientific, Bremen, Germany). The carbon and nitrogen isotopic abundance is expressed as δ^{13}C and δ^{15}N, respectively, and was calculated as follows: δ ‰ = [(Rsample/Rstandard) − 1) × 1000], where R is the ^{13}C/^{12}C or the ^{15}N/^{14}N ratio of the sample and the correspondent standard, i.e.,

Vienna Pee Dee Belemnite for C and atmospheric N_2 for N. The overall precision of the delta values was 0.1‰, as determined by repetitive measurements of standard material.

2.4. Statistical Analyses

All statistical analyses were conducted using R statistical software (RStudio 1.1.463 with R version 3.5.2). Shapiro–Wilk and Bartlett's tests were first used to test for normality and homogeneity of variances respectively, and all variables met the assumption for further variance analysis. We applied linear mixed-effects models (R package nlme) with species (*P. cembra* and *L. decidua*), tree age (sapling and adult tree), altitude (timberline and treeline), and season (mid-growing season: August; late-season: October) as fixed effects and transects as a random effect to account for variances between transects. The Wilcoxon tests for means were further conducted to analysis differences between fixed factors (age and altitude) with the ggpubr package. Since we focused on ontogenetic- or species-variations, or both, of the detected variables along altitudinal gradients, the effects of season transition were not presented in the following part.

3. Results

The levels of tissue NSCs (with the exception of shoot starch), C:N, and shoot N concentration, as well as leaf $\delta^{13}C$ and $\delta^{15}N$ differed significantly between the two species (Tables S2–S3, Figure 2). *Pinus cembra* had lower levels of NSCs than *Larix decidua*, e.g., NSC in leaves, shoots, roots of *P. cembra* was 12.12% ± 0.36%, 7.39% ± 0.19% and 5.51% ± 0.21%, but 15.27% ± 0.46%, 8.13% ± 0.29% and 7.04% ± 0.27% in *L. decidua*, respectively. Tissue C:N ratio in *P. cembra* was also lower than that in *L. decidua*, e.g., C:N in leaves, shoots, roots of *P. cembra* was 7.78 ± 0.32, 3.69 ± 0.19, 4.41 ± 0.23, but 9.96 ± 0.39, 4.51 ± 0.21, 5.55 ± 0.21 in *L. decidua*, respectively. Shoot N in *L. decidua* was 18.60 ± 0.66 mg/g, which was significantly lower than that in *P. cembra* (21.05 ± 0.74 mg/g) ($F = 7.50$, $p < 0.01$). With very few exceptions (e.g., August leaf sugars and TP, and October leaf C:N in *P. cembra* at the treeline), tree age did not influence tissue NSCs (Figure 2), total N and P (Figure 3), leaf $\delta^{13}C$ and $\delta^{15}N$ values (Figure 4), and C:N and N:P ratios (Figure 5) within each species at the same elevation. Most importantly, there were no interactions between age and altitude on concentrations of tissue NSCs (soluble sugars, starch, and NSC), nutrients (TN, TP), and the stoichiometric ratios (C:N and N:P), as well as on leaf $\delta^{13}C$ values (Tables S2–S3), with the exception of leaf $\delta^{15}N$ ($F = 5.01$, $p = 0.030$). For saplings, leaf $\delta^{15}N$ increased from −2.22‰ ± 0.31‰ (timberline) to 0.34‰ ± 0.52‰ (treeline) when combining both species ($p < 0.001$); for adult individuals, leaf $\delta^{15}N$ also increased from −1.57‰ ± 0.31‰ (timberline) to −0.22‰ ± 0.49‰ (treeline) ($p < 0.01$) (Figure 4). Here, the interaction occurred due to different magnitudes of increasing leaf $\delta^{15}N$ from timber to treeline between the age classes.

Only leaf $\delta^{15}N$ ($F = 13.31$, $p < 0.001$), root TN ($F = 13.28$, $p < 0.01$), and root C:N ($F = 6.21$, $p < 0.05$) were affected by species × altitude (Tables S2–S3). For *L. decidua*, leaf $\delta^{15}N$ was −1.25‰ ± 0.27‰ (timberline) and 1.69‰ ± 0.29‰ (treeline) ($p < 0.001$, Table S5), while for *P. cembra*, leaf $\delta^{15}N$ was −2.54‰ ± 0.28‰ (timberline) and −1.58‰ ± 0.28‰ (treeline) ($p < 0.05$, Table S4) (Figure 4). Similarly, this interaction seems to be caused by different magnitudes of increasing leaf $\delta^{15}N$ from timberline to treeline between the two species studied (Figure 4). However, root TN of the two species showed opposite trends along elevational transects: for *L. decidua*, root N increased from 11.71 ± 0.60 mg g^{-1} at timberline to 14.75 ± 1.12 mg g^{-1} ($p < 0.01$) at the alpine treeline, whereas, for *P. cembra*, root TN decreased from 14.07 ± 0.78 mg g^{-1} at timberline to 11.93 ± 0.48 mg g^{-1} ($p < 0.05$) at the alpine treeline. Consequently, root C:N in *L. decidua* slightly decreased from 5.70 ± 0.26 at the timberline to 5.40 ± 0.33 ($p > 0.05$) at the treeline, while for *P. cembra*, root C:N increased from 3.86 ± 0.31 (timberline) to 4.96 ± 0.28 (treeline) ($p < 0.05$) (Figure 5).

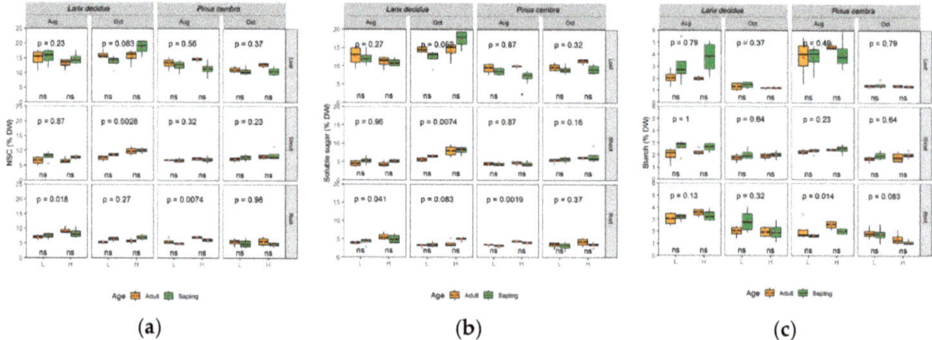

(a)   (b)   (c)

Figure 2. Non-structural carbohydrate concentration (% DW) (**a**), soluble sugar (% DW) (**b**), and starch (% DW) (**c**) (mean ± SE, *n* = 4) of different species (*Pinus cembra* L. and *Larix decidua* Mill.), age (sapling and adult tree), altitude (L for timberline and H for treeline), season (Aug for mid-growing season and October for late season) and tissues (current-year leaf, shoot, and fine root). Asterisks or "ns" at the bottom of subplot indicate significant (*p* < 0.05) or non-significant (*p* > 0.05) differences between saplings and adults within the same altitude and season, and the *p*-value at the top of subplot denote significance level by the Wilcoxon's mean test between L and H. The red points in the subplot denote mean value and grey point for outlier.

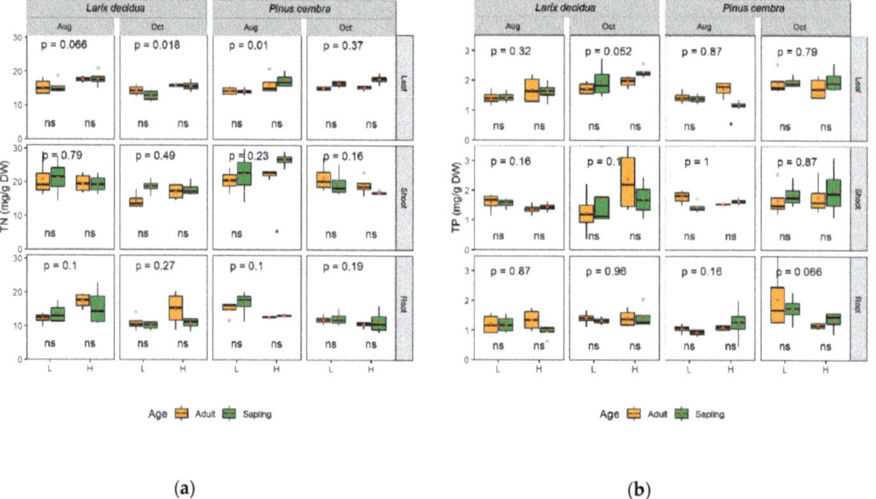

(a)   (b)

Figure 3. Total nitrogen (TN) (**a**) and phosphorus (TP) (**b**) concentrations (mg/g DW) (mean ± SE, *n* = 4) of different species (*Pinus cembra* L. and *Larix decidua* Mill.), age (sapling and adult trees), altitude (L for timberline and H for treeline), season (August for mid-growing season and October for late season), and tissues (current-year leaf, shoot, and fine root). Asterisks or "ns" at the bottom of subplot indicate significant (*p* < 0.05) or non-significant (*p* > 0.05) differences between saplings and adults within the same altitude and season, and the *p*-value at the top of subplot denote significance level by the Wilcoxon's mean test between L and H.

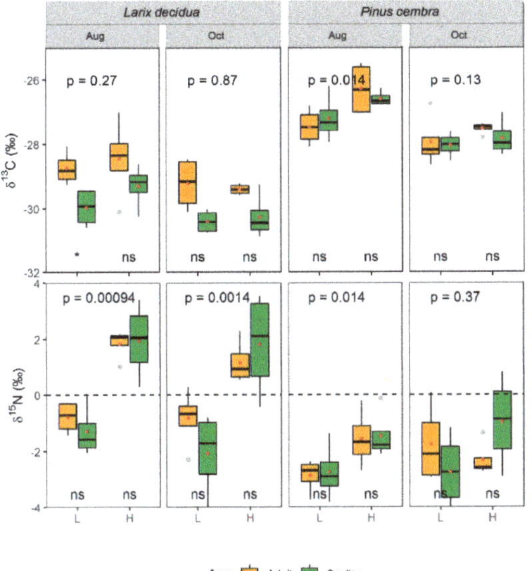

Figure 4. Leaf stable isotopes (δ^{13}C‰; δ^{15}N‰) (mean ± SE, $n = 4$) of different species (*Pinus cembra* L. and *Larix decidua* Mill.), age (sapling and adult tree), altitude (L for timberline and H for treeline), season (August for mid-growing season and October for late season). Asterisks or "ns" at the bottom of subplot indicate significant ($p < 0.05$) or non-significant ($p > 0.05$) differences between saplings and adults within the same altitude and season, and the p-value at the top of subplot denote significance level by the Wilcoxon's mean test between L and H.

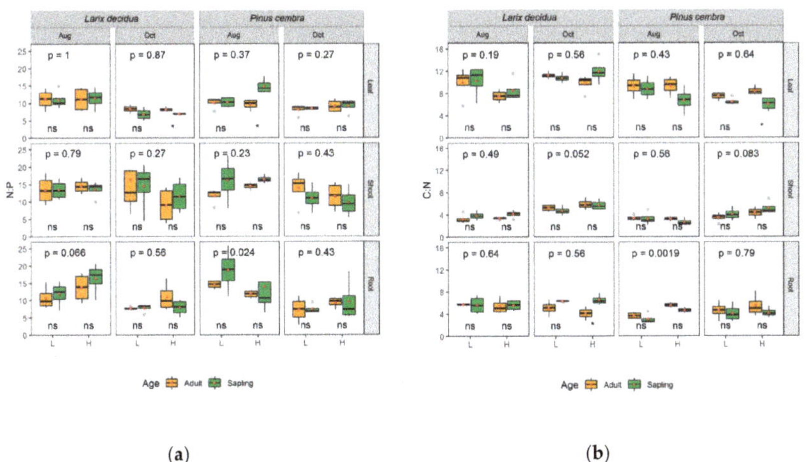

Figure 5. The ratio of N to P (**a**) and the ratio of C to N (**b**) (mean ± SE, $n = 4$) across tissues of the two species (*Pinus cembra* L. and *Larix decidua* Mill.), age (sapling and adult trees), altitude (L for timberline and H for treeline), season (August for mid-growing season and October for late season) and tissues (current-year leaf, shoot, and fine root). Asterisks or "ns" at the bottom of subplot indicate significant ($p < 0.05$) or non-significant ($p > 0.05$) differences between saplings and adults within the same altitude and season, and the p-value at the top of subplot denote significance level by the Wilcoxon's mean test between L and H.

4. Discussion

We found that there was no interaction between age and elevation on tissue NSCs (except for leaf sugars for the two species and leaf NSC for *L. decidua*) and nutrients (TN, TP) for both *P. cembra* and *L. decidua* (Tables S4 and S5). These findings suggest that saplings and adults respond consistently to combined environmental changes in terms of carbon and nutrient allocation, which does not support our hypothesis of different life stages leading to differing carbon and nutrient allocation responses along elevational changes. Niinemets [4] presented a comprehensive review which emphasized that forest trees' physiological responses to key environmental stressors and their combinations varied throughout ontogeny. The discrepancy between this review and our results may be attributed to the fact that most reviewed studies came from closed forests, where sapling and adults, due to their different ecological niches in a community, are subjected to different light, water, and temperature and nutrient regimes [38]. For elevation-driven changes in plant nutrients, Mayor et al. [39] found that declining temperatures with increasing elevation did not affect tree leaf nutrient concentrations, implying an adaptation strategy of trees growing in a harsh environment. Deciduous species tend to store N in the wood and bark of roots [30], while evergreen species store N in the youngest age class of foliage [40,41].

We found no age effects on leaf δ^{15}N for both species grown at a given altitude (timberline or treeline) (Tables S4 and S5, Figure 4), which indicates no distinct differences in nitrogen uptake and metabolism between two life stages for both species. This finding is consistent with Pardo et al. [42]. Our study found also that tree age did not affect tissue N concentration for both species (Tables S4 and S5), except for a significant age effect on leaf N in *P. cembra* (Figure S4). Previous studies [43,44], have suggested that differences in N uptake and metabolism with ontogenetic stage occur in some species and environmental conditions but not in others. Nevertheless, leaf N in both species showed an increasing trend with altitude from timberline to treeline, which corresponds with the findings from Körner [12]. There have been two main explanations for such altitudinal trends in plant N: firstly, low temperature and short growing season at higher altitude reduce growth and might consequently have an effect on leaf N contents [12]. Another cause might be the high N supply from atmospheric N deposition, due to winter snow accumulation, at the treeline [12,45] leading to higher N availability and consequently higher N uptake with increasing elevation.

Similar to leaf N content, leaf δ^{15}N was lower at the timberline than at the treeline for both species. The positive correlation between leaf N and leaf δ^{15}N was confirmed by previous studies [46–49], where leaf N was considered an index for soil N availability [50,51]. The increasing trend in leaf δ^{15}N along altitude in the present study was in line with results reported by [48] (above 1,350 m), but in contrast to other studies. For example, leaf δ^{15}N exhibited a decreasing trend with altitude in plants collected from the Kathmandu valley in Nepal [52] and from Mt. Schrankogel in Austria [53], but differences might be attributed to variations in soil ^{15}N and its drivers. For example, with increasing elevation, organic forms of N became the dominant source of N taken up by hardwood and coniferous tree species, and that variation in natural abundance foliar δ^{15}N with elevation was consistent with increasing organic N uptake [54]. Although we found similar trends along our altitudinal transect, the two species exhibited significant differences in leaf δ^{15}N, with higher leaf δ^{15}N in *L. decidua* than in *P. cembra* (Figure 4) across ontogeny and elevations. The lower leaf δ^{15}N in *P. cembra* might be associated with a higher rate of root colonization with mycorrhizal fungi [55], because mycorrhizal fungi are mainly transferring ^{15}N-depleted ammonium or amino acids to plants [56,57].

Our results showed that no significant ontogenetic differences in non-structural carbohydrates and nutrients (nitrogen and phosphorus) across leaf, shoot, and root tissues of both species grown along elevational gradients are present. However, due to the limited number of tree species growing in the alpine treeline ecotone in the Swiss Alps, we investigated only one deciduous and one evergreen tree species with four replicates. Further studies are needed to cover more tree species with more replicates for different life stages in a wider geographical range, to improve our understanding of ontogenetic effects on tree physiology.

5. Conclusions

This paper presents, to the best of our knowledge, the first field evidence to elucidate how different species integrated over different life stages cope with synthetic environmental changes (altitude) from the perspective of carbon and nutrient allocation strategies. Our results demonstrate that both deciduous and evergreen trees in their early life stage perform similarly compared to their adult individuals along an environmental gradient (from the timberline to the alpine treeline). Our results indicate that most of our knowledge of forest responses, especially shoot and root responses, to global environmental changes gained from experiments with seedlings/saplings growing in artificially controlled conditions could, to some extent, upscale to larger and mature trees growing in situ. This finding advances our understanding of plants' adaptation strategies and has considerable implications for future model-developments.

Supplementary Materials: The following are available online at http://www.mdpi.com/1999-4907/10/5/394/s1. Table S1: Description of sampling transects and plots for *Pinus cembra* L. and *Larix decidua* Mill. in the South Swiss Alps, Table S2: LMM results for the effects of species (*Pinus cembra* L. and *Larix decidua* Mill.), altitude (timberline and treeline), tree age (sapling and adult tree), sampled season (mid-growing season: August; and late-season: October) on NSCs, nutrients, and stable isotope over tissues. The interactions were retained in the models only when significant. *df*: degrees of freedom. Significant *p*-values are in bold face, Table S3: LMM results for the effects of species (*Pinus cembra* L. and *Larix decidua* Mill.), altitude (timberline and treeline), tree age (sapling and adult tree), sampled season (mid-growing season: August; and late-season: October) on the stoichiometric ratios (C:N and N:P). The interactions were retained in the models only when significant. *df*: degrees of freedom. Significant p-values are in bold face, Table S4: LMM results for the effects of altitude (timberline and treeline), tree age (sapling and adult tree), sampled season (mid-growing season: August; and late-season: October) and their interactions on NSCs, nutrients and stable isotope over tissues of *Pinus cembra* L. The interactions were retained in the models only when significant. *df*: degrees of freedom. Significant p-values are in bold face, Table S5: LMM results for the effects of altitude (timberline and treeline), tree age (sapling and adult tree), sampled season (mid-growing season: August; and late-season: October) and their interactions on NSCs, nutrients and stable isotope over tissues of *Larix decidua* Mill. The interactions were retained in the models only when significant. *df*: degrees of freedom. Significant *p*-values are in bold face.

Author Contributions: Conceptualization, M.-H.L., J.-F.L., W.-F.X., and Z.-P.J.; methodology, M.-H.L., J.-F.L., W.-F.X., and Z.-P.J.; formal analysis, J.-F.L. and Y.-Y.N.; investigation, J.-F.L. and M.-H.L.; writing—Original draft preparation, J.-F.L.; writing—Review and editing, M.-H.L., M.S., and A.G.; supervision, M.-H.L., J.-F.L., W.-F.X., Z.-P.J., M.S., and A.G.

Funding: This research was funded by the Fundamental Research Funds for the Central Non-Profit Research Institution of CAF (CAFYBB2014ZD001), the National Natural Science Foundation of China (No. 41371076 and 41371075), and the China Scholarship Council (No. 201303270003).

Acknowledgments: The authors are grateful for the laboratory assistance from Ursula Graf (WSL).

Conflicts of Interest: The authors declare no conflict of interest.

References

1. Steffen, W.; Sanderson, R.A.; Tyson, P.D.; Jäger, J.; Matson, P.A.; Moore, B., III; Oldfield, F.; Richardson, K.; Schellnhuber, H.J.; Turner, B.L.; et al. *Global Change and the Earth System: A Planet under Pressure*; Springer Science & Business Media: Heidelberg, Germany, 2006.

2. Gamfeldt, L.; Snäll, T.; Bagchi, R.; Jonsson, M.; Gustafsson, L.; Kjellander, P.; Ruiz-Jaen, M.C.; Fröberg, M.; Stendahl, J.; Philipson, C.D.; et al. Higher levels of multiple ecosystem services are found in forests with more tree species. *Nat. Commun.* **2013**, *4*, 1340. [CrossRef]

3. Wolkovich, E.M.; Cook, B.I.; Allen, J.M.; Crimmins, T.M.; Betancourt, J.L.; Travers, S.E.; Pau, S.; Regetz, J.; Davies, T.J.; Kraft, N.J.B.; et al. Warming experiments underpredict plant phenological responses to climate change. *Nature* **2012**, *485*, 494–497. [CrossRef] [PubMed]

4. Niinemets, Ü. Responses of forest trees to single and multiple environmental stresses from seedlings to mature plants: Past stress history, stress interactions, tolerance and acclimation. *For. Ecol. Manag.* **2010**, *260*, 1623–1639. [CrossRef]

5. Steppe, K.; Niinemets, Ü.; Teskey, R.O. Tree size-and age-related changes in leaf physiology and their influence on carbon gain. In *Size-and Age-Related Changes in Tree Structure and Function*; Springer: New York, NY, USA, 2011; pp. 235–253.

6. Portsmuth, A.; Niinemets, Ü.; Truus, L.; Pensa, M. Biomass allocation and growth rates in *Pinus sylvestris* are interactively modified by nitrogen and phosphorus availabilities and by tree size and age. *Can. J. For. Res.* **2005**, *35*, 2346–2359. [CrossRef]

7. Gedroc, J.J.; McConnaughay, K.D.M.; Coleman, J.S. Plasticity in Root/Shoot Partitioning: Optimal, Ontogenetic, or Both? *Funct. Ecol.* **1996**, *10*, 44–50. [CrossRef]

8. Rivas-Ubach, A.; Sardans, J.; Pérez-Trujillo, M.; Estiarte, M.; Peñuelas, J. Strong relationship between elemental stoichiometry and metabolome in plants. *Proc. Natl. Acad. Sci. USA* **2012**, *109*, 4181–4186. [CrossRef]

9. Álvarez-Yépiz, J.C.; Cueva, A.; Dovčiak, M.; Teece, M.; Yepez, E.A. Ontogenetic resource-use strategies in a rare long-lived cycad along environmental gradients. *Conserv. Physiol.* **2014**, *2*, cou034. [CrossRef]

10. Cavender-Bares, J.; Bazzaz, F.A. Changes in drought response strategies with ontogeny in *Quercus rubra*: Implications for scaling from seedlings to mature trees. *Oecologia* **2000**, *124*, 8–18. [CrossRef]

11. Bansal, S.; Germino, M.J. Variation in ecophysiological properties among conifers at an ecotonal boundary: Comparison of establishing seedlings and established adults at timberline. *J. Veg. Sci.* **2010**, *21*, 133–142. [CrossRef]

12. Körner, C. The nutritional status of plants from high altitudes. *Oecologia* **1989**, *81*, 379–391. [CrossRef] [PubMed]

13. Körner, C. The use of 'altitude' in ecological research. *Trends Ecol. Evol.* **2007**, *22*, 569–574. [CrossRef]

14. Li, M.-H.; Xiao, W.-F.; Wang, S.-G.; Cheng, G.-W.; Cherubini, P.; Cai, X.-H.; Liu, X.-L.; Wang, X.-D.; Zhu, W.-Z. Mobile carbohydrates in Himalayan treeline trees I. Evidence for carbon gain limitation but not for growth limitation. *Tree Physiol.* **2008**, *28*, 1287–1296. [CrossRef] [PubMed]

15. Marqués, L.; Camarero, J.J.; Gazol, A.; Zavala, M.A. Drought impacts on tree growth of two pine species along an altitudinal gradient and their use as early-warning signals of potential shifts in tree species distributions. *Forest Ecol. Manag.* **2016**, *381*, 157–167. [CrossRef]

16. Rigling, A.; Bigler, C.; Eilmann, B.; Feldmeyer-Christe, E.; Gimmi, U.; Ginzler, C.; Graf, U.; Mayer, P.; Vacchiano, G.; Weber, P.; et al. Driving factors of a vegetation shift from Scots pine to pubescent oak in dry Alpine forests. *Glob. Change Biol.* **2013**, *19*, 229–240. [CrossRef]

17. Peng, G.; Wu, C.; Xu, X.; Yang, D. The age-related changes of leaf structure and biochemistry in juvenile and mature subalpine fir trees (*Abies faxoniana* Rehder & E.H. Wilson.) along an altitudinal gradient. *Pol. J. Ecol.* **2012**, *60*, 311–321.

18. Zhao, C.; Chen, L.; Ma, F.; Yao, B.; Liu, J. Altitudinal differences in the leaf fitness of juvenile and mature alpine spruce trees (*Picea crassifolia*). *Tree Physiol.* **2008**, *28*, 133–141. [CrossRef] [PubMed]

19. Shi, P.; Körner, C.; Hoch, G. A test of the growth-limitation theory for alpine tree line formation in evergreen and deciduous taxa of the eastern Himalayas. *Funct. Ecol.* **2008**, *22*, 213–220. [CrossRef]

20. Fajardo, A.; Piper, F.I.; Pfund, L.; Körner, C.; Hoch, G. Variation of mobile carbon reserves in trees at the alpine treeline ecotone is under environmental control. *New Phytol.* **2012**, *195*, 794–802. [CrossRef]

21. Fajardo, A.; Piper, F.I.; Hoch, G. Similar variation in carbon storage between deciduous and evergreen treeline species across elevational gradients. *Ann. Bot.* **2013**, *112*, 623–631. [CrossRef]

22. Rabasa, S.G.; Granda, E.; Benavides, R.; Kunstler, G.; Espelta, J.M.; Ogaya, R.; Peñuelas, J.; Scherer-Lorenzen, M.; Gil, W.; Grodzki, W.; et al. Disparity in elevational shifts of European trees in response to recent climate warming. *Glob. Chang. Biol.* **2013**, *19*, 2490–2499. [CrossRef]

23. Máliš, F.; Kopecký, M.; Petřík, P.; Vladovič, J.; Merganič, J.; Vida, T. Life stage, not climate change, explains observed tree range shifts. *Glob. Chang. Biol.* **2016**, *22*, 1904–1914. [CrossRef] [PubMed]

24. Bell, D.M.; Bradford, J.B.; Lauenroth, W.K. Early indicators of change: Divergent climate envelopes between tree life stages imply range shifts in the western United States. *Glob. Ecol. Biogeogr.* **2014**, *23*, 168–180. [CrossRef]

25. Bansal, S.; Germino, M.J. Temporal variation of nonstructural carbohydrates in montane conifers: Similarities and differences among developmental stages, species and environmental conditions. *Tree Physiol.* **2009**, *29*, 559–568. [CrossRef] [PubMed]

26. Baber, O.; Slot, M.; Celis, G.; Kitajima, K. Diel patterns of leaf carbohydrate concentrations differ between seedlings and mature trees of two sympatric oak species. *Botany* **2014**, *92*, 535–540. [CrossRef]

27. Hoch, G.; Körner, C. Growth and carbon relations of tree line forming conifers at constant vs. variable low temperatures. *J. Ecol.* **2009**, *97*, 57–66. [CrossRef]

28. Gruber, A.; Pirkebner, D.; Oberhuber, W. Seasonal dynamics of mobile carbohydrate pools in phloem and xylem of two alpine timberline conifers. *Tree Physiol.* **2013**, *33*, 1076–1083. [CrossRef]

29. Zhu, W.Z.; Cao, M.; Wang, S.G.; Xiao, W.F.; Li, M.H. Seasonal dynamics of mobile carbon supply in *Quercus aquifolioides* at the upper elevational limit. *PLoS ONE* **2012**, *7*, e34213. [CrossRef]

30. Li, M.H.; Jiang, Y.; Wang, A.; Li, X.; Zhu, W.; Yan, C.F.; Du, Z.; Shi, Z.; Lei, J.; Schönbeck, L. Active summer carbon storage for winter persistence in trees at the cold alpine treeline. *Tree Physiol.* **2018**. [CrossRef]

31. Hoch, G.; Körner, C. The carbon charging of pines at the climatic treeline: A global comparison. *Oecologia* **2003**, *135*, 10–21. [CrossRef]

32. Zhu, W.Z.; Xiang, J.S.; Wang, S.G.; Li, M.H. Resprouting ability and mobile carbohydrate reserves in an oak shrubland decline with increasing elevation on the eastern edge of the Qinghai-Tibet Plateau. *For. Ecol. Manag.* **2012**, *278*, 118–126. [CrossRef]

33. Li, M.H.; Hoch, G.; Körner, C. Spatial variability of mobile carbohydrates within *Pinus cembra* trees at the alpine treeline. *Phyton-Ann. Rei. Bot. A* **2001**, *41*, 203–213.

34. Seifter, S.; Dayton, S.; Novic, B.; Muntwyler, E. The Estimation of Glycogen with the Anthrone Reagent. *Arch. Biochem.* **1950**, *25*, 191–200. [PubMed]

35. Osaki, M.; Shinano, T.; Tadano, T. Redistribution of carbon and nitrogen compounds from the shoot to the harvesting organs during maturation in field crops. *Soil Sci. Plant Nutr.* **1991**, *37*, 117–128. [CrossRef]

36. Page, A.L. *Methods of Soil Analysis. Part 2. Chemical and microbiological Properties*; American Society of Agronomy, Soil Science Society of America: Madison, WI, USA, 1982.

37. Gebauer, G.; Schulze, E.-D. Carbon and nitrogen isotope ratios in different compartments of a healthy and a declining *Picea abies* forest in the Fichtelgebirge, NE Bavaria. *Oecologia* **1991**, *87*, 198–207. [CrossRef]

38. Remy, E.; Wuyts, K.; Boeckx, P.; Ginzburg, S.; Gundersen, P.; Demey, A.; Van Den Bulcke, J.; Van Acker, J.; Verheyen, K. Strong gradients in nitrogen and carbon stocks at temperate forest edges. *For. Ecol. Manag.* **2016**, *376*, 45–58. [CrossRef]

39. Mayor, J.R.; Sanders, N.J.; Classen, A.T.; Bardgett, R.D.; Clément, J.-C.; Fajardo, A.; Lavorel, S.; Sundqvist, M.K.; Bahn, M.; Chisholm, C.; et al. Elevation alters ecosystem properties across temperate treelines globally. *Nature* **2017**, *542*, 91–95. [CrossRef] [PubMed]

40. Millard, P.; Grelet, G.-a. Nitrogen storage and remobilization by trees: Ecophysiological relevance in a changing world. *Tree Physiol.* **2010**, *30*, 1083–1095. [CrossRef]

41. Piper, F.I.; Fajardo, A. Foliar habit, tolerance to defoliation and their link to carbon and nitrogen storage. *J. Ecol.* **2014**, *102*, 1101–1111. [CrossRef]

42. Pardo, L.H.; Semaoune, P.; Schaberg, P.G.; Eagar, C.; Sebilo, M. Patterns in δ^{15}N in roots, stems, and leaves of sugar maple and American beech seedlings, saplings, and mature trees. *Biogeochemistry* **2013**, *112*, 275–291. [CrossRef]

43. Reich, A.; Ewel, J.; Nadkarni, N.; Dawson, T.; Evans, R.D. Nitrogen isotope ratios shift with plant size in tropical bromeliads. *Oecologia* **2003**, *137*, 587–590. [CrossRef]

44. Hu, Y.-L.; Yan, E.-R.; Choi, W.-J.; Salifu, F.; Tan, X.; Chen, Z.C.; Zeng, D.-H.; Chang, S. Soil nitrification and foliar δ^{15}N declined with stand age in trembling aspen and jack pine forests in northern Alberta, Canada. *Plant Soil* **2014**, *376*, 399–409. [CrossRef]

45. Hiltbrunner, E.; Schwikowski, M.; Körner, C. Inorganic nitrogen storage in alpine snow pack in the Central Alps (Switzerland). *Atmos. Environ.* **2005**, *39*, 2249–2259. [CrossRef]

46. Chen, L.; Flynn, D.F.B.; Zhang, X.; Gao, X.; Lin, L.; Luo, J.; Zhao, C. Divergent patterns of foliar δ^{13}C and δ^{15}N in *Quercus aquifolioides* with an altitudinal transect on the Tibetan Plateau: An integrated study based on multiple key leaf functional traits. *J. Plant Ecol.* **2015**, *8*, 303–312. [CrossRef]

47. Bai, E.; Boutton, T.; Liu, F.; Wu, X.B.; Archer, S.; Hallmark, C.T. Spatial variation of the stable nitrogen isotope ratio of woody plants along a topoedaphic gradient in a subtropical savanna. *Oecologia* **2009**, *159*, 493–503. [CrossRef]

48. Liu, X.; Wang, G.; Li, J.; Wang, Q. Nitrogen isotope composition characteristics of modern plants and their variations along an altitudinal gradient in Dongling Mountain in Beijing. *Sci. China Ser. D-Earth Sci.* **2010**, *53*, 128–140. [CrossRef]

49. Garten, C.T., Jr.; Miegroet, H.V. Relationships between soil nitrogen dynamics and natural ^{15}N abundance in plant foliage from Great Smoky Mountains National Park. *Can. J. Forest Res.* **1994**, *24*, 1636–1645. [CrossRef]

50. Chen, Y.; Han, W.; Tang, L.; Tang, Z.; Fang, J. Leaf nitrogen and phosphorus concentrations of woody plants differ in responses to climate, soil and plant growth form. *Ecography* **2013**, *36*, 178–184. [CrossRef]

51. Ordoñez, J.C.; Van Bodegom, P.M.; Witte, J.-P.M.; Wright, I.J.; Reich, P.B.; Aerts, R. A global study of relationships between leaf traits, climate and soil measures of nutrient fertility. *Glob. Ecol. Biogeogr.* **2009**, *18*, 137–149. [CrossRef]

52. Sah, S.; Brumme, R. Altitudinal gradients of natural abundance of stable isotopes of nitrogen and carbon in the needles and soil of a pine forest in Nepal. *J. Forest Sci.* **2003**, *49*, 19–26. [CrossRef]

53. Huber, E.; Wanek, W.; Gottfried, M.; Pauli, H.; Schweiger, P.; Arndt, S.; Reiter, K.; Richter, A. Shift in soil–plant nitrogen dynamics of an alpine–nival ecotone. *Plant Soil* **2007**, *301*, 65–76. [CrossRef]

54. Averill, C.; Finzi, A. Increasing plant use of organic nitrogen with elevation is reflected in nitrogen uptake rates and ecosystem δ^{15}N. *Ecology* **2010**, *92*, 883–891. [CrossRef]

55. Keller, G. Utilization of inorganic and organic nitrogen sources by high-subalpine ectomycorrhizal fungi of *Pinus cembra* in pure culture. *Mycol. Res.* **1996**, *100*, 989–998. [CrossRef]

56. Hobbie, E.A.; Högberg, P. Nitrogen isotopes link mycorrhizal fungi and plants to nitrogen dynamics. *New Phytol.* **2012**, *196*, 367–382. [CrossRef]

57. Hasselquist, N.J.; Metcalfe, D.B.; Inselsbacher, E.; Stangl, Z.; Oren, R.; Näsholm, T.; Högberg, P. Greater carbon allocation to mycorrhizal fungi reduces tree nitrogen uptake in a boreal forest. *Ecology* **2016**, *97*, 1012–1022. [CrossRef]

 *forests*

Article

Leaf Age Compared to Tree Age Plays a Dominant Role in Leaf δ^{13}C and δ^{15}N of Qinghai Spruce (*Picea crassifolia* Kom.)

Caijuan Li [1,2], Bo Wang [1,2], Tuo Chen [1,*], Guobao Xu [1], Minghui Wu [1,2], Guoju Wu [1] and Jinxiu Wang [1,2]

[1] State Key Laboratory of Cryospheric Science, Northwest Institute of Eco-Environment and Resources, Chinese Academy of Sciences, Lanzhou 730000, China; licaijuan@lzb.ac.cn (C.L.); wangbo900824@lzb.ac.cn (B.W.); xgb234@lzb.ac.cn (G.X.); wumh2017@lzb.ac.cn (M.W.); guojuwu@lzb.ac.cn (G.W.); nwnujinxiu_w@163.com (J.W.)
[2] University of Chinese Academy of Sciences, Beijing 100049, China
* Correspondence: chentuo@lzb.ac.cn

Received: 17 January 2019; Accepted: 31 March 2019; Published: 4 April 2019

Abstract: Leaf stable isotope compositions (δ^{13}C and δ^{15}N) are influenced by various abiotic and biotic factors. Qinghai spruce (*Picea crassifolia* Kom.) as one of the dominant tree species in Qilian Mountains plays a key role in the ecological stability of arid region in the northwest of China. However, our knowledge of the relative importance of multiple factors on leaf δ^{13}C and δ^{15}N remains incomplete. In this work, we investigated the relationships of δ^{13}C and δ^{15}N to leaf age, tree age and leaf nutrients to examine the patterns and controls of leaf δ^{13}C and δ^{15}N variation of *Picea crassifolia*. Results showed that ^{13}C and ^{15}N of current-year leaves were more enriched than older ones at each tree age level. There was no significant difference in leaf δ^{13}C values among trees of different ages, while juvenile trees (<50 years old) were ^{15}N depleted compared to middle-aged trees (50–100 years old) at each leaf age level except for 1-year-old leaves. Meanwhile, relative importance analysis has demonstrated that leaf age was one of the most important indicators for leaf δ^{13}C and δ^{15}N. Moreover, leaf N concentrations played a dominant role in the variations of δ^{13}C and δ^{15}N. Above all, these results provide valuable information on the eco-physiological responses of *P. crassifolia* in arid and semi-arid regions.

Keywords: Leaf δ^{13}C; Leaf δ^{15}N; Growth stage; Environmental factors; Relative importance

1. Introduction

As one of the most powerful tools for studying plant eco-physiology, stable isotope techniques provide fundamental insights into how plants interact with and respond to biotic or abiotic environmental factors, helping us to better understand the relationship between plants and their environment [1–3]. In particular, leaf carbon isotope composition (δ^{13}C), which reflects the balance between leaf conductance and photosynthetic rate [4], is widely used to analyze intraspecific or interspecific differences in photosynthetic and physiological characteristics [5,6], to measure the long-term water use efficiency under different environmental conditions and to reveal significant functional changes in plant metabolism and adaptation to various environmental stresses [7–9]. In addition, the natural abundance of ^{15}N in leaves or roots has been proposed as an important tracer to reflect the outcome of different processes affecting δ^{15}N compositions, thereby providing an integrative measure of terrestrial N processes [10–12].

However, to our knowledge, multiple previous studies have been conducted on large global or regional-scale variations in plant δ^{13}C or δ^{15}N values, while the understanding of δ^{13}C and δ^{15}N patterns on intermediate spatial or temporal scales is rather limited [11,12]. More importantly, variations of $\delta^{13}C$ and $\delta^{15}N$ values during different plant development and growth stages (for example, stand age class, tree age class) have been neglected. Currently, increasing attention has been paid to investigating the significant variations in stable carbon isotopes among different plant organs, such as leaves, stems, shoots, roots, or different plant species including C_3 and C_4 plants [3,13–15]. However, research about the variations in the natural abundance of ^{13}C and ^{15}N with leaf habit, phenological leaf traits or leaf age class on intermediate scales remains incomplete [10,13]. For example, a previous study demonstrated that leaf age was of special interest when exploring isotope fractionation, because younger leaves show different physiological properties and mechanisms of carbon and nitrogen assimilation compared to older leaves [13,16]. Likewise, significant variations in the $\delta^{15}N$ values with stand age were discussed. Li et al. [10] reported that leaf $\delta^{15}N$ variations at the community, plant growth form and species levels were significantly reduced with increasing stand age over shorter times and at smaller spatial scales. In addition, Vitoria et al. [17] demonstrated that there was no significant difference in leaf $\delta^{13}C$ and $\delta^{15}N$ values of evergreen and deciduous species within a site. In addition, despite tremendous progress over the past few decades in investigating what causes variation in plants $\delta^{13}C$ and $\delta^{15}N$ values, limited studies have allowed a comprehensive explanation for the relative importance of study variables on carbon and nitrogen compositions [16,17]. Therefore, it is necessary to conduct further investigation on the natural abundance of ^{13}C and ^{15}N during different plant growth stages over intermediate spatial or temporal scales.

Moreover, there are a variety of other abiotic and biotic factors that control leaf $\delta^{13}C$ and $\delta^{15}N$ values during plant development and growth [18,19]. For example, leaf $\delta^{13}C$ values changed with leaf habit, morphology, genetics and irradiance [1,16], which may reflect differences in photosynthetic water use efficiency. Current work has demonstrated that plant $\delta^{13}C$ is also influenced by various environmental factors such as precipitation, humidity, soil moisture, and air temperature [18,19]. Furthermore, leaf functional elements such as nitrogen (N) and phosphorus (P) also play a key role in $\delta^{13}C$ values through their indirect effects on photosynthetic capacity and the synthesis of proteins, DNA and RNA [20,21]. However, our knowledge on the relative role of these parameters in leaf $\delta^{13}C$ values remains incomplete. Whether and how these established relationships could hold true in specific biomes remains largely uncertain. This uncertainty is especially true for forests of different study regions that might display contrasting responses to climate [16]. Additionally, compared to plant $\delta^{13}C$, the relationship between leaf $\delta^{15}N$ and those mentioned factors has received less attention. Earlier studies have focused on the spatial or seasonal variation in plant $\delta^{15}N$ values along a specific factor gradient, but did not consider the relative importance of those variables in the variations of $\delta^{15}N$ values [11,12]. Thus, for detailed knowledge, additional empirical studies are required to address the relative effect of biotic or abiotic factors in the variations of $\delta^{13}C$ and $\delta^{15}N$ values at specific biome levels.

Qinghai spruce (*Picea crassifolia* Kom.) as a common coniferous evergreen species is widely distributed at altitudes ranging from 2300 to 3300 m in the subalpine and alpine environments of Qilian Mountains in the Northern China, in which the availability of water, nutrients and temperature is crucial for determining plant performance, abundance and distribution. It exhibits a wide tolerance to different environmental conditions and has significant ecological function in northwest China. However, limited studies have been conducted on temporal and spatial variations in the stable carbon and nitrogen isotope compositions in different aged leaves and trees of *P. crassifolia*. Therefore, it is necessary to fully understand the effects of the variables mentioned above on the $\delta^{13}C$ and $\delta^{15}N$ values of *P. crassifolia*.

We hypothesized that the $\delta^{13}C$ and $\delta^{15}N$ values of *P. crassifolia* would change with leaf age and tree age to adapt to the growth stage needs. In addition, environmental variables could contribute to the growth and eco-physiology of *P. crassifolia*. The main objectives of this study were (i) to quantify the variation in *P. crassifolia* leaf $\delta^{13}C$ and $\delta^{15}N$ values along the leaf age and tree age gradients and provide evidence of the physiological mechanisms underlying the variations in the $\delta^{13}C$ and $\delta^{15}N$

values; (ii) to investigate the relationship between leaf δ^{13}C, leaf δ^{15}N and leaf nutrients; and (iii) to explain the relative role of leaf physiological properties and leaf nutrients in the variations of δ^{13}C and δ^{15}N.

2. Materials and methods

2.1. Site Description

The research was conducted in the Shuang Longgou region (longitude 102°17′18″–102°33′42″ E; latitude 37°18′12″–37°25′18″ N) located northwest of Tianzhu County at the eastern margin of the Qilian Mountains (Figure 1). The climate is generally characterized as a semiarid continental climate with water availability being the major abiotic factor-limiting plant growth. Mean annual precipitation is less than 400 mm and mean annual air temperature in this temperate location is approximately 1.5 °C, respectively. Moreover, the dominant forest species in this study area is *P. crassifolia*, which grows naturally with no manual management, and *Juniperus przewalskii* Kom., *Betula albo-sinensis* Burk. and *Populus davidiana* Dode are the second minor contributors to the shady understory. Additionally, the shrub species are mainly dominated by *Salix cupularis*, *Rhododendron simsii* Planch., *Caragana jubata* (Pall.) Poir. and *Spiraea alpine* Pall.

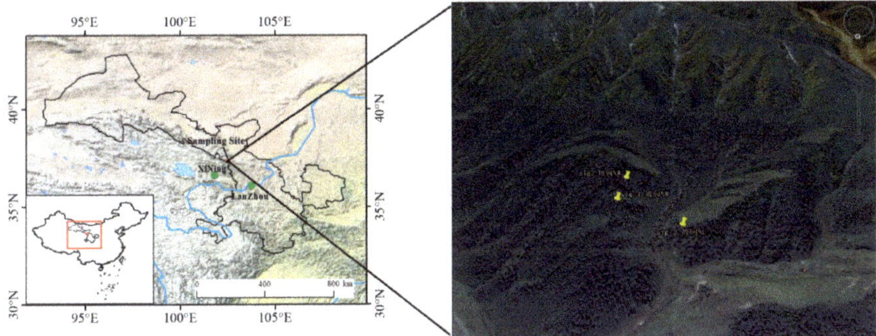

Figure 1. Geographic location of the collection sites (slg-2, slg-3, slg-4) of *Picea crassifolia* located in Shuanglonggou region.

2.2. Sample Collection

In 2015, *P. crassifolia* forests were selected from sites with similar conditions, such as topography, the composition of the undergrowth vegetation and stand age. Three plots (slg-2, slg-3, and slg-4 are shown in Figure 1) ranging from 2824 to 2914 m were established for study. Table 1 provides the general information related to the sampling sites. For each sampling plot, trees were divided into four/five groups according to the classes of diameter at breast height. Subsequently, there were five breast-height diameter classes determined and three to five trees with similar diameter in each group were chosen as the samples. The sampled *P. crassifolia* leaves developed in full light of an open canopy and were carried out from the upper third portion of the tree crown.

Moreover, considering the variation caused by differences in leaf nutrient contents at different orientations of the shoots, only leaves with a healthy appearance (avoiding damaged leaves) were cut with a pole pruner, and sampling was carried out from different orientations as much as possible. Overall, 4 years' leaves (from current year and up to 3 years-old) of each sampled tree were collected in our study. There was a clear joint on the branch between growing seasons, which aided in the determination of leaf age. We defined leaf age as 0 for the needles from the current, and then the next group was year 1 (the previous year+1) and so on. Thus, leaves aged 0–3 years were detached from the twigs before sending the samples to the laboratory for analysis. Additionally, for the estimation of tree age, tree-ring cores of selected *P. crassifolia* trees were collected at tree breast height (approximately

1.3 m above the surface) using an increment borer. Actual tree ages of sample cores were determined using dendrochronological methods [22].

Table 1. Site information for four sampling quadrats.

Site code	Elevation (m)	Slope aspect	Latitude	Longitude	MH (m)	M-dbh (cm)
slg-2	2824	NW	37°23′17.79″ N	102°30′39.47″ E	10.77	27.73
slg-3	2914	NE	37°23′02.92″ N	102°30′48.26″ E	6.53	17.09
slg-4	2828	NW	37°23′12.79″ N	102°30′44.58″ E	6.84	16.86

MH is mean tree height and M-dbh is mean diameter at breast height.

2.3. Stable Carbon and Nitrogen Isotope Analyses

The samples were washed in distilled water to remove dust particles, air dried before oven drying at 65 °C for 12 h and at 110 °C for 10 min to deactivate the enzymes, ground into a homogeneous fine powder, and sieved in the laboratory. A stable carbon and nitrogen analysis was performed in the Environmental Stable Isotope Laboratory, Institute of Environment and Sustainable Development in Agriculture, No.12, Zhongguancun South Street, Haidian District, Beijing 100081, China.

The isotopic compositions of the leaf samples were measured on an Isoprime100-EA mass spectrometer (Germany). The carbon or nitrogen isotope ratios are expressed relative to an international standard using the delta notation:

$$\delta_{\text{sample}} = (R_{\text{sample}} - R_{\text{standard}})/R_{\text{standard}}. \tag{1}$$

where δ_{sample} was defined by this relationship, R_{sample} indicated the $^{13}C/^{12}C$ or $^{15}N/^{14}N$ ratio of the sample, and R_{standard} indicated the $^{13}C/^{12}C$ or $^{15}N/^{14}N$ ratio of the standard. The international standard reference for carbon was PDB (Pee Dee Belemnite), and for nitrogen, it was an average of $^{15}N/^{14}N$ from atmospheric air [23].

2.4. Statistical Analysis

We analyzed the dataset by subdividing them into four groups based on leaf age (current year leaves, 1-year-old leaves, 2-year-old leaves, and 3-year-old leaves) and tree age (<50-year-old, 51 to 100-year-old, 101 to 150-year-old, and >150-year old), respectively. For leaf $\delta^{13}C$ and $\delta^{15}N$ values of different leaf ages and tree ages, the mean, median, standard error, and coefficient of variation (CV) were calculated, respectively. Here, the analysis was the leaf age–tree age combination. We first used two-way analysis of variance and Tukey's post hoc test to compare differences of leaf $\delta^{13}C$ and $\delta^{15}N$ values between leaf ages and tree ages. Next, the regression analysis was applied to investigate the relationship between leaf $\delta^{13}C$, leaf $\delta^{15}N$ and leaf nutrients (leaf N, P concentrations and the C:N ratios). Furthermore, we calculated the relative importance (refers to the quantification of an individual regressor's contribution to a multiple regression model) of each predictor on leaf $\delta^{13}C$ and $\delta^{15}N$ with the R package relaimpo [24].

The R package relaimpo demonstrates six different metrics for assessing the relative importance of regressors (all regressors are uncorrelated) in the model [24]. Each predictor's contribution is just the R^2 from univariate regression, and all univariate R^2-values add up to the full model R^2. R^2 represents the proportion of variation in y that is explained by the p regressors in the model. Correlation analysis was conducted using the SPSS 22.0 [25] and R 3.2.4 [26].

3. Results

3.1. Variations of P. crassifolia Leaf $\delta^{13}C$ and $\delta^{15}N$ Values with Leaf and Tree Ages

Changes of leaf $\delta^{13}C$ and $\delta^{15}N$ in relation to leaf age and tree age are shown in Figures 2 and 3. All leaf $\delta^{13}C$ values varied from −28.22‰ to −24.09‰, with a mean of −26.76‰ and a variance of

3.39%. The average carbon isotopic values from current year to 3-year-old leaves were −25.54‰, −27.15‰, −27.24‰ and −27.07‰, respectively (Figure 2A). The δ^{13}C was significantly more enriched in current year leaves than others at each tree age level ($p < 0.01$, Figure 3A), while no differences were observed among other older leaves. Meanwhile, at each leaf age level, the δ^{13}C value of mature trees (>150 years old) was lower than that of other aged trees (Figure 2A), but the difference was not significant (Figure 3A).

The mean leaf δ^{15}N of all samples was −5.91‰, with a range of −8.7‰ to −2.89‰ and a variance of 16.80%. The average nitrogen isotopic values from current year to 3-year-old leaves were −4.86‰, −6.00‰, −6.15‰ and −6.55‰, respectively (Figure 2B). Leaf δ^{15}N showed substantial variability among leaf ages and tree ages. Older leaves were more depleted in ^{15}N than current year leaves at each tree age level (Figure 3B). Moreover, there was a significant difference in δ^{15}N between middle-aged trees (50–100) and juvenile trees (<50) except for 1-year-old leaves ($p < 0.05$).

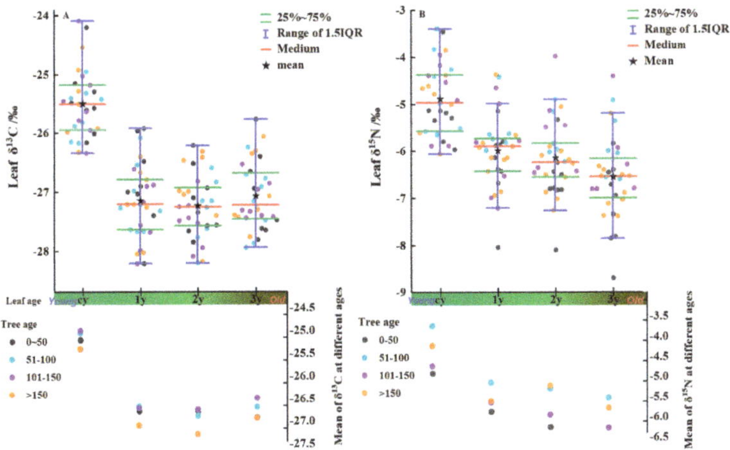

Figure 2. Box and scatter plots showing that the spatial patterns and variations of the δ^{13}C (**A**) and δ^{15}N (**B**) values changed with differently aged leaves and trees of *P. crassifolia*. Leaf ages including current year to 3-year-old leaves were represented by cy, 1y, 2y, and 3y, respectively.

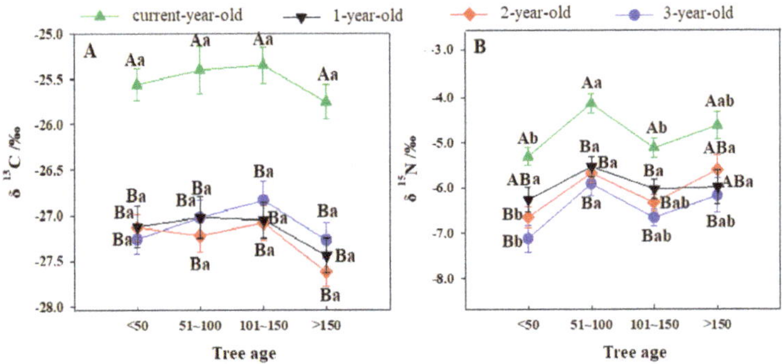

Figure 3. Differences in leaf δ^{13}C (**A**) and δ^{15}N (**B**) between 4 years' leaves collected along different tree age levels. The value was Mean± SE (standard error). Different uppercase letters represent significant differences among leaf ages at each tree age level, while different lowercase letters represent significant differences among tree ages at each leaf age level.

3.2. Relationship Between Leaf $\delta^{13}C$, $\delta^{15}N$ and N, P Concentrations as well as C:N Ratio

For all leaf samples, $\delta^{13}C$ was significantly positively correlated with leaf N and P concentrations, but negatively related to the C/N ratio ($p < 0.001$, Table 2). However, nutrient patterns did not differ among leaf ages as the leaf $\delta^{13}C$ showed no significant relationship between leaf N, leaf P concentrations and the C/N ratio except for current-year-old leaves ($p > 0.05$).

Likewise, leaf $\delta^{15}N$ showed significant positive correlation with leaf N and P concentrations but negative correlation with the C:N ratio in all leaf samples ($p < 0.001$). When examining the relationship between these leaf nutrients and $\delta^{15}N$ at each leaf age level, a significant positive correlation between leaf N concentration and $\delta^{15}N$ and a negative correlation between the C:N ratio and leaf $\delta^{15}N$ were only observed in current-year-old and 1-year-old leaves, respectively, while there was no significant relationship between leaf $\delta^{15}N$ and P concentrations at each leaf age group.

Table 2. Regression equations for leaf $\delta^{13}C$ and $\delta^{15}N$ values against leaf N and P concentrations and the C:N ratios for 4 years' leaves.

Nutrient Variables	Leaf Ages	Statistic parameters			
		$R^2(\delta^{13}C)$	Regression coefficient	$R^2(\delta^{15}N)$	Regression coefficient
Leaf N	All leaves	0.4295 ***	4.3035	0.3829 ***	3.6156
	Current-year-old	0.0894	−2.0361	0.2335 ***	2.2635
	1-year-old	0.0176	−0.6792	0.1249 *	1.3664
	2-year-old	0.0007	0.0701	0.0621	0.4335
	3-year-old	0.0617	−0.5253	0.0006	−0.0327
Leaf P	All leaves	0.4259 ***	0.7824	0.2641 ***	0.5462
	Current-year-old	0.0836	−0.5255	0.0023	0.0604
	1-year-old	0.0127	0.0444	0.0262	0.0477
	2-year-old	0.0319	0.0491	0.0015	0.0073
	3-year-old	0.0029	0.0144	0.0741	-0.0473
C:N	All leaves	0.3848 ***	−19.373	0.3597 ***	−0.0216
	Current-year-old	0.1768 *	5.008	0.2629 ***	−4.2006
	1-year-old	0.0027	1.7585	0.2446 ***	−12.641
	2-year-old	0.0198	−4.9676	0.0597	−5.8591
	3-year-old	0.0131	3.1193	0.0108	−1.8541

Note: *** $p < 0.001$,* $p < 0.05$.

3.3. Relative Importance of Leaf Age, Tree Age, Tree Height as well as Leaf Nutrients on the $\delta^{13}C$ and $\delta^{15}N$ Values

In this analysis, we assumed that the effect of leaf age may not be linear. Among the studied variables, leaf age, tree height and tree age have accounted for 18.78% variance in the model (Figure 4A). The independent effects of leaf age showed a larger contribution ($R^2 = 14.49\%$, $p < 0.05$) to total variation in leaf $\delta^{13}C$ compared with tree age and tree height. In addition, leaf nutrients such as leaf nitrogen and phosphorus concentrations were the other most important predictors for the $\delta^{13}C$ ($R^2 = 33.56\%$). In particular, leaf N concentrations explained the largest percentage of variation in leaf $\delta^{13}C$ and its effect was significant ($R^2 = 19.24\%$, $p < 0.01$).

Leaf physiological properties (leaf age, tree height and tree age) and leaf nutrients (N and P concentrations) together explained 47.29% of the variations in leaf $\delta^{15}N$ (Figure 4B). Among the study variables, leaf age was the most important predictor for the $\delta^{15}N$ ($R^2 = 15.31\%$, $p < 0.01$). Meanwhile, tree age as another physiological factor also played a key role in leaf nitrogen isotope compositions ($R^2 = 3.52\%$, $p < 0.01$). Moreover, looking at the relative contribution of each predictor in leaf nutrients, leaf nitrogen concentrations independently explained the largest percentage of variation in leaf $\delta^{15}N$ ($R^2 = 19.29\%$, $p < 0.001$).

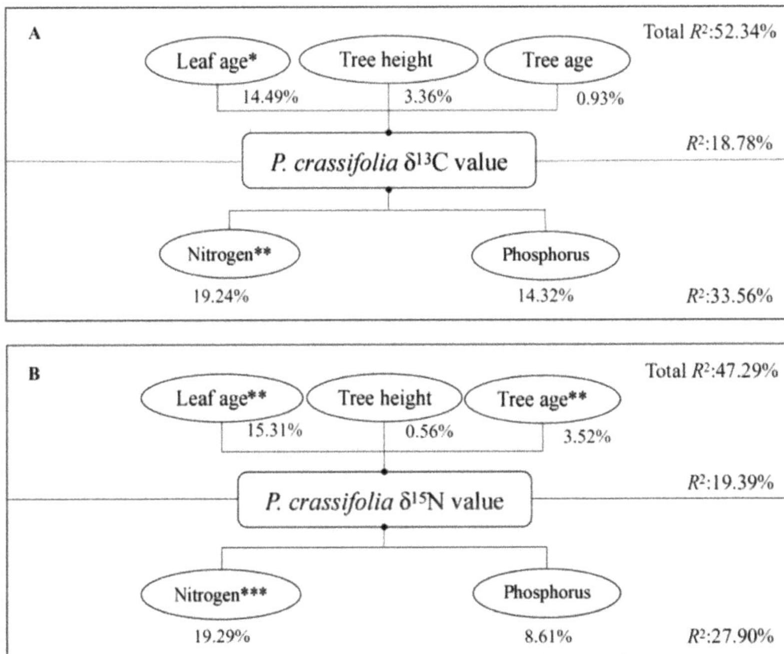

Figure 4. Variation partitioning (%, R^2) of physiological factors including leaf age, tree height and tree age; and leaf nutrients such as leaf nitrogen and phosphorus concentrations in accounting for leaf $\delta^{13}C$ (**A**) and $\delta^{15}N$ (**B**) values. The percentage values represent the proportion of the variance explained by each predictor in the model. *** $p < 0.001$, ** $p < 0.01$, * $p < 0.05$.

3.4. Relationship Between $\delta^{15}N$ and $\delta^{13}C$ in P. crassifolia Leaves

Relationship between the $\delta^{13}C$ and $\delta^{15}N$ values of all leaf ages pooled was significantly positive for *P. crassifolia* samples ($p < 0.001$, Figure 5). The correlation coefficient between them was 0.44. Different leaf ages (range from current-year leaves to 3-year-old leaves) were represented by triangles, squares, circles, and rhombuses, respectively.

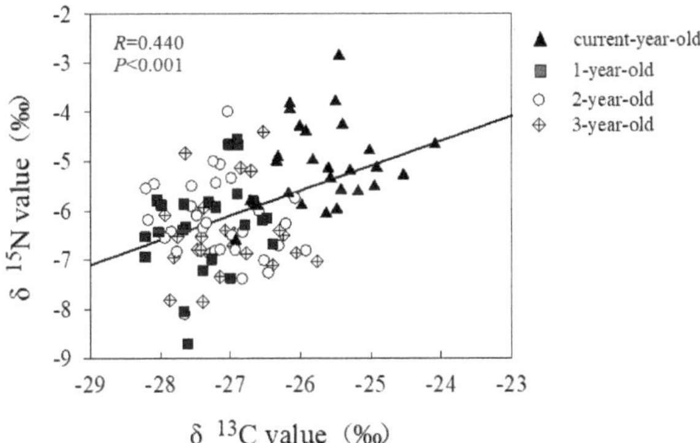

Figure 5. Relationship between leaf $\delta^{13}C$ and $\delta^{15}N$ values of *P. crassifolia*. Leaf ages (current year leaves: triangles; 1-year-old leaves: squares; 2-year-old leaves: circle; 3-year-old leaves: rhombic) are also given.

4. Discussion

4.1. Leaf $\delta^{13}C$ Changed with Leaf Age and Tree Age

P. crassifolia is widely distributed throughout the arid zone of northwest China [22]. In our work, leaf $\delta^{13}C$ varied from −28.22‰ to −24.09‰, with a mean of −26.76‰, nearly identical to that reported by Pei et al. [27] (−28.58‰ to −25.02‰) and Yun et al. [28] (−28.9‰ to −25.4‰) for *P. crassifolia* in the Qilian Mountains. Moreover, the findings were almost the same as average values reported for *Pinus tabulaeformis* in northwest China with values of −26.82‰, which fell within the range of −28.6‰ to −25.02‰ [21].

Previous studies have confirmed that the rate of photosynthesis and respiration declines with leaf aging [13]. In this study, a significant difference in $\delta^{13}C$ between current-year leaves and other old leaves was observed. The $\delta^{13}C$ values achieved a relative maximum at the current-year leaves (−25.54‰) and lower values occurred in other old leaves. Since carbon was directly assimilated from the air or remobilized from reserve carbohydrates [13,29], we assumed that the isotopic patterns reported in our study might be caused by several effects. First of all, variations in leaf $\delta^{13}C$ values were related with *P. crassifolia* interior biochemical processes. In the initial developmental stages, a 'hungry' state of intercellular CO_2 concentrations exists because plants grow relatively rapidly and need to synthesize large amounts of organic matter to meet the demands of development and construction, leading to reduced distinguishing and exclusion of $^{13}CO_2$ [18]. Thus, the values become enriched. When the exterior morphology, interior structure and physiological metabolism functions are mature, plants have the ability to adjust physiological and biochemical reactions. Therefore, they can efficiently distinguish and exclude $^{13}CO_2$, and the $\delta^{13}C$ values are expected to be depleted [30]. More importantly, photosynthetic capacity is the central process that coordinates carbon isotope discrimination, with more photosynthetically active leaves being relatively ^{13}C-enriched [16].

Second, it was associated with leaf development stage [13]. Due to the changing growth rate between different aged leaves, the allocated proportions of the structural, functional and storage components within plants varied significantly to meet the leaves' nutrient needs. Previous $^{13}CO_2$ tracer studies have reported that there were two leaf developmental stages including heterotrophic and autotrophic stages [14]. During the heterotrophic growth stage, where organic carbon was imported from elsewhere in the plant, enrichment in ^{13}C was most evident in this stage and supposedly a result of the heterotrophic carbon source for growth [13,31]. This stage was more obvious for current-year-old leaves. Furthermore, Cernusak et al. [32] discussed six hypotheses regarding the explanation for the ^{13}C enrichment of heterotrophic versus autotrophic plant organs. Based on these hypotheses, newly expanded leaves might need to synthesize large amounts of organic matter to meet the demands of development and construction, which leads to a reduction in distinguishing among sources of C and allows the leaves to obtain C from $^{13}CO_2$. Moreover, young new leaves contained more ^{13}C-enriched cellulose and import carbon from older leaves, while the old leaves had more ^{13}C-depleted lipids and lignin and export carbon to the younger leaves [33]. As a consequence of these effects, an enrichment of ^{13}C in current-year leaves was observed. During the autotrophic growth stage, carbon was assimilated and exported to other plant organs [17], which was most evident for older leaves [14]. The lighter carbon isotope was preferentially assimilated and used to produce the lipids and lignin, while the heavier carbon isotope was transported as ^{13}C-enriched sucrose to the young new leaves [13]. As a result, the old leaves were expected to be ^{13}C-depleted.

4.2. Leaf $\delta^{15}N$ Values Changed with Leaf Age and Tree Age

Leaf nitrogen isotope compositions were determined by the isotope ratio of the external nitrogen source and physiological mechanisms within the plant. However, the intra-plant variation in isotope composition was caused by multiple assimilation events, organ-specific losses of nitrogen as well as resorption and reallocation of nitrogen [12]. In our work, ^{15}N of current year leaves were more enriched compared to other mature leaves—a pattern completely similar to stable carbon isotope compositions.

Meanwhile, the result of Figure 5 indicated that these ratios shift similarly to leaf age. Nitrogen as a key nutrient to build up the photosynthetic apparatus was translocated either from the roots, storage organs or mature leaves to growing leaves [13]. Therefore, there may be variation in $\delta^{15}N$ of leaves throughout the plant depending on sink/source activity and the timing and source of remobilized and assimilated organic nitrogen [34,35]. Nitrogen remobilization was important for perennial plant survival. During growth, there was significant variation in primary N-containing compounds being remobilized in the plant [34]. Masclaux-Daubresse et al. [36] reported that N-containing compounds (like proteins, chlorophyll, etc.) could be degraded during leaf senescence and then nitrogen may be remobilized from senescing leaves to expanding leaves at the vegetative stage. For example, ^{15}N-enriched glutamine was observed as the primary transport form of organic nitrogen, which would be remobilized to developing sink leaves (receiving enriched ^{15}N-glutamine) from a source leaf (exporting enriched ^{15}N-glutamine) [34]. Consequently, an enrichment of ^{15}N in new leaves was expected. In addition, leaf proteins and in particular photosynthetic proteins of plastids were extensively degraded during senescence, providing an enormous source of nitrogen that plants could tap into to supplement the nutrition of growing organs such as new leaves and seeds [36]. Moreover, juvenile trees (<50 years old) were ^{15}N depleted compared to middle-aged trees (50–100 years old) at each leaf age level except for 1-year-old leaves in this study. This was likely attributed to the various allocated proportions of the structural, functional and storage components within the plant bodies to meet the plant's nutrient demands [37]. In addition, there was a significant difference in water potential, stomatal conductance, photosynthetic rate, and water-use efficiency between juvenile trees and other aged trees [37].

4.3. Relationships between the $\delta^{13}C$, $\delta^{15}N$ Values and Leaf Nutrients

Multiple studies have reported various correlations between the $\delta^{13}C$ and leaf nutrients [20,21,38]. In the present work, the positive relationship between $\delta^{13}C$ and N over all aged leaves together (Table 2) is in accordance with most previous studies [20,39]. Moreover, our conclusions suggest that the relative contribution of leaf N concentrations on $\delta^{13}C$ was significant ($p < 0.01$). The main cause of the positive relationships was that photosynthetic capacity increased with leaf N concentrations [20], and there was a positive correlation between leaf $\delta^{13}C$ and photosynthetic capacity [4]. However, other studies have found a negative correlation between leaf $\delta^{13}C$ and leaf N concentrations, and this was likely attributed to the presence of nitrogen-fixing species in samples such as *Caragana microphylla* [5] or an autocorrelation with water availability in a semiarid environment. In addition, in high altitude areas, low atmospheric pressure and temperature could alter the expression of the relationship between N and photosynthetic and thus, the $\delta^{13}C$-N relation [38].

Research about the relationship between leaf $\delta^{13}C$ and P concentrations is relatively limited, and the results have been inconsistent. Some studies have demonstrated the positive relationship between leaf $\delta^{13}C$ and P owing to the effect of leaf P concentrations on photosynthetic via Rubisco, while other works have found leaf $\delta^{13}C$ to be negatively related with leaf P concentration. Our study observed that leaf $\delta^{13}C$ was positively related to all aged leaves' P concentrations in simple regression (Table 2), but the effect of leaf P concentration on $\delta^{13}C$ was not significant in multiple regression (Figure 4A). This indicates that the variations in the leaf $\delta^{13}C$ values were likely caused by stomatal limitation rather than P-related changes in photosynthetic efficiency [18,20]. In addition, our findings of the negative correlation between leaf $\delta^{13}C$ and the C:N ratios is consistent with the results from multiple previous studies and suggests that *P. crassifolia* may achieve higher water use efficiency (WUE) at the expense of decreased nitrogen use efficiency (NUE) [20,38].

The uptake and discrimination of ^{15}N are also significantly related to plant N demand and assimilation capacity [1,40]. N availability in ecosystem, N re-translocation in plants, and N fractionation after plant uptake is known to influence leaf $\delta^{15}N$ [1,41]. Multiple previous studies have reported that there is a positive relationship between plant $\delta^{15}N$ and leaf N concentrations at various spatial scales [17]. In our work, leaf $\delta^{15}N$ was also positively related to leaf N concentrations. Furthermore, the result of Figure 4B suggested that leaf nitrogen concentrations play an importance role in accounting for the variations of leaf $\delta^{15}N$. Changes in environmental nitrogen demand or supply could influence whole plant and organ level nitrogen isotope discrimination [12,34]. Likewise, leaf N concentrations of current-year leaves were significantly higher than other old leaves in our study area ($p < 0.001$, Figure S1). If the nitrogen supply of current-year leaves increased, discrimination could be expected to increase. However, with important questions still remaining about the relationship of leaf N and leaf $\delta^{15}N$, more comparative data are need to evaluate the potential drivers of leaf $\delta^{15}N$ with increasing leaf N concentrations in the future [11].

5. Conclusions

In summary, the carbon and nitrogen assimilation in *P. crassifolia* leaves resulted in the same gradient of stable isotope compositions: young *P. crassifolia* leaves were more enriched in ^{13}C and ^{15}N compared with the older leaves at each tree age level. No significant difference in $\delta^{13}C$ values among different tree ages was observed at each leaf age level, while the $\delta^{15}N$ values of middle-aged (51–100 years old) were significantly more enriched than juvenile trees (<50 years old) at each leaf age level except for 1-year-old leaves. Based on the relative importance analysis, we identified that leaf age compared to tree age plays a dominant role in variation in leaf $\delta^{13}C$ and $\delta^{15}N$ values. Leaf nutrients such as leaf nitrogen concentrations are also important determinant factors for leaf $\delta^{13}C$ and $\delta^{15}N$. However, our knowledge on the mechanism and effects of these biotic and abiotic factors on leaf $\delta^{13}C$ and $\delta^{15}N$ values at large scales are still limited. Further investigation is necessary to consider combinations of different drivers and their relative importance on the $\delta^{13}C$ and $\delta^{15}N$ values.

Supplementary Materials: The following are available online at http://www.mdpi.com/1999-4907/10/4/310/s1, Figure S1: Differences in leaf N concentrations (A) and leaf P concentrations (B) between four years' leaves (from current-year-old and up to 3-year-old) collected along different tree ages levels. Different uppercase letters represent significant differences among leaf ages at each tree age level, while different lowercase letters represent significant differences among tree ages at each leaf age level.

Author Contributions: Writing—original draft preparation, writing—review and editing, data curation, formal analysis and validation, C.L.; methodology, software, visualization and supervision, B.W.; conceptualization, project administration and funding acquisition, T.C.; investigation, resources and visualization, G.X.; investigation and software, M.W.; resources, G.W.; investigation, J.W.

Funding: This research was funded by National Natural Science Foundation of China, grant numbers 31670475, 41421061.

Acknowledgments: We appreciate three anonymous reviewers and editors for their helpful comments to improve the manuscript. We thank Gaosen Zhang for helping us with leaves sampling. We also thank American Journal Experts help us to improve the language.

Conflicts of Interest: The authors declare no conflict of interest.

References

1. Dawson, T.E.; Mambelli, S.; Plamboeck, A.H.; Templer, P.H.; Tu, K.P. Stable isotopes in plant ecology. *Ann. Rev. Ecol. Syst.* **2002**, *33*, 507–559. [CrossRef]
2. Gatica, M.G.; Aranibar, J.N.; Pucheta, E. Environmental and species-specific controls on $\delta^{13}C$ and $\delta^{15}N$ in dominant woody plants from central-western Argentinian drylands. *Austral Ecology* **2017**, *42*, 533–543. [CrossRef]
3. Voronin, P.Y.; Mukhin, V.A.; Velivetskaya, T.A.; Ignat'ev, A.V.; Kuznetsov, V.V. Isotope composition of carbon and nitrogen in tissues and organs of *Betula pendula*. *Russ. J. Plant Physiol.* **2017**, *64*, 184–189. [CrossRef]
4. Farquhar, G.D.; And, J.R.E.; Hubick, K.T. Carbon isotope discrimination and photosynthesis. *Ann. Rev. Plant Physiol. Plant Mol. Biol.* **1989**, *40*, 503–537. [CrossRef]
5. Ma, F.; Liang, W.Y.; Zhou, Z.N.; Xiao, G.J.; Liu, J.L.; He, J.; Jiao, B.Z.; Xu, T.T. Spatial variation in leaf stable carbon isotope composition of three Caragana species in Northern China. *Forests* **2018**, *9*, 297. [CrossRef]
6. Krishnan, P.; Black, T.A.; Jassal, R.S.; Chen, B.; Nesic, Z. Interannual variability of the carbon balance of three different-aged Douglas-fir stands in the Pacific Northwest. *J. Geophys. Res. Biogeosci.* **2015**, *114*, 355. [CrossRef]

7. Acosta-Rangel, A.; Avila-Lovera, E.; Guzman, M.E.D.; Torres, L.; Haro, R.; Arpaia, M.L.; Focht, E.; Santiago, L.S. Evaluation of leaf carbon isotopes and functional traits in avocado reveals water-use efficient cultivars. *Agric. Ecosyst. Environ.* **2018**, *263*, 60–66. [CrossRef]

8. Aguilar-Romero, R.; Pineda-Garcia, F.; Paz, H.; Gonzalez-Rodriguez, A.; Oyama, K. Differentiation in the water-use strategies among oak species from central Mexico. *Tree Physiol.* **2017**, *37*, 915–925. [CrossRef]

9. Zhang, C.Z.; Zhang, J.B.; Zhang, H.; Zhao, J.H.; Wu, Q.C.; Zhao, Z.H.; Cai, T.Y. Mechanisms for the relationships between water-use efficiency and carbon isotope composition and specific leaf area of maize (*Zea mays*, L.) under water stress. *Plant Growth Regul.* **2015**, *77*, 233–243. [CrossRef]

10. Li, M.C.; Zhu, J.J.; Zhang, M.; Song, L.N. Foliar δ^{15}N variations with stand ages in temperate secondary forest ecosystems, Northeast China. *Scand. J. For. Res.* **2013**, *28*, 428–435. [CrossRef]

11. Craine, J.M.; Elmore, A.J.; Aidar, M.P.; Bustamante, M.; Dawson, T.E.; Hobbie, E.A.; Kahmen, A.; Mack, M.C.; McLauchlan, K.K.; Michelsen, A.; et al. Global patterns of foliar nitrogen isotopes and their relationships with climate, mycorrhizal fungi, foliar nutrient concentrations, and nitrogen availability. *New Phytol.* **2010**, *183*, 980–992. [CrossRef] [PubMed]

12. Evans, R.D. Physiological mechanisms influencing plant nitrogen isotope composition. *Trends Plant Sci.* **2001**, *6*, 121–126. [CrossRef]

13. Werth, M.; Mehltreter, K.; Briones, O.; Kazda, M. Stable carbon and nitrogen isotope compositions change with leaf age in two mangrove ferns. *Flora* **2015**, *210*, 80–86. [CrossRef]

14. Ghashghaie, J.; Badeck, F.W. Opposite carbon isotope discrimination during dark respiration in leaves versus roots—A review. *New Phytol.* **2014**, *201*, 751–769. [CrossRef] [PubMed]

15. Eley, Y.; Dawson, L.; Pedentchouk, N. Investigating the carbon isotope composition and leaf wax n-alkane concentration of C$_3$, and C$_4$, plants in Stiffkey saltmarsh, Norfolk, UK. *Org. Geochem.* **2016**, *96*, 28–42. [CrossRef]

16. Vitoria, A.P.; Vieira, T.D.O.; Camargo, P.D.B.; Santiago, L.S. Using leaf δ^{13}C and photosynthetic parameters to understand acclimation to irradiance and leaf age effects during tropical forest regeneration. *For. Ecol. Manag.* **2016**, *379*, 50–60. [CrossRef]

17. Vitoria, A.P.; Avila-Lovera, E.; Tatiane, D.O.V.; Do Couto-Santos, A.P.L.; Pereira, T.J.; Funch, L.S.; Freitas, L.; de Miranda, L.D.; Rodrigues, P.J.F.; Rezende, C.E.; et al. Isotopic composition of leaf carbon (δ^{13}C) and nitrogen (δ^{15}N) of deciduous and evergreen understorey trees in two tropical Brazilian Atlantic forests. *J. Trop. Ecol.* **2018**, *34*, 145–156. [CrossRef]

18. Sun, L.K.; Liu, W.Q.; Liu, G.X.; Chen, T.; Zhang, W.; Wu, X.K.; Zhang, G.S.; Zhang, Y.H.; Li, L.; Zhang, B.G.; et al. Temporal and spatial variations in the stable carbon isotope composition and carbon and nitrogen contents in current-season twigs of *Tamarix chinensis* Lour. and their relationships to environmental factors in the Laizhou Bay wetland in China. *Ecol. Eng.* **2016**, *90*, 417–426. [CrossRef]

19. Yang, Y.; Siegwolf, R.T.W.; Koerner, C. Species specific and environment induced variation of δ^{13}C and δ^{15}N in Alpine plants. *Front. Plant Sci.* **2015**, *6*, 423. [CrossRef]

20. Zhou, Y.C.; Cheng, X.L.; Fan, J.W.; Harris, W. Relationships between foliar carbon isotope composition and elements of C$_3$ species in grasslands of Inner Mongolia, China. *Plant Ecol.* **2016**, *217*, 883–897. [CrossRef]

21. Li, S.J.; Zhang, Y.F.; Chen, T. Relationships between foliar stable carbon isotope composition and environmental factors and leaf element contents of *Pinus tabulaeformis* in northwestern China. *Chin. J. Plant Ecol.* **2011**, *35*, 596–604. [CrossRef]

22. Wang, B.; Chen, T.; Wu, G.J.; Xu, G.B.; Zhang, Y.F.; Gao, H.N.; Zhang, Y.; Feng, Q. Qinghai spruce (*Picea crassifolia*) growth–climate response between lower and upper elevation gradient limits: A case study along a consistent slope in the mid-Qilian Mountains region. *Environ. Earth Sci.* **2016**, *75*, 236. [CrossRef]

23. Coplen, T.B. Guidelines and recommended terms for expression of stable-isotope-ratio and gas-ratio measurement results. *Rapid Commun. Mass Spectrom.* **2011**, *25*, 2538–2560. [CrossRef] [PubMed]

24. Braun, M.T.; Oswald, F.L. Exploratory regression analysis: A tool for selecting models and determining predictor importance. *Behav. Res. Methods* **2011**, *43*, 331–339. [CrossRef] [PubMed]

25. Aljandali, A. *Multivariate Methods and Forecasting with IBM®SPSS®Statistics*; Springer: Berlin/Heidelberg, Germany, 2017.

26. Grunsky, E.C. R: A data analysis and statistical programming environment—An emerging tool for the geosciences. *Comput. Geosci.* **2002**, *28*, 1219–1222. [CrossRef]

27. Pei, H.J. Spatial and Temporal Characteristics of Carbon and Nitrogen and Related Mechanism of *Sabina przewalskii.* and *Picea crassifolia* Kom. Ph.D. Thesis, Graduate University of Chinese Academy of Sciences, Beijing, China, 2012.

28. Yun, H.B.; Chen, T.; Liu, X.; Zhang, Y.F.; Xu, G.B.; Gong, D. Relationship between foliar stable carbon isotope composition and physiological factors in *Picea crassifolia* in the Qilian Mountains. *J. Glaciol. Geocryol.* **2010**, *32*, 151–156.

29. Farquhar, G.D.; O'Leary, M.H.; Berry, J.A. On the Relationship between Carbon Isotope Discrimination and the Intercellular Carbon Dioxide Concentration in Leaves. *Aust. J. Plant Physiol.* **1982**, *9*, 281–292. [CrossRef]

30. Cao, S.K.; Feng, Q.; Su, Y.H.; Chang, Z.Q.; Xi, H.Y. Research on the water use efficiency and foliar nutrient status of *Populus euphratica* and *Tamarix ramosissima* in the extreme arid region of China. *Environ. Earth Sci.* **2011**, *62*, 1597–1607. [CrossRef]

31. Terwilliger, V.J.; Kitajima, K.; Le Roux-Swarthout, D.J.; Mulkey, S.; Wright, S.J. Intrinsic water-use efficiency and heterotrophic investment in tropical leaf growth of two neotropical pioneer tree species as estimated from ^{13}C values. *New Phytol.* **2001**, *152*, 267–281. [CrossRef]

32. Cernusak, L.A.; Turner, W.B.L. Leaf nitrogen to phosphorus ratios of tropical trees: Experimental assessment of physiological and environmental controls. *New Phytol.* **2010**, *185*, 770–779. [CrossRef]

33. Ameziane, R.E.; Deleens, E.; Noctor, G.; Morot-Gaudry, J.F.; Limami, M.A. Stage of development is an important determinant in the effect of nitrate on photo-assimilate (^{13}C) partitioning in chicory (Cichorium intybus). *J. Exp. Bot.* **1997**, *48*, 25–33. [CrossRef]

34. Kalcsits, L.A.; Buschhaus, H.A.; Guy, R.D. Nitrogen isotope discrimination as an integrated measure of nitrogen fluxes, assimilation and allocation in plants. *Physiol. Plant.* **2014**, *151*, 293–304. [CrossRef]

35. Tcherkez, G. Natural ^{15}N/^{14}N isotope composition in C$_3$ leaves: Are enzymatic isotope effects informative for predicting the ^{15}N-abundance in key metabolites? *Funct. Plant Biol.* **2010**, *38*, 1–12. [CrossRef]

36. Masclaux-Daubresse, C.; Danielvedele, F.; Dechorgnat, J.; Chardon, F.; Gaufichon, L.; Suzuki, A. Nitrogen uptake, assimilation and remobilization in plants: Challenges for sustainable and productive agriculture. *Ann. Bot.* **2010**, *105*, 1141–1157. [CrossRef]

37. Donovan, L.A.; Ehleringer, J.R. Ecophysiological Differences among Juvenile and Reproductive Plants of Several Woody Species. *Oecologia* **1991**, *86*, 594–597. [CrossRef]

38. Zhou, Y.; Fan, J.; Harris, W.; Zhong, H.; Zhang, W.; Cheng, X. Relationships between C$_3$ plant foliar carbon isotope composition and element contents of grassland species at high altitudes on the Qinghai-Tibet Plateau, China. *PLoS ONE* **2013**, *8*, e60794. [CrossRef]

39. Li, C.Y.; Wu, C.C.; Duan, B.L.; Korpelainen, H.; Luukkanen, O. Age-related nutrient content and carbon isotope composition in the leaves and branches of *Quercus aquifolioides* along an altitudinal gradient. *Trees* **2009**, *23*, 1109–1121. [CrossRef]

40. Cheng, W.X.; Chen, Q.S.; Xu, Y.Q.; Han, X.G.; Li, L.H. Climate and ecosystem ^{15}N natural abundance along a transect of Inner Mongolian grasslands: Contrasting regional patterns and global patterns. *Glob. Biogeochem. Cycles* **2009**, *23*, GB2005. [CrossRef]

41. Nardoto, G.B.; Pierre, J.; Balbaud, H.; Ehleringer, J.R.; Higuchi, N.; Maria, M.; Martinell, L.A. Understanding the influences of spatial patterns on N availability within the Brazilian Amazon forest. *Ecosystem* **2008**, *11*, 1234–1246. [CrossRef]

Article

Reasons for the Extremely Small Population of putative hybrid *Sonneratia × hainanensis* W.C. Ko (Lythraceae)

Mengwen Zhang [1], Xiaobo Yang [2,*], Wenxing Long [3], Donghai Li [2] and Xiaobo Lv [1]

[1] College of Life and Pharmaceutical Sciences, Hainan University, Haikou 570228, China; zaizai_qq@163.com (M.Z.); lvxbo1980@sina.cn (X.L.)
[2] College of Ecology and Environment, Hainan University, Haikou 570228, China; dhlye@163.com
[3] College of Forestry, Hainan University, Haikou 570228, China; oklong@hainanu.edu.cn
* Correspondence: yanfengxb@hainu.edu.cn

Received: 13 April 2019; Accepted: 22 June 2019; Published: 25 June 2019

Abstract: *Sonneratia × hainanensis*, a species once endemic to Hainan Island in China, is now endangered. China's State Forestry Administration lists this species as a wild plant species with an extremely small population. Field fixed-point investigations, artificial pollination, and laboratory experiments, as well as other methods, were applied to study the reproductive system and seed germination of *S. × hainanensis* to elucidate the reasons for the endangerment of this species. The results are as follows: (1) Outcrossing index, pollen-ovule ratio, and artificial pollination showed *S. × hainanensis* has a mixed mating system and mainly focuses on outcrossing with some self-compatibility. (2) Fruit and seed placement tests showed that the fruit predators on the ground were mainly Fiddler crab and squirrel, with the predation rates being 100%. The artificially spread seeds do not germinate under natural conditions. The mean seed destruction rate and remaining rate of were 82.5% and 17.5%. (3) Seeds need to germinate under ambient light conditions, with an optimal photoperiod of 12 h. Seed germination is extremely sensitive to low temperatures because of optimum temperatures from 30 °C to 40 °C. At an optimal temperature of 35 °C, the seeds germinate under salinities ranging from 0‰ to 7.5‰, with an optimal salinity of 2.5‰, which shows the sensitivity of seed germination to salinity, with low salinity promoting germination, whereas high salinity inhibits germination. These findings indicate that the limited regeneration of *S. × hainanensis* is caused by the following: (1) Pollen limitation and inbreeding recession caused by the extremely small population of *S. × hainanensis*. (2) Seeds near parent trees are susceptible not only to high fruit drop rate, but to high predation beneath the parent trees' canopy as well. (3) Seed germination has weak adaptability to light, temperature, and salinity.

Keywords: endangered; *Sonneratia × hainanensis*; reproductive system; seed germination; light; temperature; salinity

---

1. Introduction

Research on reproductive characteristics and reproductive dynamics effectively identifies the course that endangers the plant population and the underlying mechanism [1]. Moreover, it clarifies the reproductive biological characteristics and the effects of external factors on the most critical aspect of plant reproduction, as well as the mechanism that endangers plant species [2].

At the stage of pollination, research on the reproductive system of plants is helpful for understanding their life history and causes of endangerment [3]. Recent studies have shown that corolla shape, diameter, color, odor, and other factors affect pollinator variety, visiting frequency, and visual and olfactory reactions, which reduces the possibility of hybridization and adaptability of future generations. This effect leads to a decline in the number of species [4,5]. Pollen viability and stigma receptivity differ among plant species [6]. Highly dynamic pollen and stigma are conducive to pollination [7]. Conversely, presentation of pollen viability and stigma receptivity at different times influences pollination efficiency and limits the fruiting rate [8]. Moreover, the decline in pollination quality also causes various negative effects, including reproductive failure because of pollen restrictions, decline in genetic diversity, and depression from inbreeding [9]. Studies on plant species based on different population sizes and genetic variation demonstrate that inbreeding significantly affects seed rate, seed germination, survival, and resistance to stress [10].

Seeds are an important part of the life history of endangered plants [11]. Ecological research on the seeds of endangered plants reveals better approaches to breeding seeds and seedlings, which are of great significance for population expansion and preventing population extinction [12]. Some endangered plants have less natural production and low seed quality. This phenomenon fundamentally reduces the seed germination rate and restricts population growth [13]. Other endangered plants have special fruit morphologies or structures that are not conducive to the spread of seeds or germination, which also endangers these plants [14]. The seeds of some endangered plants have distinctive features, such as congenital dormancy or substances that inhibit germination, which leads to low sprouting rates and the scarcity of seedlings under natural forests [15]. Climate change and inefficient population regeneration in modern habitats can also be attributed to the limited expansion of endangered seeds and seedlings [16]. Some studies on *Phoebe bournei* (Hemsley) Yen C. Yang J.W. and *Abies chensiensis* Tieghem showed that animal damage to seeds and seedlings contribute to their scarcity in natural populations [17]. Moreover, the short life of seeds of endangered plants, such as *Cinnamomum micranthum* (Hay.) Hay and *Cathaya argyrophylla* Chun & Kuang, also restricts their expansion [18].

Sonneratia × *hainanensis* W.C. Ko, E.Y. Chen and W.Y. Chen is an aiphyllium from the genus *Sonneratia* (*Sonneratiaceae*) [19]. This species grows on beaches near the low water line or on low tide beaches along inland rivers with low tidewater salinity [20]. The adult tree of *Sonneratia* × *hainanensis* is only eight and they are currently only found at Dongge and Touyuan in the Qinglan protection zone in Wenchang City, Hainan Province [21]. China's State Forestry Administration lists this species as a wild plant species with an extremely small population [22]. Studies on *S.* × *hainanensis* show that it has an outcross reproductive system with a long flowering phase [23]. It has flowers that can bloom almost year-round, although some *Sonneratia* species have overlapping flowering phases [24]. The outcross reproductive system, interspecific crossing sympatry, and overlapping flowering phases of *Sonneratia* provide chances for interspecific hybridization [25]. Among these species, *S.* × *hainanensis* is a diploid (22 chromosomes) hybrid, possibly between *Sonneratia alba* Smith and *Sonneratia ovata* Backer [26]. The *S.* × *hainanensis* population has a low genetic diversity, but various groups have significant genetic differentiation [27]. This phenomenon is possibly closely related to its reproductive system. However, few studies have investigated the reasons for restricted *S.* × *hainanensis* population in terms of the effects of reproductive ecology and environment on seed germination.

We conducted field fixed-point observations and laboratory experiments to study the reproductive system and seed germination of *S.* × *hainanensis*, as well as their relationship with the environment to determine the reasons for its limited natural regeneration.

2. Materials and Methods

2.1. An Overview of the Sample Plot

The sample plot was located at Qinglangang Nature Reserve of Hainan Island. *Sonneratia* plant communities in the region grow on the beach near the low water line or low tide beach along the inland river with low tidewater salinity. The area is located at 19°37′36″ N and 110°49′53″ E. Mangrove plants present abundant varieties in these communities, including 24 species of true mangroves and 20 species of semi-mangroves. *Sonneratia caseolaris* (Linn.) Engler and *Bruguiera sexangula* (Loureiro) Poiret are the dominant mangrove species. In China, *Sonneratia* includes six species, namely, *S. caseolaris*, *S. alba*,

Sonneratia apetala Buchanan-Hamilton, *S. ovata*, *Sonneratia* × *gulngai* N. C. Duke & B.R. Jackes, and *S.* × *hainanensis*. *S. apetala* is an introduced exotic plant with a developed population, whereas, *S.* × *hainanensis* is a seriously endangered plant species. The geographical location is shown in Figure 1.

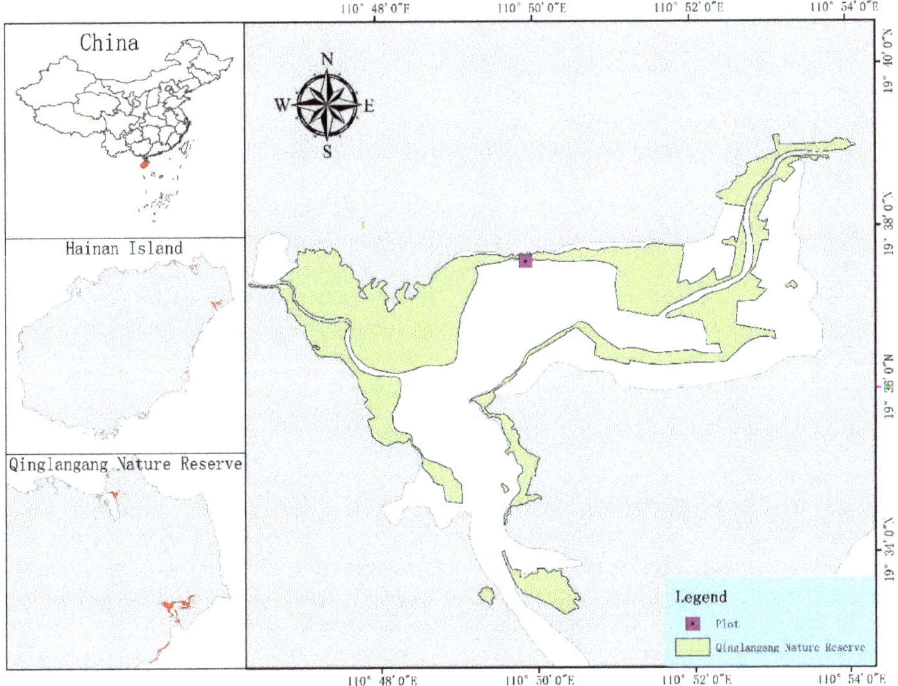

Figure 1. Geographic distribution of the fixed plot.

2.2. Study Object

We selected 8 trees of different *S.* × *hainanensis* strains from the bottomland of an enclosed sea harbor in Paigang Village, Dongge Town, Wenchang City, Hainan Province for the study. All trees were at least 100 years old, with strong trunks and branches. They were about 8 m high, with a diameter at breast height of 100 cm, and a crown breadth of 15 m × 15 m. The fruit type of *S.* × *hainanensis* was spherical berry with a diameter of 50–80 mm. The seeds were sickle-shaped, V-shaped, or irregular, with an average length of 10 mm and the outer seed coat was brown. There were 2 cotyledons, elliptical or oblong, 2–3 cm long, and 1–3cm wide.

2.3. Determination of the Reproductive System

2.3.1. Estimating the Outcrossing Index (OCI)

Inflorescence diameter, flower size, and flowering behavior were measured and the assay of the reproductive system was specifically conducted according to the standards proposed by Dafni [28].

2.3.2. Pollen to Ovule (P/O) ratio

Ten buds just coming into bloom with undehisced anthers were randomly selected and fixed in formalin-acetic acid-alcohol (FAA). The number of single pollen grains was counted according to the method by Cruden, and the number of ovules was measured by paraffin section. The pollen to ovule

(P/O) ratio of each flower was calculated by dividing the number of pollen grains by the number of ovules [29].

2.4. Determination of the Reproductive System

2.4.1. Changes in Stigma Receptivity and Pollen Vitality with Time

A total of 50 flowers were selected at the start and end of the period of high stigma receptivity. All stamens were carefully removed using tweezers before the blooming and isolated in a bag to prevent self-pollination. Fresh pollen grains were collected from other flowers. Flowers within 1 day to 5 days after pollination were isolated in a bag. Ten replicates were carried out during each flowering period.

A total of 50 flowers were selected at the start and end periods of pollen vitality. All stamens were carefully removed using tweezers before blooming. At the flowering day, the stigmas were pollinated with pollen from other flowers on the day of anther dehiscence and at 1 day to 5 days after dehiscence. Then, the stigmas were subjected to bag isolation. Ten repetitions were carried out for each pollen sample. Pollen germination rate and stigma receptivity were calculated according to the fruit setting rate during the preliminary stage. An ANOVA was used to compare the differences between pollen germination rate and stigma receptivity at different periods.

2.4.2. Artificial Pollination and Bagging Experiments

During the bud stage, one strain and 280 buds of *S.* × *hainanensis* were selected for seven pollination treatments: A, natural control (natural hybrid); B, bagging with castration without pollination (apomixia); C, net isolation with castration (anemophily pollination); D, no bagging; E, no bagging with castration (bagging before flowering and pollen from the same flower was given after flowering); F, artificial geitonogamy (mutual pollination of No. 1 and No. 10, No. 2 and No. 9, and so on for 10 buds); and G, artificial xenogamy (pollen source was from another strain of *S.* × *hainanensis* 100 m away from the experiment strain). The E, F, and G treatments involved artificial pollination, whereas A, B, C, and D were performed for comparison. The fruit dropping time, number of fruits, fruiting rate, and seed setting rate of single fruits were determined.

2.5. Seed Germination and Seedling Survival Experiment

2.5.1. Dynamic Changes in *S.* × *hainanensis* Seeds under Natural Conditions

Three parental trees were randomly selected, and there were no *S.* × *hainanensis* trees within 50 m around the three mother trees. In order to calculate the number of fruits dropped, at the beginning of the fruit drop, every 5 days all the fruits dropped were collected and brought back to the laboratory for germination experiments until fruit dropping ceased.

One small quadrat of 1 m × 1 m was set in four regions of 1 m, 3 m, 5 m, and 10 m away from each parent tree. Thirty mature fruits covered with wire netting (mesh size was 50 mm × 50 mm) were naturally placed in each quadrat. A seed germination experiment was set up, not far from each quadrat. Eighty seeds were scattered on the ground, and the seeds were covered with a plastic mesh (mesh size was 1 mm × 1 mm). There were 3 replicates around a parental tree. The number of fruit predated by animals and seed germination rate at different distances from the parent trees were counted.

2.5.2. The Relationship between *S.* × *hainanensis* Seed Germination and Environmental Factors

Germination tests were performed indoors. Each test included 50 full seeds with three replicates. The seeds were first disinfected with 0.1% potassium permanganate solution for 5 min, rinsed with distilled water, and then sown into 11 cm Petri dishes padded with filter paper. Distilled water (for light and temperature experiments) and saline (for salinity experiments) were added until the seeds were immersed in solution. Seed germination was observed daily after sowing and the liquid was changed once daily. The germinated seeds were transferred into another dish. After 5 d, 10 strains were selected from each dish to measure the radicle length and perfectness ratio. Germination rate, germination potential, radicle length, and radicle perfectness ratio were calculated using a germination standard of a radicle length of 3 mm and an experimental time of 20 d.

The light–dark cycles were set to 4/20, 8/16, 12/12, 16/8, and 24-hour light/0 hours darkness. The temperature was set to 28 °C during the day and 23 °C during the night. Relative air humidity (75%) and light intensity (700 lx) were constant during treatment, and complete darkness was used as the control. All tests were divided into six treatment groups.

Seeds were sown in culture dishes and placed in an incubator set to different constant temperatures (15, 20, 25, 30, 35, 40, and 45 °C). The glass door was closed to allow the seeds to germinate under natural indoor light.

Ten groups with different salinities (0‰, 2.5‰, 5‰, 7.5‰, 10‰, 12.5‰, 15‰, 20‰, 25‰, and 30‰) were prepared, and saline with the salinity of seawater (18.4‰ to 19.2‰) was used as the control. All tests were divided into 11 groups, and all germination experiments were conducted under indoor natural light in an incubator at a constant 35 °C. The seawater was replaced daily.

2.6. Data Analysis Method

Mean and standard error (SE) of three replicates were calculated. Data on all germination rates, germination viability, radicle lengths, and radicle perfectness ratio were analyzed for differences among different treatments by analysis of variance. If the difference was significant at $p < 0.05$, a Duncan test was employed to determine the potential source of the difference. All statistical analyses were performed with SPSS, version 16.0 (SPSS Inc., Chicago, Illinois, U.S.A.). Statistical significance was defined as $p < 0.05$.

3. Results

3.1. Determination of the Reproductive System

S. × hainanensis has a bell-shaped calyx tube. The top of the flower was about 4.5 cm to 5.5 cm from the base. The flower diameter was measured from the bell-shaped central part, which was approximately 2.4 cm to 3.2 cm. Flowers with diameters greater than 6 mm were still scored 3. In the field experiments, if the anther dehiscence of *S. × hainanensis* flowers presented no time interval from stigma receptivity, it was scored as 0. If the stigma grew faster than stamens and kept ahead of stamens, it was scored as 1. Thus, the OCI of *S. × hainanensis* was 4, which indicates that the sexual reproductive system of *S. × hainanensis* is partially self-compatible, outcrossing, and needs pollinators.

Table 1 shows the P/O ratio values of *S. × hainanensis*. According to Cruden, P/O ratio values between 244.7 and 2588.0 indicate facultative outcrossing. Thus, the sexual reproduction system of *S. × hainanensis* is dominated by facultative outcrossing (Table 1).

Table 1. Pollen to ovule (P/O) ratio of *Sonneratia × hainanensis*.

Items of Observation	Number of Pollen in Each Simple Flower (P)	Number of Ovules in Each Simple Flower (O)	P/O Ratio	Types of Breeding System
Results	≈12500 ± 2000	34 ± 10	354	Xenogamy

3.2. Exploring Dysgenesis in S. × hainanensis

3.2.1. Pollen Germination Rate and Stigma Receptivity

Under natural conditions, the pollen germination rate and stigma receptivity of single *S. × hainanensis* flowers is usually consistent with time (Figure 2). The pollen germination rate was highest (100% or close to 100%) on the first and second day of flowering. The pollen germination rate declined greatly by the third day and was lowest (20%) by the fifth day. Stigma receptivity peaked on the day of flowering, and 100% of the pollinated flowers bore fruits. On the fifth day of flowering, the entire style withered and stigma receptivity (5%) was almost lost.

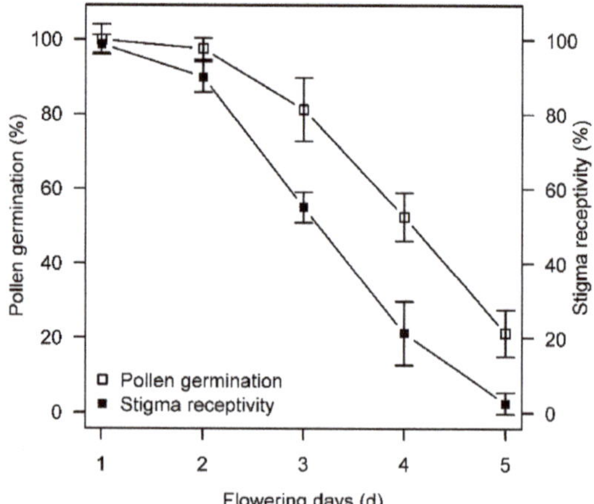

Figure 2. Variations in pollen germination rate and stigma receptivity with flower age in *S.* × *hainanensis*.

3.2.2. Artificial Pollination Experiment

After bagging with castration (apomixia), the *S.* × *hainanensis* ovaries grew slightly larger around the 10th day, and then stopped changing. After one month, the ovaries withered and dropped. The fruit setting rate was 0 (Table 2). Under natural conditions, the fruiting rate of *S.* × *hainanensis* was 45%. The fruiting rate without bagging with castration was very low (17.5%). The fruiting rate (10%) under net isolation with castration was lower than that under natural pollination. However, under artificial pollination, the fruiting rate in all treatment conditions (42.5% to 62.5%) increased remarkably.

Table 2. Fruit setting rate and seed setting number under different pollination methods.

Pollination Method	Number of Flowers	Fruit Dropping Time (d)	Number of Fruits	Fruiting Rate (%)	Seed Setting Number in a Single Flower
Net isolation with castration	40	7	4	10	16.25
Bagging with castration	40	0	0	0	0
No bagging treatment	40	7	7	17.5	19
Natural pollination	40	7	18	45	30.4
Artificial self-pollination	40	15	17	42.5	24.2
Artificial geitonogamy	40	15	20	50	34.6
Artificial xenogamy	40	18	25	62.5	56.2

3.3. Relationships among Seed Germination, Survival of Seedlings, and Environmental Factors

3.3.1. Seed Germination Rate at Different Fruit Dropping Times

The artificial control experiment for the seed germination rate of dropped fruits at different times indicated that all indices significantly increase with increasing fruit dropping times (Table 3). At the fruit dropping time of 5–10 d, the seed germination rate was 0, and the seed germination rate was less than 10% at 20–25 d. When the fruit dropping time was 30 d, the number of fruit falling (84.67%) and the seed germination rate (46.55%) reached the maximum. When the fruit drop time was 35 d, the fruit was no longer dropped. A chi square test showed that the seed germination rate was highly correlated with fruit dropping time ($\chi^2 = 63.29$, $p < 0.01$).This indicates that under natural conditions, the early dropping of *S.* × *hainanensis* fruits leads to the incapacity of immature seeds to germinate, which could result in source limitation of seed.

Table 3. Fruit dropping amount and seed germination rate of *S.* × *hainanensis* at different fruit dropping times.

Dropping Time	5 d	10 d	15 d	20 d	25 d	30 d	35d
Fruit dropping amount	14.67 ± 4.51f	25.67 ± 4.04e	39.33 ± 9.24d	58.00 ± 8.00c	70.33 ± 4.51ab	84.67 ± 6.43a	0 ± 0g
Seed germination rate (%)	0 ± 0d	0 ± 0d	2 ± 0bc	9 ± 0b	43 ± 3a	45 ± 3a	0 ± 0d

Different letters indicate significant difference ($p < 0.05$).

3.3.2. Dynamic Observation and Verification of Seeds under Natural Conditions

It can be seen from Table 4 that under natural conditions, the fruit predation rate and seed germination rate at different distances are 100% and 0. The seed destruction rate (by animals and disease) and seed remaining rate were about 82.5% and 17.5%, respectively. Through field observations, it was found that crabs, squirrels, etc., were the main predators. In addition, the remaining seeds could not germinate. This indicates that the seed germination of *S.* × *hainanensis* under natural conditions is not only affected by animal predation, but also by habitat factors.

Table 4. Dynamic changes of fruit and seeds of *S.* × *hainanensis* under natural conditions.

Sampling Point	Fruit Predation Rate (%)	Seed Germination Rate (%)	Seed Damage Rate (%)	Seed Remaining Rate (%)
1 m	100	0	83 ± 10	17 ± 10
3 m	100	0	85 ± 0	15 ± 0
5 m	100	0	80 ± 6	20 ± 6
10 m	100	0	82 ± 13	18 ± 13
Average	100	0	82.5 ± 2.08	17.5 ± 2.08

3.3.3. Effects of Light Cycle Duration on *S.* × *hainanensis* Seed Sprouting

The seeds in the dark control group did not germinate. The germination rate, germination potential, radicle length, and radicle perfectness ratio were zero. Germination rate (Figure 3a), germination potential (Figure 3b), and radicle length (Figure 3c) demonstrated unimodal curves with increasing light cycle duration. Germination rate, germination potential, and radicle length differed under different light cycle durations (Figure 3). The germination rate (68.67%) and germination potential (58.67%) peaked at 12 h. The radicle length and radicle perfectness ratio did not statistically differ among different groups, except for the control group and the group at 4 h (Figure 3d).

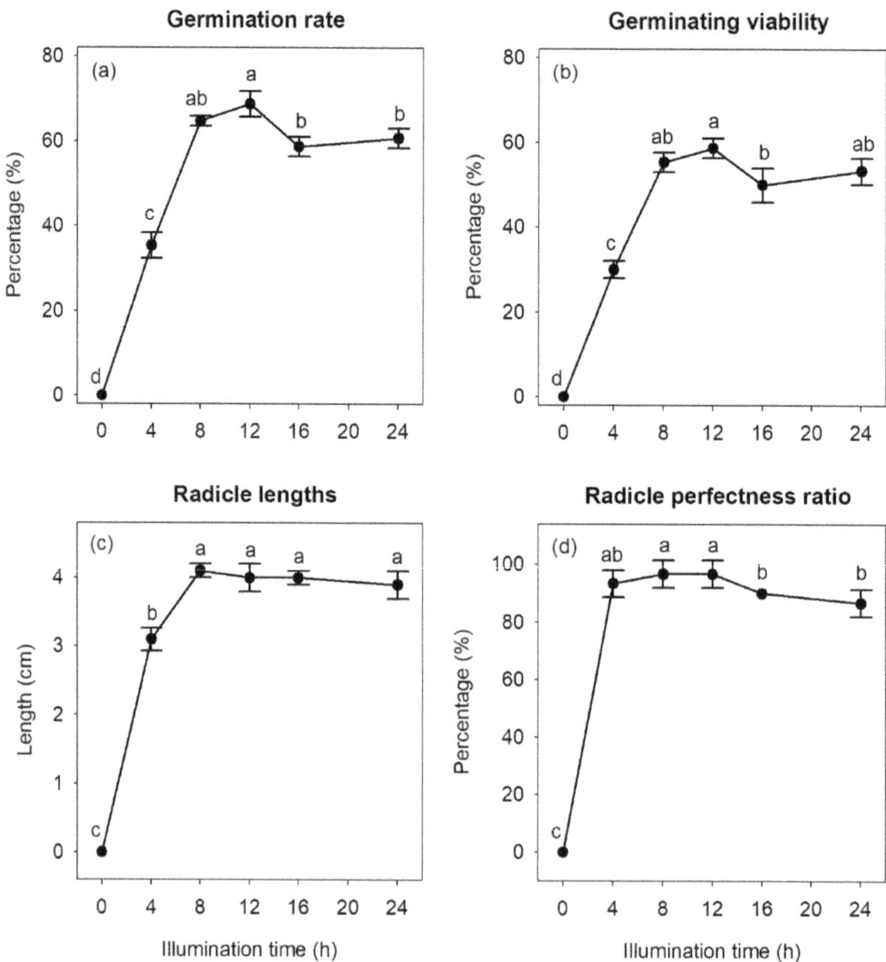

Figure 3. Effects of different illumination times on indicators. Different letters indicate significant difference ($p < 0.05$).

3.3.4. Effect of Temperature on *S.* × *hainanensis* Sprouting

The *S.* × *hainanensis* sprouting rate, germination potential, and radicle length exhibited remarkable differences under different temperatures (Figure 4). As the temperature increased, the germination rate, germination potential, and radicle length of all groups showed unimodal curve changes (Figure 4a–c). The germination rate and germination potential increased from 30 °C to 35 °C. The germination rate (90.67%) and the germination potential (78.00%) peaked at 35 °C. The germination potential reached its minimum at 15 °C and the germination rate reached its minimum at 45 °C. As the temperature increased, the radicle length and the radicle perfectness ratio also presented unimodal curve changes (as shown in Figure 4c,d). The radicle length (4.83 cm) and the radicle perfectness ratio (100%) peaked at 35 °C.

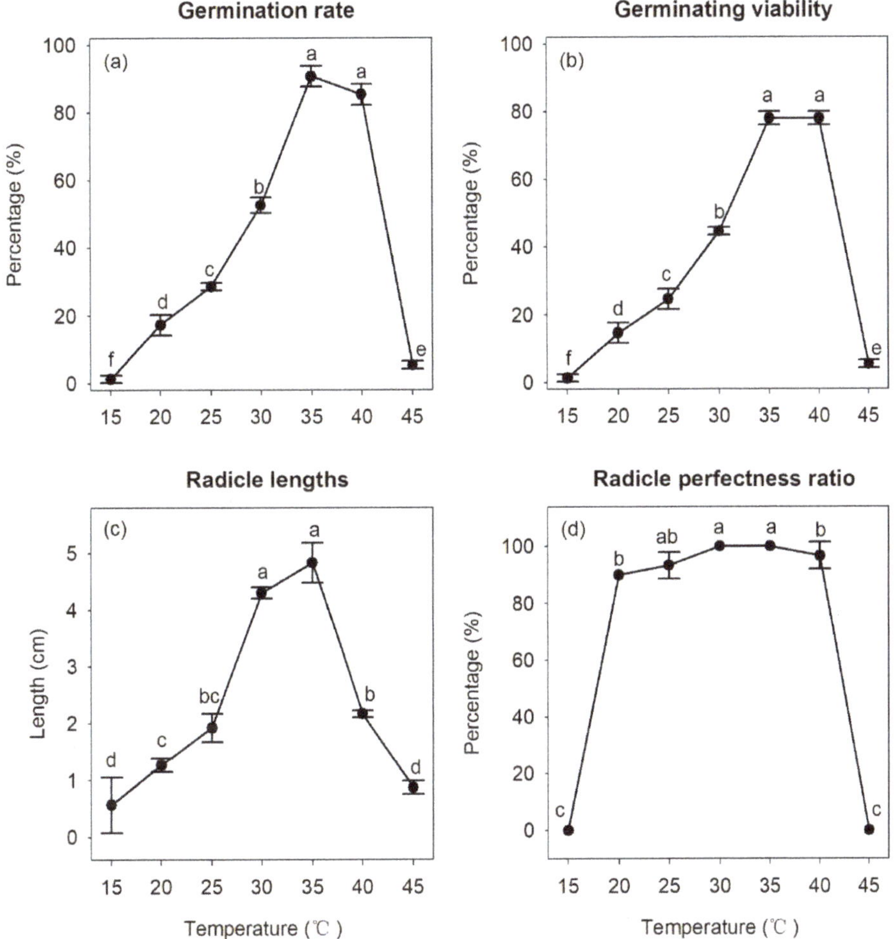

Figure 4. Effects of different temperatures on indicators. Different letters indicate significant difference ($p < 0.05$).

3.3.5. Effect of Salinity on *S.* × *hainanensis* Seed Germination

As salinity increased, the germination rate, germination potential, radicle length, and intact rate of each group significantly decreased (Figure 5). The germination rate (82.67%) and the germination potential (75.33%) peaked at 0‰ salinity (Figure 5a,b). The difference was not obvious compared to the germination rate (81.33%) and germination potential (72.67%) at 2.5‰ salinity. However, the radicle growth of the latter was more robust and the radicle perfectness ratio was ideal (100%; Figure 5d); radicle length under low salinity (<7.5‰) almost doubled under high salinity (>10‰) (Figure 5c). The appropriate germination salinity of *S.* × *hainanensis* varied from 0‰ to 7‰.

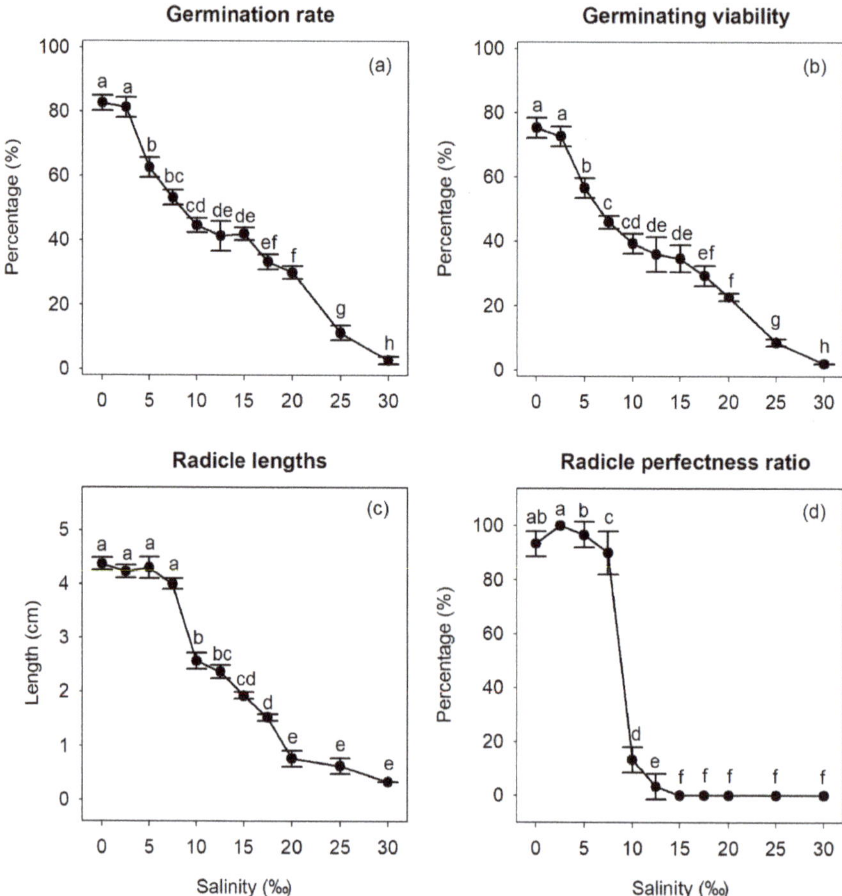

Figure 5. Effects of different salinities on indicators. Different letters indicate significant difference ($p < 0.05$).

4. Discussion

4.1. Reproductive System of S. × hainanensis

The understanding of the plant reproductive system is a prerequisite for understanding the life history of plants, and is also the basic background knowledge required for other related research [30]. Due to the different gender systems of plants, they often exhibit diverse types of reproductive systems [31]. Different gender systems have different effects on their hybridization rate, pollination mechanism, and breeding system [32].

According to the standards by Dafni, further experiments can verify that the sexual reproductive system of *S. × hainanensis* is partially self-compatible, outcrossing, and needs pollinators [28]. Based on the P/O ratio by Cruden, its reproductive system is facultative outcrossing. Cross-pollination, especially xenogamy, has a high maturation rate [29]. The fruiting rate of artificial pollination is higher than the other pollination approaches. Thus, *S. × hainanensis* flowers have a mixed mating system. Selfing may mainly come from geitonogamous pollination. Moreover, the study by Tomlinson showed that geitonogamous plants have an outcross reproductive system [24]. The differences in these studies may be attributed to the narrowed population of *S. × hainanensis*, which led to a higher chance of selfing and inbreeding. During evolution, *S. × hainanensis* gradually changed from obligate outcrossing to facultative outcrossing with self-compatible tendencies. This reproductive system possibly evolved into a stable mixed mating system.

Research shows that small populations weaken the attraction of insects for pollination [33]. As a result, individual plants in small populations only acquire limited pollen with inferior quality compared to large populations, which leads to the pollen restriction on the reproductive success of small populations [33]. According to population genetics, small populations tend to undergo genetic drift, which reduces genetic diversity. This process is often accompanied by the accumulation of harmful mutations and it enhances inbreeding [34]. When populations shrink, the amount of pollen and its quality during pollination are greatly affected, which undermines reproduction [34]. According to research, only eight strains of *S.* × *hainanensis* are naturally distributed, with remarkable distances from each other. Under natural conditions, *S.* × *hainanensis* mainly conducts self-pollination. The artificial pollination experiment shows that artificially improving pollination notably improves the fruit setting rate of *S.* × *hainanensis*. The seed germination rate from the outcrossing population is much higher than from selfing. Therefore, pollen restriction and inbreeding depression can be attributed to the low seed germination rate of small populations, which is consistent with the finding of Li on *Lumnitzera littorea* (Jack) Voigt [35].

4.2. Effects of Natural Environment on S. × hainanensis Seed Germination

In the wild environment, in addition to abiotic factors (light, temperature, salinity), seed germination was also affected by biological factors (animals, diseases) [36]. The study of *Taxus chinensis var. mairi* has showed that only very small portions of the seed rain supplied the seed bank, and most of the falling seed did not replenish the soil seed bank due to predation, human disturbance, and environmental factors [37]. Seeds of *Phoebe bournei* were not only susceptible to soil pathogens, but to high predation beneath parent trees' canopy as well, which resulted in poor field seed germination [38].

In this experiment, the fruit drop of *S.* × *hainanensis* was serious during fruiting stages, and only the seeds of mature stage had higher germination rate (about 40%), and the seed germination rate in other periods was very low (less than 10%), so most of the fruits were wasted. Because the fruit of *S.* × *hainanensis* was sweet, it was welcomed by animals such as squirrels and Fiddler crabs. Most of the seeds were destroyed and could not be germinated. Therefore, seed predation by animals was also one of the reasons that the seed germination of *S.* × *hainanensis* was limited.

4.3. Effects of Simulated Environmental Factors on S. × hainanensis Seed Germination

Seed germination is the weakest phase for plants to resist environmental stress. Furthermore, factors that adversely affect seed germination directly influence the generation and supplementation of new individual plants into the population, as well as its stability [39]. In this research, the major obstacles to the regeneration of *S.* × *hainanensis* populations were closely related to seed germination and indispensable environmental factors that influence germination.

4.3.1. Effect of Light on S. × hainanensis Seed Germination

Light is indispensable for the seed germination of certain plants and different plants have different light requirements for seed germination [40]. Studies have shown that the seeds of the endangered plant, *Garcinia paucinervis* Chun & F.C. How, can germinate in the presence or absence of light, indicating that light is not a necessary condition for its germination [41]. The seed of *Lumnitzera littorea* is a light-requiring seed [42]. Under darkness and constant temperature, the newly collected *S.* × *hainanensis* seeds did not completely germinate, and increased light promoted seed germination and radicle growth. Exposing the seeds to light for 12 h per day was optimal for germination. After the spreading of seeds, the weak air and water permeability of sludge under the forest results in insufficient light acquisition. Thus, most seeds only receive insufficient light, which causes an obvious decline in sprouting rate. Seedlings under the forests are scarce. Actual investigation reveals no *S.* × *hainanensis* seedling under the forest. Hence, the insufficient light in forests limits *S.* × *hainanensis* seed germination, which is consistent with the finding of Liao on *S. caseolaris* [43].

4.3.2. Effect of Temperature on *S.* × *hainanensis* Seed Germination

Temperature is one of the key factors during seed germination. However, the response of different endangered plant seeds to temperature is also different [44]. The endangered plant *Cercidiphyllum japonicum* Siebold & Zuccarini has a low seed germination rate below 5 °C, and there is no significant difference between 10 °C and 30 °C [40]. The seed germination of most endangered plants, such as *Dracaena cambodiana* Pierre ex Gagnepain, *Sinia rhodoleuca* (Diels) M.C.E. Amaral and *Lumnitzera littorea*, are very sensitive to temperature, and 25 °C is optimal. At the same time, seeds cannot germination when the temperature is below 15 °C, and if it is higher than 30 °C [45–47]. Under laboratory conditions, the suitable temperature for *S.* × *hainanensis* seed germination was between 30 °C and 40 °C, with an optimum temperature of 35 °C. This condition indicates that *S.* × *hainanensis* seeds are sensitive to low temperatures, but are able to resist temperature changes within a small range. In 2008, because of extreme cold in southern China, all seedlings of *S.* × *hainanensis* were dead in the mangrove forest nature protection area of Dongzhaigang [48]. Therefore, temperature is also one of the factors limiting the seed germination of *S.* × *hainanensis*.

4.3.3. Effect of Salinity on *S.* × *hainanensis* Seed Germination

The optimum salinity for *S.* × *hainanensis* seed was between 0‰ and 7.5‰. Within this range, the germination potential, germination rate, radicle length, and radicle perfectness ratio, which reflect the sprouting ability, are better than those under other salinities. Further increases in salinity caused a drop in germination potential, germination rate, radicle length, and radicle perfectness ratio, which demonstrates that low salinity is conducive to *S.* × *hainanensis* seed germination, whereas high salinity inhibits germination. This result is consistent with findings that most halophytic plants have higher germination rates at low salinities [49–51].

Generally, seed germination at low salinity only slightly differs from that at 0‰ salinity. Increasing salinity gradually inhibits germination, but the seeds of most plants regain their vigor after being transferred into freshwater, which increases their accumulated germination rate [52]. For example, placing seeds that cannot germinate under high salinity into fresh water partially restores their germination rate, which shows that high salinity triggers dormancy in *S.* × *hainanensis* seeds. Exposing the seeds to optimal conditions allows them to germinate. This may be an important adaptation mechanism for *S.* × *hainanensis* in saline environments. Studies in China on the seed germination and seedling growth of other mangrove trees such as *S. apetala* under saline stress revealed that seed germination is better under low salinity than under high salinity [53]. This observation is partially consistent with the findings in this experiment. The *S.* × *hainanensis* seeds exhibited the highest germination index under 0‰ salinity and the low salinity value (2.5‰). However, radicle growth was more robust and the radicle perfectness ratio was higher under low salinity. This observation implies that *S.* × *hainanensis* seed germination requires stimulation under a certain salinity to allow the radicle to grow more robustly, more adaptable to the saline environment, and to improve the survival rate of seedlings. In addition, the seawater salinity control demonstrates that salinity limits the seed germination of *S.* × *hainanensis*.

5. Conclusions

Ecological analysis of the reproductive system, seeds, and seedlings indicated that the following factors limit the regeneration of *S.* × *hainanensis*: (1) Pollen limitation and inbreeding recession caused by the extremely small population of *S.* × *hainanensis*. (2) Seeds near parent trees are susceptible not only to high fruit drop rate, but to high predation beneath parent trees' canopy as well. (3) Seed germination has weak adaptability to light, temperature, and salinity.

To protect *S. × hainanensis*, we recommend implementing artificial xenogamy during the flowering stage to improve pollination efficiency, the fruit setting rate, and the fruiting, as well as building a small simulation greenhouse and collecting seeds in time. The populations can be expanded through indoor seedling breeding and domestication to factors such as salinity.

Author Contributions: Conceptualization, M.Z. and X.Y.; data curation, M.Z. and X.L.; funding acquisition, D.L.; investigation, M.Z. and X.L.; methodology, M.Z.; project administration, X.Y.; resources, W.L.; writing—original draft, M.Z.; writing – review and editing, M.Z. and X.Y.

Funding: This work was financially supported by the China National key R & D Program during the 13th Five-Year Plan Period (2016YFC0503100).

Acknowledgments: We thank LetPub (https://www.letpub.com.cn/) for editing this manuscript.

Conflicts of Interest: The authors declare no conflict of interest.

References

1. Arroyo, J.; Thompson, J.D. Plant reproductive ecology and evolution in a changing Mediterranean climate. *Plant Boil.* **2018**, *20*, 3–7. [CrossRef]
2. Sheng, M.Y.; Shen, C.Z.; Chen, X.; Tian, X.J. Resource status and protection countermeasures of endangered wild plants in China. *Chin. J. Nat.* **2011**, *33*, 455–467.
3. Liu, K.Q.; Deng, H.P. Floral biology and breeding system of endangered plant *Scutellaria tsinyunensis* endemic to Chongqing, China. *Bull. Bot. Res.* **2011**, *31*, 403–407.
4. Sigrist, M.R.; Sazima, M. Phenology, reproductive biology and diversity of buzzing bees of sympatric Dichorisandra species (Commelinaceae): breeding system and performance of pollinators. *Plant Syst. Evol.* **2014**, *301*, 1005–1015. [CrossRef]
5. Tsuji, K.; Ohgushi, T. Florivory indirectly decreases the plant reproductive output through changes in pollinator attraction. *Ecol. Evol.* **2018**, *8*, 2993–3001. [CrossRef]
6. Waelti, M.O.; Muhlemann, J.K.; Widmer, A.; Schiestl, F.P. Floral odour and reproductive isolation in two species of Silene. *J. Revolut. Biol.* **2008**, *21*, 111–121. [CrossRef]
7. Gross, C.L.; Bartier, F.V.; Mulligan, D.R. Floral structure, breeding system and fruit-set in the threatened sub-shrub *Tetratheca juncea* Smith (Tremandraceae). *Ann. Bot.* **2003**, *92*, 771–777. [CrossRef]
8. Hamston, T.J.; Wilson, R.J.; De Vere, N.; Rich, T.C.G.; Stevens, J.R.; Cresswell, J.E. Breeding system and spatial isolation from congeners strongly constrain seed set in an insect–pollinated apomictic tree: Sorbus subcuneata (Rosaceae). *Sci. Rep.* **2017**, *7*, 45122. [CrossRef]
9. Sawyer, N.W. Reproductive Ecology of *Trillium recurvatum* (*Trilliaceae*) in Wisconsin. *Am. Midl. Nat.* **2010**, *163*, 146–160. [CrossRef]
10. Opedal, Ø.H.; Armbruster, W.S.; Pélabon, C. Inbreeding effects in a mixed–mating vine: effects of mating history, pollen competition and stress on the cost of inbreeding. *AoB PLANTS* **2015**, *7*, 133. [CrossRef]
11. Takebayashi, N.; Morrell, P.L. Is self–fertilization an evolutionary dead end? Revisiting an old hypothesis with genetic theories and a macro–evolutionary approach. *Am. J. Bot.* **2001**, *88*, 1143–1150. [CrossRef] [PubMed]
12. Bebawi, F.F.; Campbell, S.D.; Mayer, R.J. Seed ecology of Captain Cook tree [*Cascabela thevetia* (L.) Lippold] – germination and longevity. *Rangel. J.* **2017**, *39*, 307. [CrossRef]
13. Oakley, C.G.; Ågren, J.; Schemske, D.W. Heterosis and outbreeding depression in crosses between natural populations of Arabidopsis thaliana. *Heredity* **2015**, *115*, 73–82. [CrossRef] [PubMed]
14. Prill, N.; Bullock, J.M.; Van Dam, N.M.; Leimu, R. Loss of heterosis and family–dependent inbreeding depression in plant performance and resistance against multiple herbivores under drought stress. *J. Ecol.* **2014**, *102*, 1497–1505. [CrossRef]
15. Jiménez-Alfaro, B.; Silveira, F.A.; Fidelis, A.; Poschlod, P.; Commander, L.E. Seed germination traits can contribute better to plant community ecology. *J. Veg. Sci.* **2016**, *27*, 637–645. [CrossRef]
16. Phartyal, S.S.; Rosbakh, S.; Poschlod, P. Seed germination ecology in Trapa natans L., a widely distributed freshwater macrophyte. *Aquat. Bot.* **2018**, *147*, 18–23. [CrossRef]
17. Zhang, S.X. A preliminary study on cutting propagation techniques for old plants of Cinnamomum micranthum. *J. Fujian Sci. Tech.* **2000**, *27*, 69–71.

18. Xie, Z.Q.; Chen, W.L. The endangered causes and protection measures of *Cathaya argyrophylla*, an endemic to China. *Acta Phytoecol. Sin.* **2003**, *27*, 661–666.

19. Gao, Y.Z. *Sonneratiaceae: The Flora of China*; Science Press: Beijing, China, 1983; pp. 114–116.

20. Lin, P.; Lu, C.Y. Mangrove community in Hainan Island. *J. Xiamen Univ.* **1985**, *24*, 1117–1119.

21. Li, H.S.; Chen, G.Z. Biological characteristics and protection of Chinese endemic plant *S.* × *hainanensis*. *J. Guangdong Educ. Inst.* **2003**, *23*, 48–51.

22. Zhang, Z.; Guo, Y.; He, J.-S.; Tang, Z. Conservation status of wild plant species with extremely small populations in China. *Biodivers. Sci.* **2018**, *26*, 572–577. [CrossRef]

23. Duke, N.C. A mangrove hybrid, *Sonneratia* × *gulngai* (Sonneratiaceae) from north-eastern Australia. *Austrobaileya* **1984**, *2*, 103–105.

24. Tomlinson, P.B. *The botany of Mangroves*; Cambridge University Press: Cambridge, UK, 1986.

25. Zhang, Y.L.; Wang, K.F.; Li, Z. Study on pollen morphology(Sonneratiaceae) in China and Its paleoecological significance. *Mar. Geol. Quat. Geol.* **1997**, *17*, 2993–3001.

26. Renchao, Z. Natural Hybridization and Speciation in Sonneratia. Ph.D. Thesis, Zhongshan University, Guangzhou, China, 2006.

27. Li, H.S.; Chen, G.Z.; Si, S.H. ISSR study on genetic diversity of Hainan Sanghai. *Acta Sci. Nat. Univ. Sunyatseni* **2004**, *43*, 68–70.

28. Dafni, A. *Pollination Ecology*; Oxford University Press: New York, NY, USA, 1992.

29. Cruden, R.W. Pollen-ovule ratios: A conservative indicator of reproductive systems in flowering plants. *Evolution* **1977**, *31*, 32–46. [CrossRef] [PubMed]

30. Holsinger, K.E. Pollination biology and the evolution of mating systems in flowering plants. *Evol. Biol.* **1996**, *29*, 107–149.

31. Morgan, J.W. Effects of population size on seed production and germinability in an endangered, fragmented grassland plant. *Conserv. Boil.* **1999**, *13*, 266–273. [CrossRef]

32. Bie, P.F.; Tang, T.; Hu, J.Y.; Jiang, W. Flowering phenology and breeding system of an endangered and rare species Urophysa rockii(Ranunculaceae). *Acta Ecol. Sin.* **2018**, *38*, 3899–3908.

33. Lande, R. Risk of population extinction from fixation of new deleterious mutations. *Evolution* **1995**, *48*, 1460–1469. [CrossRef]

34. Kéry, M.; Matthies, D.; Spillmann, H.H. Reduced fecundity and offspring performance in small populations of the declining grassland plants *Primula veris* and *Gentiana lutea*. *J. Ecol.* **2000**, *88*, 17–30. [CrossRef]

35. Zhang, Y.; Li, Y.H.; Zhang, X.N.; Yang, Y. Flower phenology and breeding system of endangered mangrove *Lumnitzera littorea* (Jack.) Voigt. *Chin. J. Appl. Env. Biol.* **2017**, *23*, 0077–0081.

36. Zhang, J.H.; Liu, B.W. Patterns of seed predation and removal of Mongolian oak by rodents. *Acta Ecol. Sin.* **2014**, *34*, 1201–1211.

37. Yue, H.J.; Tong, C.; Zhu, J.M.; Huang, J.F. Seed rain and soil seed bank of endangered *Taxus chinensis* var *mairei* in Fujian, China. *Acta Ecol. Sin.* **2010**, *30*, 4389–4400.

38. Wu, D.R.; Wang, B.S. Seed and seedl ing ecology of the endangered *Phoebe bournei* (Lauraceae). *Acta Ecol. Sin.* **1997**, *11*, 1752–1757.

39. Manfred, J.; Lesley, P.; Birgitte, S. Habitat specificity, seed germination and experimental translocation of the endangered herb *Brachycome muelleri* (Asteraceae). *Biol. Conserv.* **2004**, *116*, 251–267.

40. Finch-Savage, W.E.; Leubner-Metzger, G. Seed dormancy and the control of germination. *New Phytol.* **2006**, *171*, 501–523. [CrossRef]

41. Zhang, J.J.; Chai, S.F.; Wei, X.; Lv, S.H.; Wu, S.H. Germination characteristics of the seed of a Rare and endangered plant, *Garcinia paucinervis* Australia. *Sci. Silvae Sin.* **2018**, *54*, 175–178.

42. Yang, Y.; Zhong, C.R.; Li, Y.H.; Zhang, Y. The morphological structure and germination characters of seed of endangered *Mangrove Lumnitzera littorea* (Jack.) Voigt. *Mol. Plant Breed.* **2016**, *14*, 2851–2858.

43. Liao, B.W.; Zheng, D.Z.; Zheng, S.F.; Li, Y. Seed germination conditions of *Sonneratia caseolaris* of mangrove. *J. Cent. South For. Univ.* **1997**, *17*, 25–27.

44. Xu, X.L. Effects of Cold Storage and Temperature on Seed Germination Characteristics of Two Common Plants in Alpine Meadow. Ph.D. Thesis, Lanzhou University, Lanzhou, China, 2007.

45. Li, W.L.; Zhang, X.P.; Hao, C.Y.; Zhang, H.; Holsinger, K.E. Characteristics of seed germination of the rare plant Cercidiphyllum japonicum. *Acta Ecol. Sin.* **2008**, *28*, 107–149.

46. Chai, S.F.; Jiang, J.S.; Wei, X.; Wang, M.L.; Li, L.; Qi, X.X. Seed germination characteristics of endangered plant Sinia rhodoleuca. *Chin. J. Ecol.* **2010**, *29*, 233–237.
47. Zheng, D.J.; Wu, Y.J.; Yun, Y.; Jiang, D.Q.; Chen, X.; Zhang, Z.L. Seed germination and its environment adaptability of endangered tree Dracaena cambodiana. *J. Trop. Subtrop. Bot.* **2016**, *24*, 71–79.
48. Chen, L.Z.; Wang, W.Q.; Zhang, Y.H.; Huang, L.; Zhao, C.L.; Yang, S.C. Damage to mangroves from extreme cold in early 2008 in sourthern China. *Chin. J. Plant Ecol.* **2010**, *34*, 186–194.
49. You, H.-M. Adaptability of mangrove Kandelia obovata seedlings to salinity–waterlogging. *Ying yong sheng tai xue bao = J. Appl. Ecol.* **2015**, *26*, 675–680.
50. Tang, M.; Li, K.; Xiang, H.Y.; Dong, X.; Jin, H.X.; Wang, Y.; Yang, H.J.; Zhang, Z.X. Research on ecological, physiological and morphological adaptability of two mangrove species to salt stress. *Ecol. Sci.* **2014**, *33*, 513–519.
51. Zhang, Y.; Ye, Y.; Lu, C.Y. Seed germination and seedling growth of mangrove *Excoecaria agallocha* under different salinities. *J. Xiamen Univ.* **2010**, *49*, 145–148.
52. Khan, M.A.; Gul, B.; Weber, D.J. Germination responses to *Salicornia rubra* to temperature and salinity. *J. Arid Env.* **2000**, *45*, 207–214. [CrossRef]
53. Li, Y.; Zheng, D.Z.; Liao, B.W.; Zheng, S.F.; Song, X.Y. Effect of salinity and temperature on seed germination of mangrove *Sonneratia apetala Buch. Ham. For. Res.* **1997**, *10*, 137–142.

MDPI
St. Alban-Anlage 66
4052 Basel
Switzerland
Tel. +41 61 683 77 34
Fax +41 61 302 89 18
www.mdpi.com

Forests Editorial Office
E-mail: forests@mdpi.com
www.mdpi.com/journal/forests